International Association of Geodesy Symposia

Chris Rizos, Series Editor
Pascal Willis, Series Associate Editor

International Association of Geodesy Symposia

Chris Rizos, Series Editor
Pascal Willis, Series Associate Editor

Symposium 104: Sea SurfaceTopography and the GeoidSymposium
Symposium 105: Earth Rotation and Coordinate Reference Frames
Symposium 106: Determination of the Geoid: Present and Future
Symposium 107: Kinematic Systems in Geodesy, Surveying, and Remote Sensing
Symposium 108: Application of Geodesy to Engineering
Symposium 109: Permanent Satellite Tracking Networks for Geodesy and Geodynamics
Symposium 110: From Mars to Greenland: Charting Gravity with Space and Airborne Instruments
Symposium 111: Recent Geodetic and Gravimetric Research in Latin America
Symposium 112: Geodesy and Physics of the Earth: Geodetic Contributions to Geodynamics
Symposium 113: Gravity and Geoid
Symposium 114: Geodetic Theory Today
Symposium 115: GPS Trends in Precise Terrestrial, Airborne, and Spaceborne Applications
Symposium 116: Global Gravity Field and Its Temporal Variations
Symposium 117: Gravity, Geoid and Marine Geodesy
Symposium 118: Advances in Positioning and Reference Frames
Symposium 119: Geodesy on the Move
Symposium 120: Towards an Integrated Global Geodetic Observation System (IGGOS)
Symposium 121: Geodesy Beyond 2000: The Challenges of the First Decade
Symposium 122: IV Hotine-Marussi Symposium on Mathematical Geodesy
Symposium 123: Gravity, Geoid and Geodynamics 2000
Symposium 124: Vertical Reference Systems
Symposium 125: Vistas for Geodesy in the New Millennium
Symposium 126: Satellite Altimetry for Geodesy, Geophysics and Oceanography
Symposium 127: V Hotine Marussi Symposium on Mathematical Geodesy
Symposium 128: A Window on the Future of Geodesy
Symposium 129: Gravity, Geoid and Space Missions
Symposium 130: Dynamic Planet - Monitoring and Understanding...
Symposium 131: Geodetic Deformation Monitoring: From Geophysical to Engineering Roles
Symposium 132: VI Hotine-Marussi Symposium on Theoretical and Computational Geodesy
Symposium 133: Observing our Changing Earth
Symposium 134: Geodetic Reference Frames
Symposium 135: Gravity, Geoid and Earth Observation
Symposium 136: Geodesy for Planet Earth
Symposium 137: VII Hotine-Marussi Symposium on Mathematical Geodesy
Symposium 138: Reference Frames for Applications in Geosciences
Symposium 139: Earth on the Edge: Science for a sustainable Planet
Symposium 140: The 1st International Workshop on the Quality of Geodetic Observation and Monitoring Systems (QuGOMS'11)
Symposium 141: Gravity, Geoid and Height systems (GGHS2012)
Symposium 142: VIII Hotine-Marussi Symposium on Mathematical Geodesy
Symposium 143: Scientific Assembly of the International Association of Geodesy, 150 Years
Symposium 144: 3rd International Gravity Field Service (IGFS)
Symposium 145: International Symposium on Geodesy for Earthquake and Natural Hazards (GENAH)

More information about this series at http://www.springer.com/series/1345

REFAG 2014

Proceedings of the IAG Commission 1 Symposium Kirchberg, Luxembourg, 13–17 October, 2014

Edited by

Tonie van Dam

Springer

Volume Editor
Tonie van Dam
University of Luxembourg
Faculty of Science, Technology, and Communication
Luxembourg
Luxembourg

Series Editor
Chris Rizos
School of Surveying & Geospatial Engineering
University of New South Wales
Sydney
Australia

Associate Editor
Pascal Willis
Institut national de l'Information
géographique et forestière
Service de la Recherche et de l'Enseignement
France

ISSN 0939-9585 ISSN 2197-9359 (electronic)
International Association of Geodesy Symposia
ISBN 978-3-319-45628-7 ISBN 978-3-319-45629-4 (eBook)
DOI 10.1007/978-3-319-45629-4

Library of Congress Control Number: 2016963341

Printed on acid-free paper

This Springer imprint is published by Springer Nature
The registered company is Springer International Publishing AG
The registered company address is: Gewerbestrasse 11, 6330 Cham, Switzerland

Preface

Reference Frames for Applications in Geoscience (2014 IAG Commission 1 Symposium: Luxembourg, 13–17 October, 2014)

Accurate reference frames are fundamentally important for Earth science studies, satellite navigation, many applications relying on geo-spatial information, and increasingly for demanding commercial applications such as agriculture, aviation, construction, public safety, and transportation.

In October, approximately 100 geodesists from 24 countries met to discuss the role of global and regional reference frames in science and practice. The international symposium, REFAG2014, took place at the Melia Hotel in Kirchberg, Luxembourg. This symposium provided a forum where the developers of the International Terrestrial Reference Frame (ITRF), the scientific users, and government sponsors of observing systems could discuss the most important questions. The symposium featured sessions on six topics: the theory and concepts in developing reference frames, the spacebased geodetic techniques required to generate the reference frames, the theory to tie the terrestrial reference frames to the celestial reference frame and Earth rotation parameters, regional reference frames, the importance of reference frames in the geosciences, and how georeferencing is used in practice.

The meeting consisted of 60 oral presentations and 18 posters, with 10 invited presentations over 4.5 days. Invited speakers were chosen for their expertise, but also for their ability to look beyond current practices and requirements and to visualize needs for the future. The invited speakers represented the groups working at the cutting edge to develop algorithms to combine all the geodetic observations into a frame. They also represented some of the best scientists using geodetic data for geodynamic and climate studies.

This volume contains the proceedings of selected papers from sessions organized under the six themes listed below. In contrast to most workshops or symposia, here we decided to arrange the sessions around the most pressing questions regarding reference frames that need to be addressed at this point in time.

Session 1: Theory and Concepts
Associate Editor: Zuheir Altamimi

- Are present reference frame theory and concepts adequate for current measurement accuracy and all types of station position time variations?
- How well do the concepts match current practice (e.g., UTC/Terrestrial time versus geocentric time)? Are changes needed?
- What about the mismatch between long-term linear theoretical frames and the reality of constantly moving stations? Is a new conceptual framework needed?
- To what extent are fundamental physical standards limiting, e.g., the present inaccuracy of the constants G and GmE, relativistic formulation, the time-variable gravity field and the present inaccuracy of geocenter and geocenter motion?

Session 2: Geodetic Measurement Techniques
Associate Editor: Tom Herring

- How can the strengths and weaknesses of current technologies be better meshed to improve the ITRF?
- What are the present leading geophysical model limitations and how can they be improved? How do we ensure consistency of geophysical model application among different techniques?
- How can we best cope with the rapidly expanding volume of geodetic data (esp. GNSS data)?
- Should more synergistic and multi-technique analysis approaches be promoted for the future?
- How do we precisely quantify the limiting factors of the legacy network and optimally deploy nextgeneration systems?

Session 3: Regional Reference Frames
Associate Editor: Richard Wonnacott

- Is there a future role for sub-global reference frames of high accuracy?
- If the main justification for regional frames is to remove large-scale tectonic effects, how well can this really be done? What about residual motions?
- How should national and regional frame efforts be best coordinated?

Session 4: Celestial to Terrestrial Frame Transformations
Associate Editor: Johannes Boehm

- Can the value of EOPs as monitors of global variations of the Earth system be improved (e.g., better accuracy, higher temporal resolution)?
- Is additional, non-geodetic data needed to achieve greater insight or value? If so, what types?
- Is the current array of observing systems evolving towards an optional delivery of EOP results or are changes needed?
- Is the interaction between the geodetic EOP community and groups engaged in modelling large-scale terrestrial fluids adequate? Can the benefits to both sides be improved?
- How can the current ICRF implementation made consistent with the ITRF and EOPs and what is the current level of consistency?

Session 5: Usage and Applications of Reference Frames in Geosciences
Associate Editor: Tonie van Dam

- In what ways are geoscience and geophysical applications under-served by the present geodetic system? Are large-scale shifts in emphasis needed or beneficial?
- What are the major questions of geoscience that are not currently well addressed by the geodetic observing systems? What is the required accuracy of all types of geodetic products needed?
- How best to integrate geodetic and gravimetric systems?
- How to maintain a suitable balance between societal needs (e.g., natural hazard monitoring) and basic scientific goals?

Session 6: Georeferencing in Practice
Associate Editor: Mikael Lilje

- Are emerging commercial geodetic applications (e.g., GIS) approaching high accuracy requirements?
- How best to synchronize commercial requirements with basic research needs? How can conflicts of interest be minimized?
- How can the basic infrastructure observing systems that have traditionally been funded by national research bodies be sustained, especially as commercial applications grow in importance?
- Are large-scale changes in the national, regional, and global geodetic frameworks needed? If so, what sort?

I would like to express my gratitude to all those who have contributed to the success of REFAG2014: the conveners who devoted their time to organizing interesting and timely sessions all the meeting participants, my post-docs and PhD students who helped me to organize and run the meeting, and Jim Ray and Zuheir Altamimi whose advice I sought throughout the organization of this meeting. I also thank the Luxembourg Fonds National de la Recherche, the University of Luxembourg, the European Geosciences Union, and Leica for their financial support of the symposium

I thank the associate editors of these proceedings who played a leading role in the peer review process and all reviewers who contributed to reviewing the papers contained in this volume. Finally, my sincere thanks also goes to the IAG proceedings editor Dr. Pascal Willis whose reminders kept the review and publication processes on schedule.

Luxembourg, Luxembourg Tonie van Dam

Contents

Part I

Theory and Concepts

The Helmert Transformation Approach in Network Densification Revisited

C. Kotsakis, A. Vatalis, and F. Sansò

Abstract

The mapping problem of an adjusted network from its initial frame to another target frame through the Helmert transformation (HT) is discussed in this paper. We present an optimal solution which can be easily computed by a closed-form expression in terms of appropriate corrections to the standard HT solution that is often used in geodetic practice. Its advantage is the minimization of the propagated noise from the initial network coordinates to their estimated values in the target frame, both at the reference and non-reference stations. This is accomplished by an additional filtering step within the transformation procedure which exploits the known covariance structure of the underlying network in both frames. The presented approach is a suitable tool for aligning an existing network solution to a secular frame such as the ITRF and, as shown in the paper, it can be unequivocally related to the constrained network adjustment directly in the target frame. Nevertheless, any unmodeled non-secular signals hidden in the initial coordinates will be affected by the aforementioned filtering step, and thus the frame alignment methodology presented herein is not tuned for geophysical loading studies with respect to a secular reference frame.

Keywords

Constrained adjustment • Frame alignment • Geodetic networks • Helmert transformation • Least squares estimation • Reference frames

1 Introduction

The main prerequisite for the alignment of a geodetic network to ITRF (Altamimi et al. 2011), or to any other global frame of interest, is to process the network measurements along with a subset of reference stations with well known positions in the desired frame. Using the prior information of the reference stations, there are mainly two alternative strategies to express the network coordinates in the desired frame:

- Constraining, either stochastically or absolutely, the coordinates of the reference stations to their known values during the network adjustment – hereafter called the *constrained network adjustment (CNA) approach*, or
- Performing the network adjustment in an unknown or weakly defined frame (e.g. free-net solution) and then fitting the computed solution to the desired frame using a set of Helmert transformation parameters derived from the available reference stations – hereafter called the *Helmert transformation (HT) approach*.

Both of these approaches have been widely used in practice for geodetic network densifications at global, regional and local scales (e.g. Gurtner et al. 1997; Altamimi 2003; Bruyninx et al. 2013). From a theoretical perspective, the first approach is the optimal densification strategy in the

C. Kotsakis (✉)
Department of Geodesy and Surveying, Aristotle University of Thessaloniki, Thessaloniki, Greece
e-mail: kotsaki@topo.auth.gr

A. Vatalis • F. Sansò
DICA - Politecnico di Milano, c/o Polo Regionale di Como, Como, Italy

T. van Dam (ed.), *REFAG 2014*, International Association of Geodesy Symposia 146,
DOI 10.1007/1345_2015_204

sense that the estimated coordinates in the desired frame will have the smallest error variances amongst any other linear unbiased methodology using the same network data (provided that the inverse covariance matrix of the reference stations coordinates is used as a weight matrix in the constrained adjustment). The HT approach, on the other hand, is able to preserve the network geometry as defined by the actual measurements – something that is not ensured in the CNA approach – through a least-squares fit to the desired frame using the Helmert transformation either in its full form or in an abridged form by omitting some of its original parameters. Its weakness is related to the so-called network configuration effect which can produce biases to the transformed coordinates, especially in regional and local networks (Altamimi 2003). This effect stems from the ill-conditioning in the least-squares adjustment of the HT model over reference stations with limited (non-global) spatial support, and it leads to highly correlated estimates of the transformation parameters and overly reduced accuracy for the transformed coordinates mostly at the non-reference stations.

Following the recent study by Kotsakis et al. (2014), the aim of this paper is to retrace the HT approach for network densification and to give a revised formulation which improves the accuracy of the estimated coordinates in the target frame. Compared to the usual frame transformation methodology, the presented scheme contains a noise filtering step that reduces the propagated random errors from the initial coordinates to their transformed values by exploiting the known stochastic characteristics of the underlying network. Despite the accessibility of the full covariance (CV) matrices of the initial and the reference coordinates, this extra step is absent from the determination of Helmert-transformed coordinates in geodetic networks. However, its implementation is easy and it can provide an effective "regularization" tool that may compensate, to some extent, the ill-conditioning caused by the network configuration effect in the transformation solution. To further support our findings, a useful relationship is derived between the CNA and the HT estimators for the station coordinates in the target frame. Besides its theoretical elegance, such a result is particularly useful as not only does it identify the conditions under which the two densification schemes give identical results, but it also justifies the frequently suggested use of an abridged HT model in frame alignment problems.

The viewpoint of this study lies on the optimal mapping of a network solution to a target frame which is realized by prior coordinates (and their CV matrix) in a subset of the network stations. The aforementioned noise filtering step is an essential part of this procedure, yet it could damp useful hidden information of geodynamical interest within the initial solution that will not be properly transferred for further scientific inference in the target frame (e.g. study of loading displacements). Therefore, the presented approach is a suitable tool for aligning a network solution to a secular frame such as the ITRF, but it will not rightly reproduce unmodeled non-secular signals in the transformed coordinates. This should be strongly emphasized in order to avoid any confusion to the reader regarding the applicability of our proposed optimal estimator, and it will be further underlined in following sections of the paper.

2 HT-Based Network Mapping to a Target Reference Frame

The geodetic formulation of the Helmert transformation is commonly expressed as

$$\mathbf{X} = \mathbf{X}' + \mathbf{G}\,\boldsymbol{\theta} \tag{1}$$

where $\mathbf{X}'$ and $\mathbf{X}$ are the Cartesian coordinate vectors of a set of stations with respect to an initial and a target frame, respectively. The vector $\boldsymbol{\theta}$ contains the frame transformation parameters whereas the matrix $\mathbf{G}$ originates from the simplified Jacobian of the nonlinear similarity transformation under sufficiently small orientation and scale differences between the involved frames. The above well known linearized model provides the basis for our following analysis in this section.

At first, let us recall the standard HT approach in network densification problems which is implemented in two steps as follows. Initially, a least-squares estimate of the transformation parameters is obtained from a group of reference stations with known noisy coordinates in both frames. We consider the fully weighted case where the CV matrices of both coordinate sets, $\boldsymbol{\Sigma}_{\mathbf{X}}$ and $\boldsymbol{\Sigma}_{\mathbf{X}'}$, are used in the estimation process according to the standard formula:

$$\begin{aligned}\widehat{\boldsymbol{\theta}} = \left(\mathbf{G}^T\left(\boldsymbol{\Sigma}_{\mathbf{X}} + \boldsymbol{\Sigma}_{\mathbf{X}'}\right)^{-1}\mathbf{G}\right)^{-1} \times \\ \times\, \mathbf{G}^T\left(\boldsymbol{\Sigma}_{\mathbf{X}} + \boldsymbol{\Sigma}_{\mathbf{X}'}\right)^{-1}\left(\mathbf{X} - \mathbf{X}'\right)\end{aligned} \tag{2}$$

Subsequently, the estimated Helmert parameters are employed to transform the coordinates of the reference and non-reference stations (denoted by $\mathbf{X}'$ and $\mathbf{Z}'$ respectively) from their initial frame to the target frame:

$$\begin{bmatrix}\widehat{\mathbf{x}}^{\text{st}} \\ \widehat{\mathbf{z}}^{\text{st}}\end{bmatrix} = \begin{bmatrix}\mathbf{X}' \\ \mathbf{Z}'\end{bmatrix} + \begin{bmatrix}\mathbf{G} \\ \tilde{\mathbf{G}}\end{bmatrix}\widehat{\boldsymbol{\theta}} \tag{3}$$

where the superscript 'st' indicates that the computed coordinates are obtained by the standard HT approach. The matrices $\mathbf{G}$ and $\tilde{\mathbf{G}}$ denote the Helmert transformation matrices for the reference and non-reference stations, respectively. Note that the coordinate vectors $\mathbf{X}'$ and $\mathbf{Z}'$ are always correlated with each other as they are jointly obtained by a least-squares network adjustment in some initial frame (which does not need to be further specified at this point).

The solution in Eq. (3) does not arise from a particular estimation principle but it is merely deduced by the forward implementation of the HT model. Although it preserves the network geometry as defined by the adjusted measurements in the initial frame, this solution lacks the property of minimizing the propagated data noise to the transformed coordinates in the target frame. In fact, the random errors of the coordinate vectors $\mathbf{X}'$ and $\mathbf{Z}'$ will be fully absorbed into the result of Eq. (3) which means that the standard approach does not provide full optimal control upon the transformation solution or, by using frame terminology, the realization of the target frame through Eq. (3) does not produce station coordinates with optimal accuracy level.

Following a more rigorous approach, the HT-based mapping of a geodetic network to a target frame can be formulated by the combined system of observation equations

$$\mathbf{X} = \mathbf{x} + \mathbf{v_X}\,, \qquad \mathbf{v_X} \sim (\mathbf{0},\ \mathbf{\Sigma_X}) \tag{4}$$

$$\mathbf{X}' = \mathbf{x} - \mathbf{G}\,\boldsymbol{\theta} + \mathbf{v_{X'}}\,, \qquad \mathbf{v_{X'}} \sim (\mathbf{0},\ \mathbf{\Sigma_{X'}}) \tag{5}$$

$$\mathbf{Z}' = \mathbf{z} - \tilde{\mathbf{G}}\,\boldsymbol{\theta} + \mathbf{v_{Z'}}\,, \qquad \mathbf{v_{Z'}} \sim (\mathbf{0},\ \mathbf{\Sigma_{Z'}}) \tag{6}$$

in conjunction with the data weight matrix

$$\mathbf{P} = \begin{bmatrix} \mathbf{\Sigma_X} & \mathbf{0} & \mathbf{0} \\ \mathbf{0} & \mathbf{\Sigma_{X'}} & \mathbf{\Sigma_{X'Z'}} \\ \mathbf{0} & \mathbf{\Sigma_{Z'X'}} & \mathbf{\Sigma_{Z'}} \end{bmatrix}^{-1} \tag{7}$$

The latter considers all statistical information that is commonly available in network densification problems, whereas the vectors $\mathbf{x}$ and $\mathbf{z}$ correspond to the true coordinates of the reference and non-reference stations in the target frame. The analytic form of the weighted least-squares solution of the previous system is given in Kotsakis et al. (2014). Therein it was shown that the adjusted Helmert parameters remain the same as in Eq. (2), a fact that is expected since the inclusion of the non-reference stations into the adjustment procedure does not contribute any additional information for the determination of those parameters. On the other hand, the least-squares solution for the network coordinates in the target frame differs from the classic expression in Eq. (3) as follows (Kotsakis et al. 2014).

$$\begin{bmatrix} \hat{\mathbf{x}} \\ \hat{\mathbf{z}} \end{bmatrix} = \begin{bmatrix} \hat{\mathbf{x}}^{st} \\ \hat{\mathbf{z}}^{st} \end{bmatrix} + \begin{bmatrix} \mathbf{\Sigma_{X'}} \\ \mathbf{\Sigma_{Z'X'}} \end{bmatrix} (\mathbf{\Sigma_X} + \mathbf{\Sigma_{X'}})^{-1} \left(\mathbf{X} - \hat{\mathbf{x}}^{st}\right) \tag{8}$$

The above formula gives the optimal HT-based estimator in terms of additive corrections to the standard HT-based estimator for network densification purposes. Both of these estimators produce a network solution that refers to the same target frame, namely to the one realized by the prior coordinates $\mathbf{X}$ of the reference stations. Their difference is that Eq. (8) leads to station coordinates with smaller error variances compared to the standard estimates $\hat{\mathbf{x}}^{st}$ and $\hat{\mathbf{z}}^{st}$, as it has been shown in Kotsakis et al. (2014).

Evidently, Eqs. (3) and (8) give identical results under the conditions $\mathbf{\Sigma_{X'}} = \mathbf{0}$ and $\mathbf{\Sigma_{Z'X'}} = \mathbf{0}$ none of which ever applies in practice, at least for cases of adjusted networks that need to be transformed to another frame. Furthermore, if the prior coordinates of the reference stations are assumed errorless ($\mathbf{\Sigma_X} = \mathbf{0}$) then Eq. (8) will reproduce their values ($\hat{\mathbf{x}} = \mathbf{X}$) in accordance to the rationale of the constrained network adjustment directly in the desired frame. In general, though, the CNA and HT densification schemes do not give the same estimated coordinates at the non-reference stations. This is because each scheme defines the target frame at a different stage during the network analysis, that is, either in tandem with the adjustment of the network measurements (CNA approach) or after the adjustment of the network measurements in an arbitrary initial frame (HT approach). Their actual differences are discussed and evaluated in more detail in the following section.

Let us stress that Eq. (8) suppresses the noise of the initial solution ($\mathbf{X}'$, $\mathbf{Z}'$) provided that all relevant CV matrices are correct or, at least, reliable. This noise filtering step is missing from the standard HT solution in Eq. (3), thus making the transformed coordinates ($\hat{\mathbf{x}}^{st}$, $\hat{\mathbf{z}}^{st}$) to have larger error variances compared to the result of Eq. (8). On the other hand, in frame alignments for generating coordinate time series in support of geodynamical investigations (e.g. Tregoning and van Dam 2005; Bevis and Brown 2014) such a filtering step may not be desirable as it could weaken the signal information hidden in the initial solution. This is especially true when $\mathbf{\Sigma_X} \ll \mathbf{\Sigma_{X'}}$, in which case the transformed network will be forced to "follow" the secular character of the target frame (e.g. ITRF) thus obscuring any non-secular signals originating by unmodeled geophysical loading effects (at least at the reference stations). However, the importance of Eq. (8) remains in the sense of being an effective tool for the combination of independent overlapping networks using their full covariance information in their respective frames – see also Kotsakis et al. (2014) where the more general case of inter-correlated overlapping networks is additionally treated.

3 Comparison of the CNA and HT-Based Estimators in Network Densification

For the purpose of this study, it is instructive to relate the CNA and HT densification strategies when using the same set of reference stations in each estimation scheme. To compare them in an analytic way we consider the normal equations

(NEQ) originating from the data processing in the underlying network (before adding any datum constraints and after eliminating any auxiliary parameters from the adjustment procedure)

$$\mathbf{N}\begin{bmatrix}\mathbf{x}-\mathbf{x}_o\\ \mathbf{z}-\mathbf{z}_o\end{bmatrix}=\mathbf{u} \quad (9)$$

where $\mathbf{x}_o$ and $\mathbf{z}_o$ are the approximate coordinates for the reference and non-reference stations that are used in the linearization of the NEQ system.

Restricting our attention to GNSS networks, the above system is generally invertible as it already contains the auxiliary datum information from the adopted IGS orbits. An initial "free" network solution can therefore be obtained as

$$\begin{bmatrix}\mathbf{X}'\\ \mathbf{Z}'\end{bmatrix}=\begin{bmatrix}\mathbf{x}_o\\ \mathbf{z}_o\end{bmatrix}+\mathbf{N}^{-1}\mathbf{u} \quad (10)$$

which is given in a reference frame that could be far from the ITRF since it is realized only at the precision level of the IGS orbits (a few cm). This solution can be transformed to any desired frame, e.g. ITRF, using the HT scheme that was described in the previous section.

Alternatively, a constrained solution directly in the desired frame can be obtained via the same prior information for the reference stations ($\mathbf{X}$, $\boldsymbol{\Sigma}_\mathbf{X}$) according to the well known least-squares adjustment formula

$$\begin{bmatrix}\widehat{\mathbf{x}}^c\\ \widehat{\mathbf{z}}^c\end{bmatrix}=\begin{bmatrix}\mathbf{x}_o\\ \mathbf{z}_o\end{bmatrix}+\left(\mathbf{N}+\mathbf{H}^T\boldsymbol{\Sigma}_\mathbf{X}^{-1}\mathbf{H}\right)^{-1}\times \\ \times\left(\mathbf{u}+\mathbf{H}^T\boldsymbol{\Sigma}_\mathbf{X}^{-1}(\mathbf{X}-\mathbf{x}_o)\right) \quad (11)$$

where the auxiliary matrix $\mathbf{H}$ has the partitioned form $[\mathbf{I} \mid \mathbf{0}]$ in accordance to the NEQ partitioning in Eq. (9). The above solution differs from the HT solution of Eq. (8) according to the general formula (see proof in the next section)

$$\begin{bmatrix}\widehat{\mathbf{x}}\\ \widehat{\mathbf{z}}\end{bmatrix}-\begin{bmatrix}\widehat{\mathbf{x}}^c\\ \widehat{\mathbf{z}}^c\end{bmatrix}=\left(\mathbf{N}+\mathbf{H}^T\boldsymbol{\Sigma}_\mathbf{X}^{-1}\mathbf{H}\right)^{-1}\mathbf{N}\,\mathbf{E}^T\widehat{\boldsymbol{\theta}} \quad (12)$$

where $\widehat{\boldsymbol{\theta}}$ is given by Eq. (2) and $\mathbf{E}$ is the Helmert transformation matrix for the entire network (including reference and non-reference stations)

$$\mathbf{E}=\left[\mathbf{G}^T\ \tilde{\mathbf{G}}^T\right] \quad (13)$$

The difference of the two densification solutions depends on the frame transformation model that is employed in the HT approach. This is not a trivial realization and, actually, it can explain the fact that the full (7-parameter) Helmert model might not always be the best choice to obtain a well-expressed network solution in the desired frame. In fact, Eq. (12) shows that the CNA and HT-based estimators give the same result under the condition $\mathbf{NE}^T=\mathbf{0}$, which implies that the transformation model should contain only the parameters that correspond to the datum defect of the underlying network (e.g. Blaha 1971; Sillard and Boucher 2001).

In the case of GNSS networks the NEQ system in Eq. (9) is generally invertible ($\mathbf{NE}^T\neq\mathbf{0}$), thus causing an discrepancy between the constrained solution of Eq. (11) and the transformed solution of Eq. (8) for any choice of the transformation model. However, if a reduced model is employed in the HT approach, e.g. shift-only model, then the difference of the two solutions will be dictated by (a linear combination of) the columns of the matrix $\mathbf{NE}^T$ that correspond only to the selected model parameters. Hence their consistency may be improved if the omitted parameters correspond to numerically large columns of the aforementioned matrix. Typically, such frame parameters in GNSS networks are the rotation angles and the scale factor, which could cause apparent biases to the transformed coordinates in the desired frame relative to the weighted CNA solution.

To provide a quick example of the differences among the previous densification schemes, we used daily sinex files from regional subnetworks that are regularly processed by local analysis centers of the EUREF Permanent Network (EPN). Using the constraint-free NEQ of each subnetwork, we computed and compared the ITRF2008 coordinates obtained by the *standard HT approach* (*7 parameters*), the *optimal HT approach* (*7 parameters*) and the *weighted CNA approach*, based on the same reference stations in each case. The prior coordinates of the reference stations and their full CV matrix were extracted from the ITRF2008-TRF-IGS sinex file (http://itrf.ensg.ign.fr/ITRF_solutions/2008/ITRF2008_files.php) and were reduced by their known velocities to the current epoch of the daily solutions. Some representative results from two different subnetworks are given in Fig. 1. The apparent biases in the standard HT solution due to the network configuration effect are clearly visible, and they amount to several mm in both horizontal and vertical components. The optimal HT solution, on the other hand, seems to provide a much better agreement with the weighted CNA solution over all stations in every case.

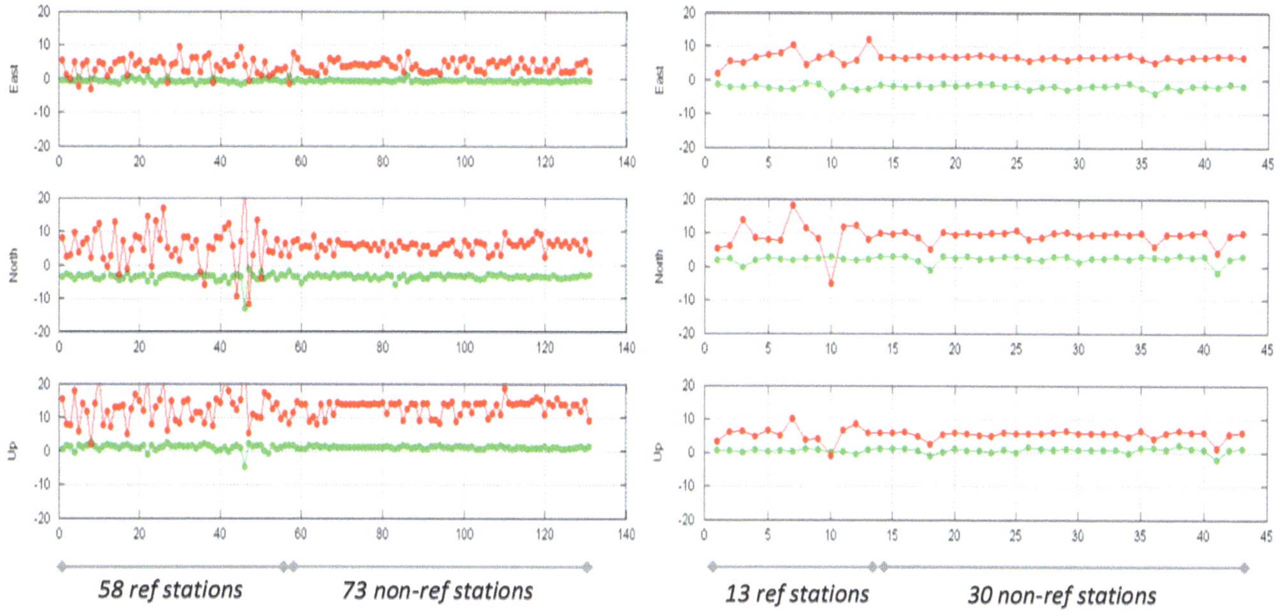

Fig. 1 Differences (in mm) of the ITRF2008 coordinates obtained by the weighted CNA solution vs. the optimal HT solution (*green line*) and the standard HT solution (*red line*). The used sinex files refer to the 6th day of GPS week 1809 from the MUT and SUT (*left and right plots* respectively) local analysis centers of the EPN network

4 The Analytic Proof of Eq. (12)

Starting from Eq. (8), and using the auxiliary matrix $\mathbf{H} = [\mathbf{I} \mid \mathbf{0}]$, the optimal HT solution can be equivalently expressed as

$$\begin{bmatrix}\widehat{\mathbf{x}}\\ \widehat{\mathbf{z}}\end{bmatrix} = \begin{bmatrix}\widehat{\mathbf{x}}^{st}\\ \widehat{\mathbf{z}}^{st}\end{bmatrix} + \begin{bmatrix}\boldsymbol{\Sigma}_{X'} & \boldsymbol{\Sigma}_{X'Z'}\\ \boldsymbol{\Sigma}_{Z'X'} & \boldsymbol{\Sigma}_{Z'}\end{bmatrix} \times \mathbf{H}^T\left(\boldsymbol{\Sigma}_X + \mathbf{H}\begin{bmatrix}\boldsymbol{\Sigma}_{X'} & \boldsymbol{\Sigma}_{X'Z'}\\ \boldsymbol{\Sigma}_{Z'X'} & \boldsymbol{\Sigma}_{Z'}\end{bmatrix}\mathbf{H}^T\right)^{-1} \times \left(\mathbf{X} - \widehat{\mathbf{x}}^{st}\right) \qquad (14)$$

Taking into account Eq. (3) and the matrix notation from Eq. (13), the previous equation becomes

$$\begin{bmatrix}\widehat{\mathbf{x}}\\ \widehat{\mathbf{z}}\end{bmatrix} = \begin{bmatrix}\mathbf{X}'\\ \mathbf{Z}'\end{bmatrix} + \mathbf{E}^T\widehat{\boldsymbol{\theta}} + \begin{bmatrix}\boldsymbol{\Sigma}_{X'} & \boldsymbol{\Sigma}_{X'Z'}\\ \boldsymbol{\Sigma}_{Z'X'} & \boldsymbol{\Sigma}_{Z'}\end{bmatrix} \times \mathbf{H}^T\left(\boldsymbol{\Sigma}_X + \mathbf{H}\begin{bmatrix}\boldsymbol{\Sigma}_{X'} & \boldsymbol{\Sigma}_{X'Z'}\\ \boldsymbol{\Sigma}_{Z'X'} & \boldsymbol{\Sigma}_{Z'}\end{bmatrix}\mathbf{H}^T\right)^{-1} \times \left(\mathbf{X} - \mathbf{X}' - \mathbf{H}\mathbf{E}^T\widehat{\boldsymbol{\theta}}\right) \qquad (15)$$

For the sake of convenience we introduce the auxiliary symbols:

$$\widehat{\boldsymbol{\xi}} = \begin{bmatrix}\widehat{\mathbf{x}} - \mathbf{x}_o\\ \widehat{\mathbf{z}} - \mathbf{z}_o\end{bmatrix},\quad \boldsymbol{\xi}' = \begin{bmatrix}\mathbf{X}' - \mathbf{x}_o\\ \mathbf{Z}' - \mathbf{z}_o\end{bmatrix},\quad \boldsymbol{\Sigma}_{\xi'} = \begin{bmatrix}\boldsymbol{\Sigma}_{X'} & \boldsymbol{\Sigma}_{X'Z'}\\ \boldsymbol{\Sigma}_{Z'X'} & \boldsymbol{\Sigma}_{Z'}\end{bmatrix} \qquad (16)$$

and thus Eq. (15) takes the form

$$\widehat{\boldsymbol{\xi}} = \boldsymbol{\xi}' + \mathbf{E}^T\widehat{\boldsymbol{\theta}} + \boldsymbol{\Sigma}_{\xi'}\mathbf{H}^T\left(\boldsymbol{\Sigma}_X + \mathbf{H}\,\boldsymbol{\Sigma}_{\xi'}\mathbf{H}^T\right)^{-1} \times \left(\mathbf{X} - \mathbf{H}\boldsymbol{\xi}' - \mathbf{x}_o - \mathbf{H}\mathbf{E}^T\widehat{\boldsymbol{\theta}}\right) \qquad (17)$$

Using the well known matrix identity (Schaffrin 1983, p. 34, Eq. (A12))

$$\mathbf{DC}\left(\mathbf{A} + \mathbf{BDC}\right)^{-1} = \left(\mathbf{I} + \mathbf{DCA}^{-1}\mathbf{B}\right)^{-1}\mathbf{DCA}^{-1} \qquad (18)$$

we have

$$\boldsymbol{\Sigma}_{\xi'}\,\mathbf{H}^T\left(\boldsymbol{\Sigma}_X + \mathbf{H}\,\boldsymbol{\Sigma}_{\xi'}\mathbf{H}^T\right)^{-1} = \left(\mathbf{I} + \boldsymbol{\Sigma}_{\xi'}\mathbf{H}^T\boldsymbol{\Sigma}_X^{-1}\mathbf{H}\right)^{-1} \times \boldsymbol{\Sigma}_{\xi'}\mathbf{H}^T\boldsymbol{\Sigma}_X^{-1} \qquad (19)$$

which allows us to express Eq. (17) as

$$\hat{\boldsymbol{\xi}} = \boldsymbol{\xi}' + \mathbf{E}^T\hat{\boldsymbol{\theta}} + \left(\mathbf{I} + \boldsymbol{\Sigma}_{\xi'}\mathbf{H}^T\boldsymbol{\Sigma}_{\mathbf{X}}^{-1}\mathbf{H}\right)^{-1}\boldsymbol{\Sigma}_{\xi'}\mathbf{H}^T\boldsymbol{\Sigma}_{\mathbf{X}}^{-1} \times \left(\mathbf{X} - \mathbf{H}\boldsymbol{\xi}' - \mathbf{x}_o - \mathbf{H}\mathbf{E}^T\hat{\boldsymbol{\theta}}\right) \tag{20}$$

Multiplying both sides of the last equation by the matrix $\mathbf{I} + \boldsymbol{\Sigma}_{\xi'}\mathbf{H}^T\boldsymbol{\Sigma}_{\mathbf{X}}^{-1}\mathbf{H}$ and after some simple algebraic manipulations, we get

$$\left(\mathbf{I} + \boldsymbol{\Sigma}_{\xi'}\mathbf{H}^T\boldsymbol{\Sigma}_{\mathbf{X}}^{-1}\mathbf{H}\right)\hat{\boldsymbol{\xi}} = \boldsymbol{\xi}' + \mathbf{E}^T\hat{\boldsymbol{\theta}} + \boldsymbol{\Sigma}_{\xi'}\mathbf{H}^T\boldsymbol{\Sigma}_{\mathbf{X}}^{-1}\ (\mathbf{X} - \mathbf{x}_o) \tag{21}$$

Let us now introduce the auxiliary vector

$$\hat{\boldsymbol{\xi}}^c = \begin{bmatrix} \hat{\mathbf{x}}^c - \mathbf{x}_o \\ \hat{\mathbf{z}}^c - \mathbf{z}_o \end{bmatrix} \tag{22}$$

which corresponds to the constrained solution directly in the desired frame using the same reference stations with the HT solution. Based on Eq. (11) we obviously have

$$\mathbf{H}^T\boldsymbol{\Sigma}_{\mathbf{X}}^{-1}\ (\mathbf{X} - \mathbf{x}_o) = \left(\mathbf{N} + \mathbf{H}^T\boldsymbol{\Sigma}_{\mathbf{X}}^{-1}\mathbf{H}\right)\hat{\boldsymbol{\xi}}^c - \mathbf{u} \tag{23}$$

By substituting the last expression into Eq. (21) and after performing straightforward operations, we get

$$\left(\mathbf{I} + \boldsymbol{\Sigma}_{\xi'}\mathbf{H}^T\boldsymbol{\Sigma}_{\mathbf{X}}^{-1}\mathbf{H}\right)\hat{\boldsymbol{\xi}} = \boldsymbol{\xi}' + \mathbf{E}^T\hat{\boldsymbol{\theta}} + \left(\mathbf{I} + \boldsymbol{\Sigma}_{\xi'}\mathbf{H}^T\boldsymbol{\Sigma}_{\mathbf{X}}^{-1}\mathbf{H}\right)\hat{\boldsymbol{\xi}}^c + \left(\boldsymbol{\Sigma}_{\xi'}\mathbf{N} - \mathbf{I}\right)\hat{\boldsymbol{\xi}}^c - \boldsymbol{\Sigma}_{\xi'}\mathbf{u} \tag{24}$$

Considering that the network solution $\boldsymbol{\xi}'$ in the initial frame is obtained by a free-net adjustment (see Eq. (10)) we have $\mathbf{N}\,\boldsymbol{\xi}' = \mathbf{u}$ and $\boldsymbol{\Sigma}_{\xi'} = \mathbf{N}^{-1}$, and thus the last equation yields

$$\left(\mathbf{I} + \boldsymbol{\Sigma}_{\xi'}\mathbf{H}^T\boldsymbol{\Sigma}_{\mathbf{X}}^{-1}\mathbf{H}\right)\hat{\boldsymbol{\xi}} = \mathbf{E}^T\hat{\boldsymbol{\theta}} + \left(\mathbf{I} + \boldsymbol{\Sigma}_{\xi'}\mathbf{H}^T\boldsymbol{\Sigma}_{\mathbf{X}}^{-1}\mathbf{H}\right)\hat{\boldsymbol{\xi}}^c \tag{25}$$

or equivalently

$$\begin{aligned} \hat{\boldsymbol{\xi}} &= \hat{\boldsymbol{\xi}}^c + \left(\mathbf{I} + \boldsymbol{\Sigma}_{\xi'}\mathbf{H}^T\boldsymbol{\Sigma}_{\mathbf{X}}^{-1}\mathbf{H}\right)^{-1}\mathbf{E}^T\hat{\boldsymbol{\theta}} \\ &= \hat{\boldsymbol{\xi}}^c + \left(\mathbf{N}^{-1}\mathbf{N} + \mathbf{N}^{-1}\mathbf{H}^T\boldsymbol{\Sigma}_{\mathbf{X}}^{-1}\mathbf{H}\right)^{-1}\mathbf{E}^T\hat{\boldsymbol{\theta}} \\ &= \hat{\boldsymbol{\xi}}^c + \left(\mathbf{N} + \mathbf{H}^T\boldsymbol{\Sigma}_{\mathbf{X}}^{-1}\mathbf{H}\right)^{-1}\mathbf{N}\,\mathbf{E}^T\hat{\boldsymbol{\theta}} \end{aligned} \tag{26}$$

Taking into account Eqs. (16) and (22) we finally have

$$\begin{bmatrix} \hat{\mathbf{x}} \\ \hat{\mathbf{z}} \end{bmatrix} = \begin{bmatrix} \hat{\mathbf{x}}^c \\ \hat{\mathbf{z}}^c \end{bmatrix} + \left(\mathbf{N} + \mathbf{H}^T\boldsymbol{\Sigma}_{\mathbf{X}}^{-1}\mathbf{H}\right)^{-1}\mathbf{N}\,\mathbf{E}^T\hat{\boldsymbol{\theta}} \tag{27}$$

which concludes our proof.

5 Conclusions

The mapping problem of a network solution to a target frame through the HT approach was discussed in this paper. It has been shown that the optimal coordinates in the target frame (in the sense of minimum estimation error variance) can be computed by a closed-form expression in terms of appropriate corrections to the standard HT estimator which is commonly used in practice. Our revised estimator is easy to implement and it does not require any additional matrix inversion other than the one already used by the classic stepwise solution of Eqs. (2) and (3). Furthermore, it was shown that its difference with the weighted CNA estimator depends on the chosen transformation model and, in particular, on the linear combination of the columns of the matrix $\mathbf{NE}^T$; see Eq. (12). This is a useful result as it implies the equivalency of the two general approaches for network densification, regardless of the number of the used reference stations, as long as the frame transformation model ($\boldsymbol{\theta}$, $\mathbf{E}$) employs only the parameters related to the rank defect of the underlying network.

The advantage of the optimal HT approach presented here is the minimization of the propagated noise from the initial network solution to the estimated coordinates in the target frame. This is accomplished by a filtering step within the transformation procedure which exploits the network's known covariance structure in both frames; see Eq. (8). It is again noted that any unmodeled non-secular signals in the initial coordinates will be affected by such filtering during their transfer to the target frame. Hence, the methodology described in this paper is not tuned in the analysis of loading signals with respect to a secular reference frame, although the aforementioned noise reduction is a critical and worth-considering aspect in support of geophysical signal detection in ITRF-aligned coordinate time series.

Acknowledgements The authors would like to acknowledge the valuable comments provided by Z. Altamimi, G. Blewitt and two anonymous reviewers, which have significantly improved the contents of this paper.

References

Altamimi Z (2003) Discussion on how to express a regional GPS solution in the ITRF, EUREF publication no. 12. Verlag des Bundesamtes für Kartographie und Geodäsie, Frankfurt am Main, pp 162–167

Altamimi Z, Collilieux X, Métivier L (2011) ITRF2008: an improved solution of the international terrestrial reference frame. J Geod 85:457–473

Bevis M, Brown A (2014) Trajectory models and reference frames for crustal motion geodesy. J Geod 88:283–311

Blaha G (1971) Inner adjustment constraints with emphasis on range observations, OSU report no. 148. Department of Geodetic Science, The Ohio State University, Columbus, Ohio

Bruyninx C, Altamimi Z, Caporali A, Kenyeres A, Lidberg M, Stangl G (2013) Guidelines for EUREF densifications (ver. 5). IAG subcommission for the European Reference Frame. ftp://epncb.oma.be/pub/general/Guidelines_for_EUREF_Densifications.pdf

Gurtner W, Boucher C, Bruyninx C, van der Marel H (1997) The use of the IGS/EUREF permanent network for EUREF densification campaigns, EUREF publication no. 6. Veröffentlichungen der Bayerischen Kommission für die Internationale Erdmessung, Bayerischen Akademie der Wissenschaften, Munich, pp 50–51

Kotsakis C, Vatalis A, Sanso F (2014) On the importance of intra-frame and inter-frame covariances in frame transformation theory. J Geod 88:1187–1201

Schaffrin B (1983) Model choice and adjustment techniques in the presence of prior information, OSU report no. 351. Department of Geodetic Science and Surveying, Ohio State University, Columbus, Ohio

Sillard P, Boucher C (2001) A review of algebraic constraints in terrestrial reference frame datum definition. J Geod 75:63–73

Tregoning P, van Dam T (2005) Effects of atmospheric pressure loading and seven-parameter transformations on estimates of geocenter motion and station heights from space geodetic observations. J Geophys Res 110:B03408. doi:10.1029/2004JB003334

A Study on the Impact of Reference Frame Implementation Strategy on GNSS Time Series for Regional Network Analysis

Miltiadis Chatzinikos and Athanasios Dermanis

Abstract

Coodinate time series from a regional GNSS network in Greece covering a period of 5 years are used in order to study the effect of the choice of the reference system on their linear, nonlinear and spectral characteristics. The standard solution where the reference system is defined by removing the translational rank defect in GNSS data by partial inner constraints is compared with two different solutions. The first is obtained by a posteriori aligning the network at each epoch to the IGS08 reference system through the coordinates of common points by means of a similarity (Helmert) transformation. The second solution is achieved by a regional stacking where the original standard coordinate time series are best fit to a linear in time coordinate model and the derived similarity transformation parameters are used to convert the standard solution into coordinate time series where variations due to reference system instability are removed. The analysis shows that the stacking solution leads to better results more suitable for regional geodynamic studies free from effects, which are not reflecting actual temporal variations in the shape of the network.

Keywords

Coordinate time series • GNSS network • Helmert transformation • Reference system • Stacking

1 Introduction

Temporal variations of network station coordinates obtained by GNSS or other space techniques are the sum of three components: (a) true variation in the shape of the relevant geodetic network, which is the geophysical signal of actual interest, (b) effect of remaining systematic observational or modeling errors and (c) temporal variation of the reference system to which coordinates refer. Ideally, coordinates should refer at all epochs to the same reference system, but this requirement is not only unfeasible but also void of meaning in the case of a deformable network. Indeed while the choice of reference system for a rigid network at a single epoch automatically defines a reference system for all epochs, in the case of a deformable network the reference system must be chosen for each observation epoch. The reference system is optimal if its motion in inertial space best reflects the motion of the network as a whole, by minimizing the remaining apparent motions of the stations (Altamimi and Dermanis 2012, 2013). This allows the separation of the inertial variation of the network point positions into a common part that represents earth orbit, rotation and a minimal remaining part representing network deformation. The means by which the temporal evolution of the reference system is chosen (its initial epoch placement being a purely conventional matter) has a direct effect on the resulting coordinate time series. This fact is not always very clear when various types of analysis are applied on them, e.g. the extraction of linear trend and additional nonlinear periodic or quasi-periodic terms. Within a GNSS separate

M. Chatzinikos (✉) • A. Dermanis
Department of Geodesy and Surveying, Aristotle University of Thessaloniki, Thessaloniki, Greece
e-mail: mchatzin@topo.auth.gr

T. van Dam (ed.), *REFAG 2014*, International Association of Geodesy Symposia 146,
DOI 10.1007/1345_2015_206

network adjustment for each epoch, the coefficient matrix of the normal equations is surprisingly numerical invertible, a fact which reflects no lack in the three components of the reference system definition (origin, orientations, scale). However, an analysis of its eigenvalues demonstrates a strong weakness in the definition of the origin. This is a known fact in GNSS analysis, since the detection of geocenter motion is very weak (see e.g. Rebischung 2014, for a theoretical explanation). For the normal equations matrix of each weekly solution of our regional network the ratios $a_i = \lambda_i/\lambda_{max}$ of the eigenvalues with respect to the larger one are very small of the order of 10^{-4} for only the three smaller ones. It is easy to verify that this is due to a translation defect by checking that the well-known basis of the null space of the design matrix corresponding to a translational network defect belongs to the span of the eigenvectors with close to zero eigenvalues. More details on how to give physical explanation to numerically detected defects will be given in a future work. For this reason, it is advisable in the EUREF guidelines (Bruyninx et al. 2012, Sect. 4.3) to treat the matrix as having a rank defect of three and use a proper set of minimal constraints. It is interesting to note that the a_i values for $i = 4, \ldots, 20$ are also small of the order of 10^{-2} before ascending to 10^{-1} values and larger.

In the present study a geodetic network of 17 selected GNSS stations (Fig. 1) in the tectonically active area of Greece are analyzed for geodynamic purposes. The network includes four stations which are also official EUREF stations and they can be used for the connection with the IGS08 reference frame (Rebischung et al. 2011). Daily GPS data were processed using Bernese GPS software v5.0 (Dach et al. 2007) following the standard method. Absolute antenna phase center corrections (IGS08.atx) were applied and two different ambiguity resolution strategies were used depending on the length of baselines (SIGMA:<200 km, QIF: >200 km). IGS final orbits (sp3), which are also known as

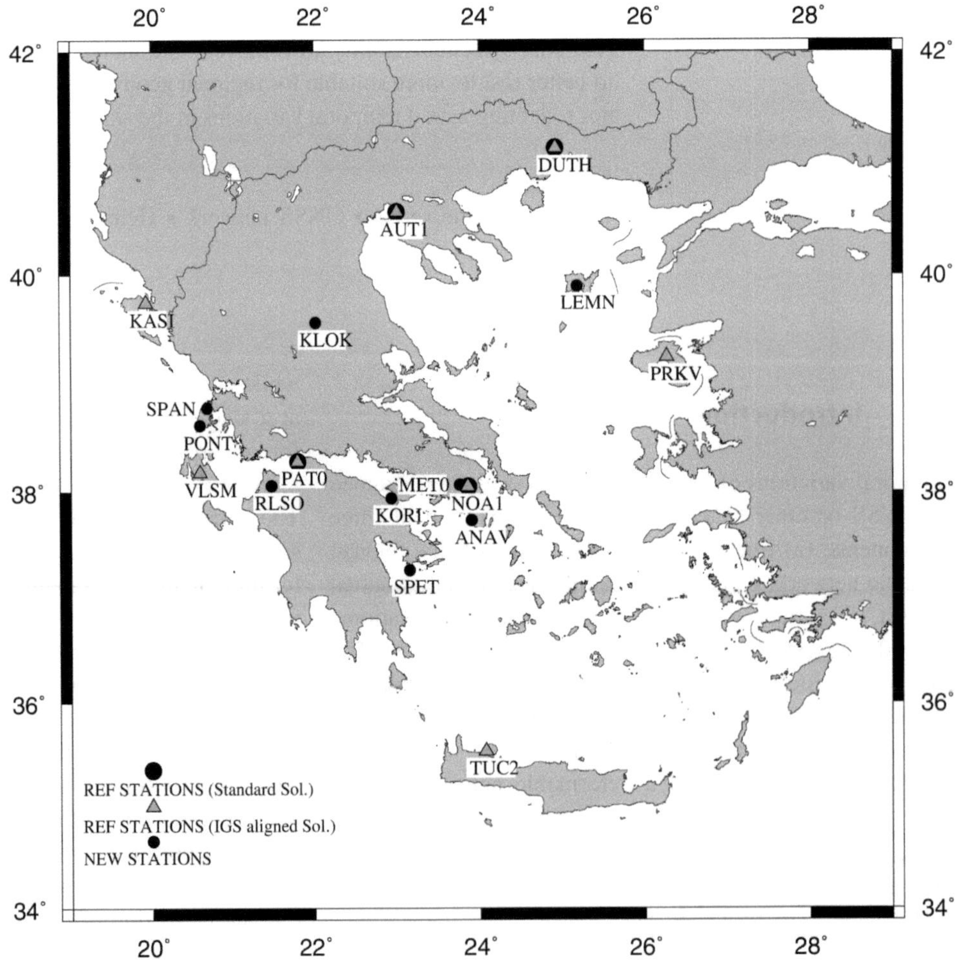

Fig. 1 The 17 GNSS network and the stations used in standard and IGS aligned solutions

"fiducial free orbits", and the associated earth orientation parameters where held fixed. In addition IGS station coordinates were minimally constrained by imposing a no-net translation condition. Each one of the final daily solutions was derived using the ionosphere free linear combination, estimating daily coordinates and 12 tropospheric parameters.

In order to study the effect of the choice of reference system on coordinate time series it is essential that all coordinate time series under comparison have the same network shape at each single epoch. This automatically excludes the use of data from stations external to the network, as it is usually done in order to align a regional network to a global one. Use of such data will lead to combined network solutions which will have different network shape at single epochs depending on how many and which particular external stations were chosen.

Within the above restrictions we study three different choices of reference system definition:

1. The first choice is to make the weekly coordinates of the four EUREF stations as close as possible to their coordinates obtained by linear interpolation based on their IGS08 initial coordinates and constant velocities. Let us note that these interpolated coordinates neglect the non-negligible nonlinear terms, which are certainly present in the weekly solutions. We will refer to this choice as the "standard solution".
2. The second choice, which we call "regional stacking solution", is an intrinsic definition of the reference system based on the stacking of the previously obtained coordinate time series. The optimal reference system is chosen through the usual 14 inner constraints for initial translation, rotation, scale and their corresponding rates. Such algebraically derived inner constraints strongly depend on the approximate values used, which in our case were zero values for the velocities and the mean of all weekly coordinates for the initial epoch coordinates for each station. This approach benefits from the well-known fact that stacking produces coordinates series with minimal variation (see for e.g. Blewitt et al. 2013; Bevis and Brown 2014) and this is the reason why it is an indispensable part of any global network analysis such as the ITRF (Altamimi et al. 2007, 2011). When applied to regional networks it provides a means for focusing on regional geodynamic studies, where only the variation of the local network shape is important, while its motion as whole with respect to a global reference system (i.e. with respect to the rest of the earth) is of secondary importance. In addition to the removal of the actual total network motion, the regional stacking approach has the advantage of removing any artifacts of a similar nature, which do not reflect real geophysical processes but are the result of instabilities in the reference system realization within GNSS data analysis, caused by systematic and modeling errors.
3. For a third choice we sought a better alignment to the IGS08. To achieve this, we utilized existing weekly coordinates of eight network stations which in some sense refer to IGS08, because they are the results of the weekly solutions of a combined network with 108 regional stations and 17 optimal distributed EUREF stations in the wider surrounding area (Chatzinikos 2013; Chatzinikos et al. 2013). The original coordinates of each weekly solution were aligned to the corresponding coordinates of the combined solution of the eight common stations through a best fit by a Helmert (7-parameter similarity) transformation. We will refer to this choice as the "IGS aligned solution".

Apart from the stacking approach which is well-known to give optimal results the other two choices we believe that reflect standard scenarios for producing coordinate time series.

By the way let us note that the alignment of a regional network to a global solution through an overlapping subnetwork does not automatically guarantee "inheritance" of the reference system of the global network. A suboptimal solution to this theoretically very interesting problem has been given by Kotsakis et al. (2014). A more complete solution has been presented by Dermanis (2013).

2 Standard, IGS Aligned and Stacking Solution for Coordinate Time Series

Standard solution: Following the EUREF guidelines (Bruyninx et al. 2012, Sect. 4.3) the standard solution is obtained by adjusting the GNSS data of every epoch t_k (weekly solutions) with additional partial (translational) inner constraints on the coordinates $\boldsymbol{x}_i(t_k)$ of the four stations (i = AUT1, DUTH, NOA1, PATO) which belong to the EUREF network.

The partial inner constraints used in every epoch t_k are the translational constraints,

$$\sum\nolimits_{i=1}^{4} \left\{x_i\left(t_k\right) - \boldsymbol{x}_i^{IGS08}\left(t_k\right)\right\} = 0 \qquad (1)$$

where $x_{i_0}^{IGS08}(t_k)$ are the linearly interpolated values

$$x_i^{IGS08}(t_k) = x_{i_0}^{IGS08} + (t_k - t_0)\, v_i^{IGS08}, \quad i = 1, 2, 3, 4, \tag{2}$$

$x_{i_0}^{IGS08}$ and v_i^{IGS08} being the initial coordinates and velocities, respectively, in the IGS08 reference frame.

IGS aligned solution: In the IGS aligned solution for each epoch t_k, the coordinates $x_i = x_i(t_k)$ of eight stations from the standard solution are aligned to their counterparts $x_i^{IGS08} = x_i^{IGS08}(t_k)$ through a linearized similarity (Helmert) transformation,

$$x_i^{IGS08} = x_i + \lambda x_i + [x_i \times]\,\theta + d + e_i, \quad i = 1, 2, \ldots, 8, \tag{3}$$

where λ, $\theta = \left[\theta_X\ \theta_Y\ \theta_Z\right]^T$, $d = \left[d_X\ d_Y\ d_Z\right]^T$ are the scale parameter, the three rotation angles and the three translation components, respectively, of the original non-linear similarity transformation $\tilde{x} = (1+\lambda)\, R(\theta)\, x + d$ and e_i the corresponding residuals expressing differences in the shape of the sub-network of the used points. The optimal estimates of the transformation parameters $\widehat{\lambda}, \widehat{\theta}, \widehat{d}$ are obtained by a least squares fit $\left(\sum_i e_i^T e_i = \min\right)$ and are then used to convert the reference coordinates x_i (i = 1, ...,17) into a new coordinate time series $\tilde{x}_i = x_i + \widehat{\lambda} x_i + [x_i \times]\,\widehat{\theta} + \widehat{d}$.

Regional stacking solution: Regional stacking is based on the same linear transformation model $x_i(t_k) = \overline{x}_i(t_k) + \lambda \overline{x}_i(t_k) + [\overline{x}_i(t_k)\times]\,\theta + d + e_i$ where now $x_i(t_k)$ are the coordinate time series of the standard solution, while the coordinates $\overline{x}_i(t_k)$ are forced to obey a linear-in-time model $\overline{x}_i = x_{0i} + (t - t_0)\, v_i$ with reference epoch coordinates $x_{0i} = \overline{x}_i(t_0)$ and constant velocities v_i. Replacing and neglecting second order terms lead to the observation equations

$$\begin{aligned} x_i(t_k) = x_{0i} + (t_k - t_0)\, v_i + \lambda x_{0i} + [x_{0i}\times]\,\theta \\ + d + e_i, \quad i = 1, \ldots, 17, \end{aligned} \tag{4}$$

where e_i are in this case discrepancies in the $x_i(t_k)$ sequence of network shapes from a linear model. The usual least square solution with 14 inner constraints (Altamimi and Dermanis 2012, 2013) provides estimates of $\widehat{\lambda}, \widehat{\theta}, \widehat{d}$, $(\widehat{x}_{0i}, \widehat{v}_i)$ which are used in the inverse transformation of the $x_i(t_k)$ coordinates into a new "stacked" coordinate time series $x_i{}'(t_k) = x_i(t_k) - \widehat{\lambda} x_i(t_k) - [x_i(t_k)\times]\,\widehat{\theta} - \widehat{d}$.

Instead of the usual East, North and Vertical coordinates, we replace the East and North directions, which unlike the vertical one have no geophysical significance, with two new perpendicular horizontal directions, different for each station, which correspond to the maximum and minimum coordinate variation. To do this the east E_k and north N_k coordinates of a station at the various epochs t_k are considered independently of their time of occurrence as samples with means $m_E = \Sigma_k E_k/M$, $m_N = \Sigma_k N_k/M$, M being the number of epochs and dispersion matrix

$$\begin{aligned} S &= \begin{bmatrix} S_E^2 & S_{EN} \\ S_{EN} & S_N^2 \end{bmatrix} \\ &= \frac{1}{M}\sum_k \begin{bmatrix} (E_k - m_E)^2 & (E_k - m_E)(N_k - m_N) \\ (N_k - m_N)(E_k - m_E) & (N_k - m_N)^2 \end{bmatrix}. \end{aligned} \tag{5}$$

Diagonalization of the dispersion matrix ($\lambda_1 \geq \lambda_2$)

$$S = R(-\theta)\begin{bmatrix} \lambda_1 & 0 \\ 0 & \lambda_2 \end{bmatrix} R(\theta), \tag{6}$$

reveals the angle θ of the direction of maximal dispersion and the horizontal coordinates are transformed into ones along the directions of maximal and minimal coordinate variability according to

$$\begin{bmatrix} H_{max} \\ H_{min} \end{bmatrix} = R(\theta)\begin{bmatrix} E \\ N \end{bmatrix} = \begin{bmatrix} \cos\theta & \sin\theta \\ -\sin\theta & \cos\theta \end{bmatrix}\begin{bmatrix} E \\ N \end{bmatrix}. \tag{7}$$

The replacement of the East–west directions with the ones of maximum-minimum variations may have the disadvantage that different directions are used for each station, but on the other hand, the concentration of horizontal position variability in the maximal component has the advantage of magnification of the relevant physical process for better comparison of different solutions. In any case we compare the coordinate time series from the different solutions for each station separately and not ones belonging to different stations.

One may object to the use of identity weight matrices in the above least squares best fits, instead of the standard use of the inverse of the error covariance matrix, which give optimal results, i.e., best (= minimum least square error), linear, uniformly (whatever the true parameter values) unbiased estimates, for the unknown parameters and all the linear functions, according to the Gauss-Markov theorem.

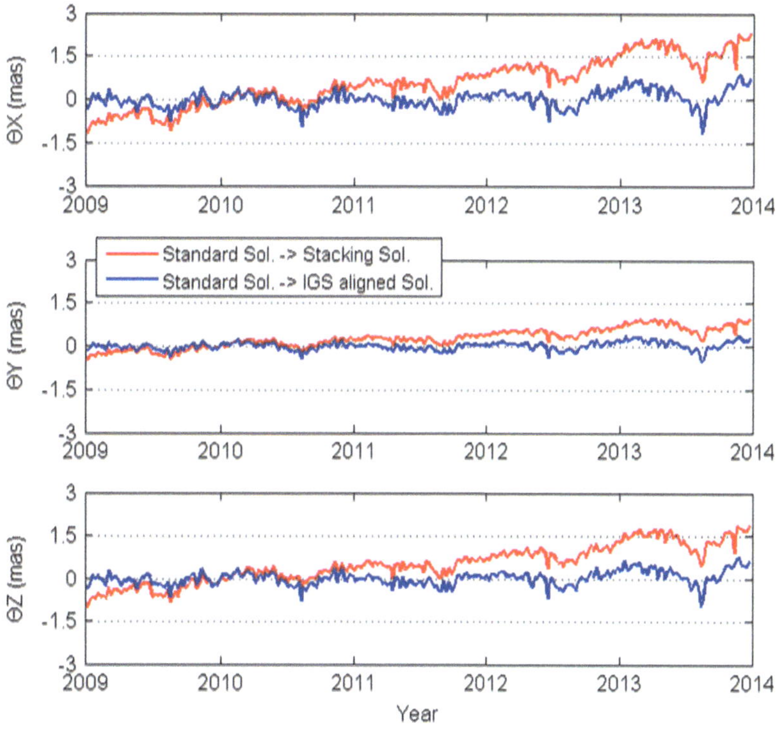

Fig. 2 Time series of the three rotation angles transforming the standard solution to the IGS aligned solution (*blue*) and to the stacking solution (*red*) (1 mas ≈ 31 mm on the earth surface)

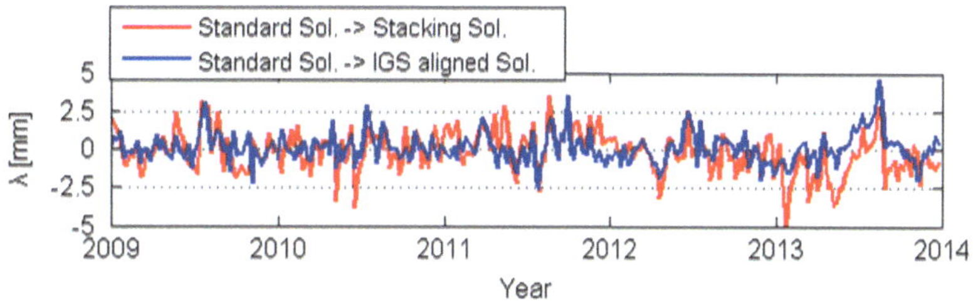

Fig. 3 Time series of the scale transforming the standard and the IGS aligned solutions to the stacking solution (*red*)

Leaving apart the problem that the input covariance matrices if correctly computed would be singular, the reason for our unit weight matrix choice is the fact that the prerequisites of the Gauss-Markov theorem do not apply here. Errors are not zero-mean due to existing biases and their true covariance matrices are unknown; the ones provided by various GNSS software are too optimistic to be taken seriously. For the stacking problem in particular, the post-linear error residuals have certainly non-zero means, namely the significant non-linear temporal variations, which are due to underlying geophysical processes and are thus of a deterministic rather than of a stochastic nature.

For the same reason we consciously abstain from any error analysis that would provide signal-to-noise ratios for the estimated parameters. To compute the covariance matrices of parameter estimates through the law of covariance propagation, we need the true (or at least good estimates) of the input error covariances, which are in reality unknown. Realistic quality assessment in GNSS data analysis remains a challenging problem.

3 Comparison of the Derived Solutions

The time series of rotation transformation parameters θ_X, θ_Y, θ_Z and those for scale λ, from the standard solution to the IGS aligned and stacking solutions, are depicted in Figs. 2 and 3 respectively. Rotation angles are the parameters of largest significance, with scale being less significant, while translation values remain below the noise level. Rotation angles demonstrate a significant linear trend for the stacking solution, which is then removed from the modified coordinate series. The larger values occur for θ_X and are about three times larger than the smaller ones of θ_Y. The derived rotation time series are larger for the stacking solution, though the relevant signals maintain the same form. In addition to a linear trend, they demonstrate periodic fluctuations, which are better understood in the corresponding spectra of the nonlinear variations depicted in Fig. 4. In order to be comparable to the spectra of station coordinate time series, the original angular time series in

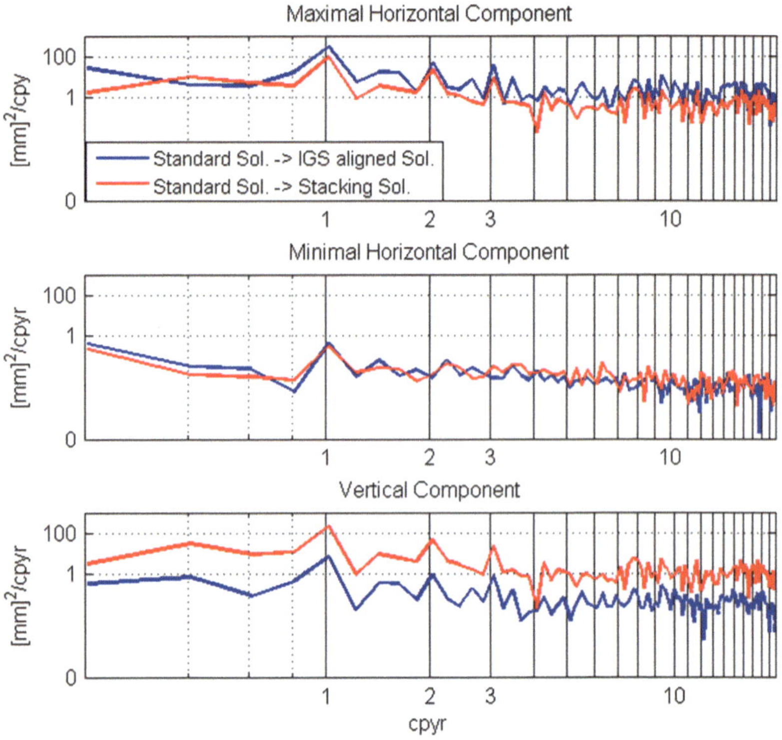

Fig. 4 Spectra of the displacements caused by the three rotation angle time series depicted in Fig. 2, after linear trend removal, for the IGS aligned solution (*blue*) and the stacking solution (*red*)

radians have been converted to equivalent displacements in vertical, maximal horizontal and minimal horizontal directions on the earth surface, by multiplying with the distance of the network barycentre from the corresponding rotation axis.

Both of the stacking and the IGS aligned solutions demonstrate an almost annual period (360 days), an almost semi-annual period (179 days) and an additional 4-month period (119 days), mostly seen in the vertical and the maximal horizontal direction, while the signal is negligibly weak in the minimal horizontal component. These signals are stronger in the stacking solution for the vertical component and somewhat smaller for the maximal horizontal component.

For each type of solution $\boldsymbol{x}_i(t_k) = \widehat{\boldsymbol{x}}_{0i} + (t_k - t_0)\widehat{\boldsymbol{v}}_i + \widehat{\boldsymbol{\varepsilon}}_i(t_k)$, as a measure of the non-linear coordinate variability in each station i after removal of the best fitted linear trend $(\widehat{\boldsymbol{x}}_{0i} + (t_k - t_0)\widehat{\boldsymbol{v}}_i)$, we use the "rms" $\widehat{\sigma}_i$ computed by $\widehat{\sigma}_i^{\,2} = \left(\sum_k \widehat{\boldsymbol{\varepsilon}}_i^{\,2}(t_k)\right)/f$, where $\widehat{\boldsymbol{\varepsilon}}_i$ are the estimated post linear residuals and f the degrees of freedom. The results for the three solutions under comparison are depicted in Fig. 5 for all stations. As expected, the largest rms appears in the vertical and the lower one in the minimal horizontal component, with the vertical variation being more than double the (maximal) horizontal one. Both the IGS aligned and stacking solutions perform better that the standard one, except for the four EUREF stations constrained in the standard solution, where as expected there is a smaller repeatability for the reference solution. In general, the IGS aligned solution gives results, which strongly depend on the distribution of the stations used for the determination of the 7-transformation parameters, while the stacking solution does not require such a choice, and gives a more homogeneous treatment of the whole network. The stacking solution produces coordinate time series with systematically smaller residuals than the IGS aligned one.

In order to see the effect of the solution on particular stations we have chosen two characteristic stations, one with the smaller tectonic activity (station KLOK in central Greece) and one with the largest one (station SPAN in the island of Lefkas). The post-linear residuals $\widehat{\boldsymbol{\varepsilon}}_i(t_k)$ of the three solutions for station KLOK are depicted in Fig. 6. Both the IGS aligned and stacking solutions are much different from the standard solution, as far as the horizontal components are concerned, while all three are similar for the vertical component. In all cases, the post-linear residuals in the stacking solution demonstrate smaller variation. The corresponding spectra are depicted in Fig. 7. Only the annual signal for the vertical solution is significant and of similar power in all three solutions. A much less powerful annual signal appears in the maximal horizontal component, which is even weaker for the stacking solution.

The post-linear residuals of the three solutions for the most tectonically active station SPAN are depicted in Fig. 8. All signals appear very similar with the stacking solution having smaller variability. The corresponding spectra are depicted in Fig. 9. Only an annual significant signal is seen for both the vertical and the maximal horizontal components, with almost the same power for all three solutions.

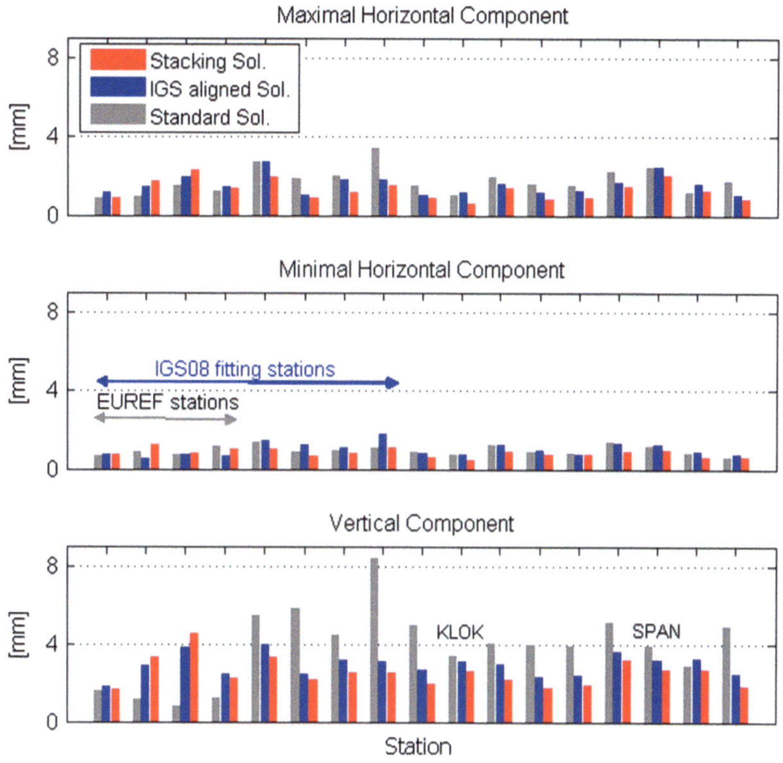

Fig. 5 Rms of post-linear residuals after removal of linear trend for the standard, IGS08 aligned and stacking solutions

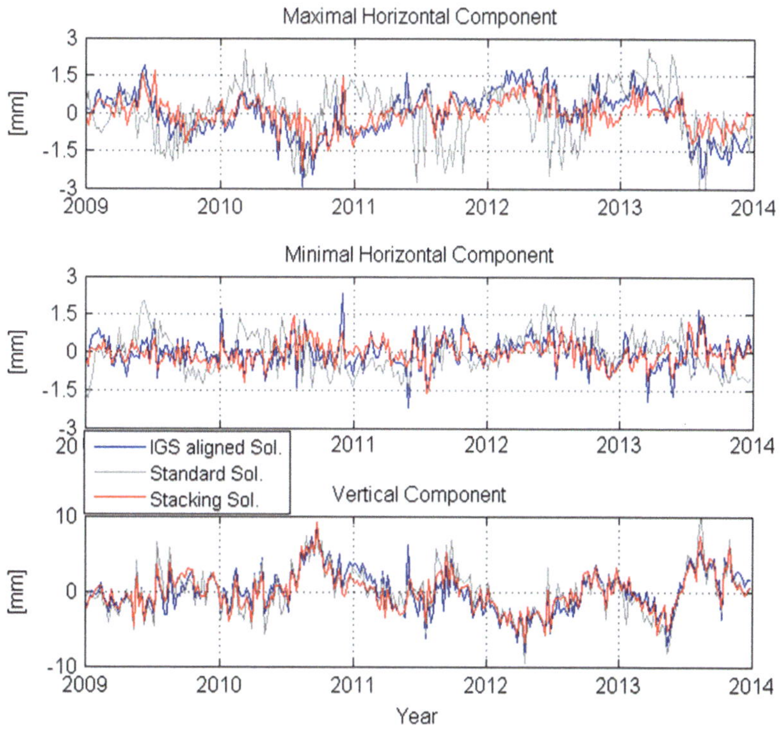

Fig. 6 Post-linear residuals of the station KLOK for the three solutions: standard (*grey*), IGS aligned (*blue*) and stacking (*red*)

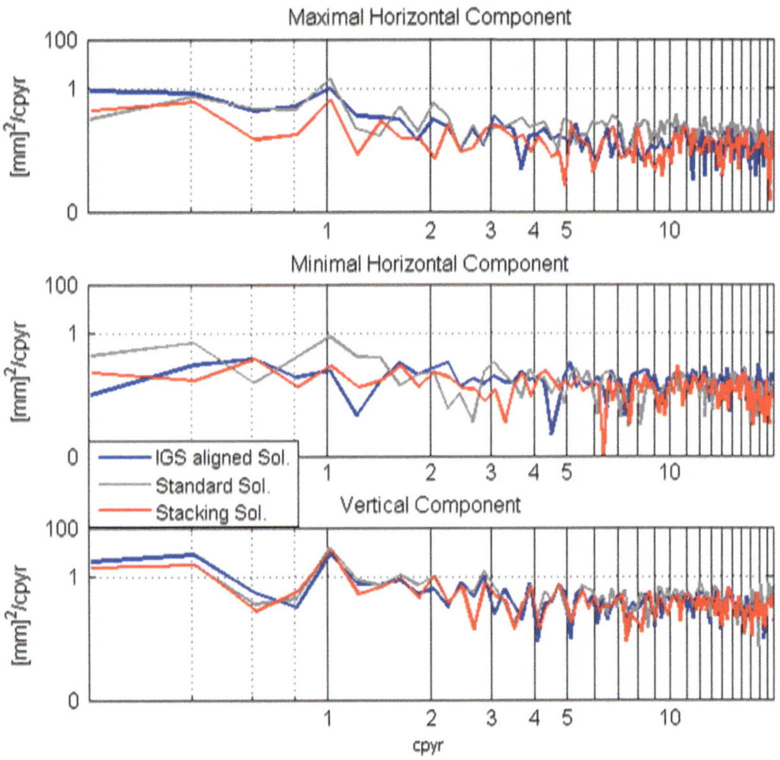

Fig. 7 Spectra of the post-linear residuals of the station KLOK depicted in Fig. 5 for the three solutions: standard (*grey*), IGS aligned (*blue*) and stacking (*red*)

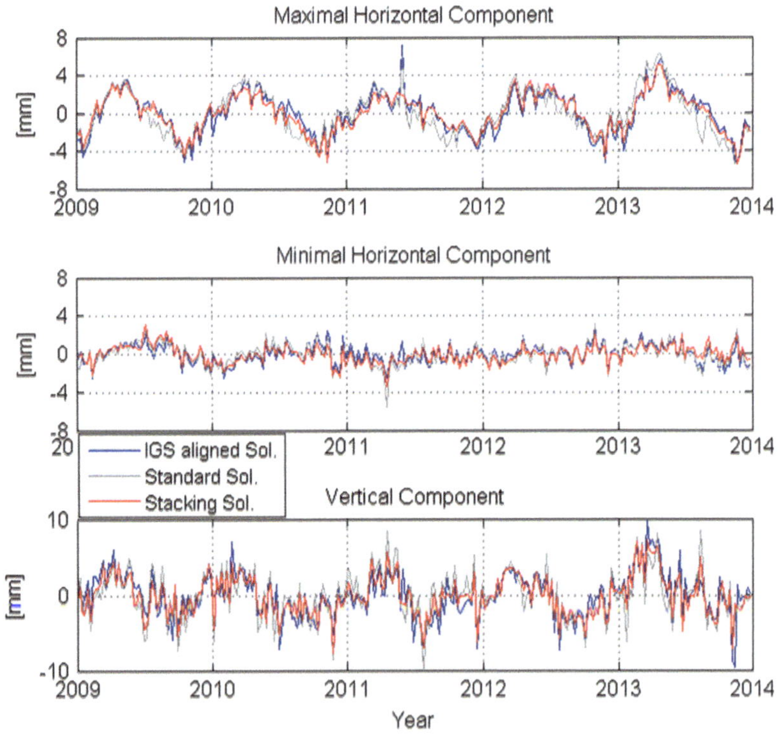

Fig. 8 Post-linear residuals of the station SPAN for the three solutions: standard (*grey*), IGS aligned (*blue*) and stacking (*red*)

4 Conclusions

It has been demonstrated that the form of coordinate time series used in local tectonic studies strongly depends on the chosen reference system. The standard solution according to EUREF guidelines suffers from apparent variations which are due to temporal instability of the used reference system and do not correspond to actual variations in network shape. To overcome this problem two different approaches have been studied. The first is a Helmert transformation fit to a global network (in this case the IGS08) and the second

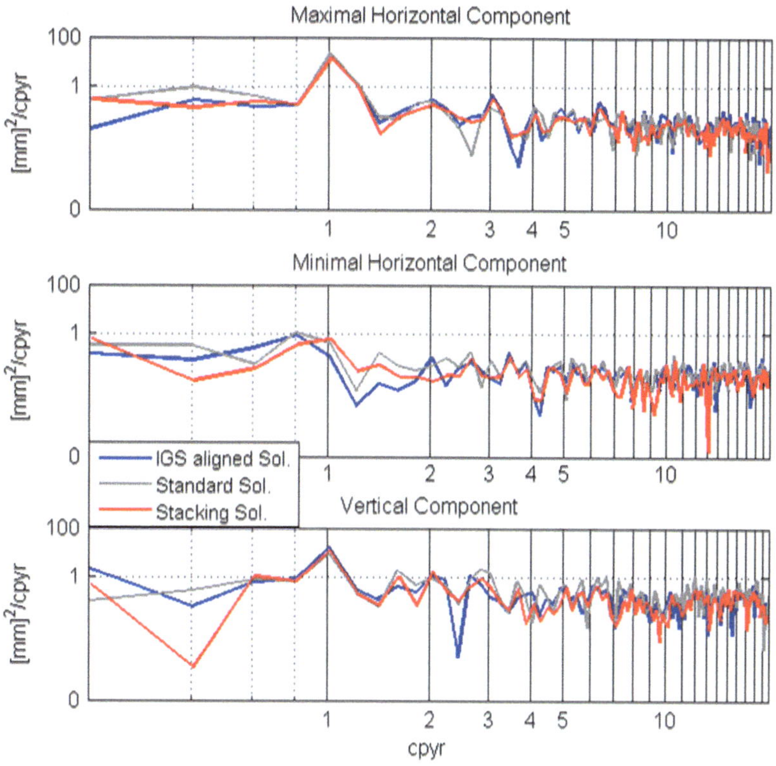

Fig. 9 Spectra of the post-linear residuals of the station SPAN depicted in Fig. 7 for the three solutions: standard (*grey*), IGS aligned (*blue*) and stacking (*red*)

a regional stacking similar to the one applied in the ITRF solution (Altamimi et al. 2007). The stacking approach demonstrates an ability to remove coordinate variations, which may be due to either the total motion of the region under study or to the effect of systematic observational errors causing temporal reference system instability.

The superiority of the stacking strategy is obviously due to the fact that the applied transformations at each epoch seek conformity to a linear-in-time model with simultaneously estimated parameters. Thus the optimal linear-in-time model is chosen among all possible ones. On the contrary the other techniques try to fit the transformation parameters to rather prescribed linear models, those implicit in the IGS08 coordinates of the chosen network stations for best fit. The standard solution has also the disadvantage that only translations are allowed assuming that scaling orientation is provided by the GNSS observables.

These results are more or less expected since stacking is well-known to provide smoother coordinate time series in general. Here however stacking is applied to a regional network and for this reason it removes, in addition to the usual effect of instabilities in reference system definition, additional effects that may correspond to either the motion of the network as a whole, or similar artifacts caused by systematic and modeling errors in GNSS data analysis.

References

Altamimi Z, Dermanis A (2012) The choice of reference system in ITRF formulation. In: Sneeuw N et al (eds) VII Hotine-Marussi symposium on mathematical geodesy, vol 137, International Association of Geodesy, symposia. Springer, Berlin, pp 329–334

Altamimi Z, Dermanis A (2013) Theoretical foundations of ITRF determination. The algebraic and the kinematic approach. In: Katsampalos KV, Rossikopoulos D, Spatalas S, Tokmakidis D (eds) On measurements of lands and constructions. Volume in honor of Prof. D. Vlachos, Publication of the School of Rural & Surveying Engineering, Aristotle University of Thessaloniki, pp 331–359. http://der.topo.auth.gr

Altamimi Z, Collilieux X, Legrand J, Garayt B, Boucher C (2007) ITRF2005: a new release of the International Terrestrial Reference Frame based on time series of station positions and Earth Orientation Parameters. J Geophys Res 112:B09401

Altamimi Z, Collilieux X, Métivier L (2011) ITRF2008: an improved solution of the international terrestrial reference frame. J Geod 85:457–473

Bevis M, Brown A (2014) Trajectory models and reference frames for crustal motion geodesy. J Geod 88(3):283–311

Blewitt G, Kreemer C, Hammond WC, Goldfarb JM (2013) Terrestrial reference frame NA12 for crustal deformation studies in North America. J Geodyn 72:11–24

Bruyninx C, Altamimi Z, Caporali A, Kenyeres A, Lidberg M, Stangl G, Torres J (2012) Guidelines for EUREF densifications. (ftp://epncb.oma.be/pub/general/Guidelines_for_EUREF_Densifications.pdf)

Chatzinikos M (2013) Study of the earth's crust displacements in the area of Greece analyzing GNSS data. Ph.D. thesis, School of Rural and Surveying Engineering, Aristotle University of Thessaloniki, Greece

Chatzinikos M, Fotiou A, Pikridas C, Rossikopoulos D (2013) The realization of a semi-kinematic datum in Greece including a new velocity model. In: Proceedings of the IAG scientific assembly, 1–6 Sept, Potsdam, Germany, IAG symposia 143. Springer, Berlin

Dach R, Hugentobler U, Fridez P, Meindl M (2007) Bernese GPS software version 5.0. Astronomical Institute, University of Bern, Bern

Dermanis A (2013) Merging local networks to the ITRF. A coordinate free approach (almost). VIII Hotine-Marussi symposium, Rome, Italy, 17–21 June 2013

Kotsakis C, Vatalis A, Sanso F (2014) On the importance of intra-frame and inter-frame covariances in frame transformation theory. J Geod 88(12):1187–1201

Rebischung P (2014) Can GNSS contribute to improving the ITRF definition? Ph.D. thesis, Observatory of Paris, France

Rebischung P, Griffiths J, Ray J, Schmid R, Collilieux X, Garayt B (2011) IGS08: the IGS realization of ITRF2008. GPS Solutions 1–12. doi:10.1007/s10291-011-0248-2

Validation of Components of Local Ties

Susanne Glaser, Mathias Fritsche, Krzysztof Sośnica, Carlos Javier Rodríguez-Solano, Kan Wang, Rolf Dach, Urs Hugentobler, Markus Rothacher, and Reinhard Dietrich

Abstract

Local ties (LTs) at co-located sites are currently used to align different space geodetic techniques for the determination of a global terrestrial reference frame (TRF). However, the currently available LT measurements are typically characterized by an inhomogeneous accuracy, which may cause inconsistencies within the TRF and limit the final TRF accuracy. An alternative strategy is a combination of common parameter types to which the individual geodetic techniques are sensitive. In this study, we combine Global Navigation Satellite Systems (GNSS) and Satellite Laser Ranging (SLR) data without using LTs but by combining the common pole coordinates and by adding proper datum constraints. In addition, we constrain the velocities at co-located sites to be the same for all markers. This allows an independent validation of measured LT components. Our data are based on a homogeneous reprocessing of GPS+GLONASS and SLR to LAGEOS-1 and LAGEOS-2 over 17 years in the time span of 1994–2010. A preliminary analysis including the elimination of outliers and the selection of core datum stations was performed based on the station position time series of the single-technique solutions. Applying our combination approach, the north and height components of the LTs can be directly derived from our combined coordinate solution. The differences of the measured and the estimated LTs remain below 1 cm for 96% in the north component and for 50% in the height component of all co-located sites.

Keywords

GNSS • Inter-technique combination • Local ties • SLR

S. Glaser (✉)
Institut für Planetare Geodäsie, Technische Universität Dresden, Dresden, Germany

Present address: Institut für Geodäsie und Geoinformationstechnik, Technische Universität Berlin, Berlin, Germany
e-mail: susanne.glaser@tu-berlin.de

M. Fritsche
GFZ German Research Centre for Geosciences, Potsdam, Germany

K. Sośnica • R. Dach
Astronomisches Institut, Universität Bern, Bern, Switzerland

Present address of K. Sosnica: Institute of Geodesy and Geoinformatics, Wroclaw University of Environmental and Life Sciences, Wroclaw, Poland

C.J. Rodríguez-Solano • U. Hugentobler
Institut für Astronomische und Physikalische Geodäsie, Technische Universität München, Munich, Germany

K. Wang • M. Rothacher
Institut für Geodäsie und Photogrammetrie, Eidgenössische Technische Hochschule Zürich, Zurich, Switzerland

R. Dietrich
Institut für Planetare Geodäsie, Technische Universität Dresden, Dresden, Germany

T. van Dam (ed.), *REFAG 2014*, International Association of Geodesy Symposia 146,
DOI 10.1007/1345_2015_190

1 Introduction

The International Terrestrial Reference Frame ITRF2008 (Altamimi et al. 2011) is the current realization of the International Terrestrial Reference System. For its determination, solutions of the space geodetic techniques Global Navigation Satellite System (GNSS), Satellite Laser Ranging (SLR), Very Long Baseline Interferometry (VLBI) and Doppler Orbitography and Radiopositioning Integrated by Satellite (DORIS) have been combined in order to get an optimal solution in terms of accuracy, stability and availability of the TRF. Therefore, the IAG (International Association of Geodesy) Services, namely the International GNSS Service (IGS), the International Laser Ranging Service (ILRS), the International VLBI Service (IVS), and the International DORIS Service (IDS), provide station positions and Earth orientation parameters (EOP) which were tied together using additional information or through a direct combination (e.g. for EOP). The vector between the reference points of the instruments of the different techniques at co-located sites has to be exactly known. These so-called local ties (LTs) can be obtained from local terrestrial and GNSS measurements. The LTs play a key role in the combination of the techniques and affect directly the accuracy of the ITRF, since any discrepancies or inaccuracies propagate into the estimated coordinates and may cause distortions within the combined network (Thaller et al. 2011). For instance, according to Altamimi et al. (2011) such inaccuracies range between few mm and up to almost 10 cm between the GPS receiver and the VLBI antenna at the station Kashima, Japan. In almost all cases the discrepancies exceed the precision of the single-technique solutions. Since the data base of the LTs is very inhomogeneous without providing the full variance-covariance information in general, the evaluation of the accuracy is difficult (Ray and Altamimi 2005). Further problems arise in the fact that the LT stations are globally not well distributed and that there are only a few LTs between VLBI and SLR (Krügel and Angermann 2005). Therefore, the IERS (International Earth Rotation and Reference System Service) Working Group on Site Survey and Co-location[1] was established to deal with specific questions about this topic.

It should be emphasized that the vector closure between stations at co-located sites is also affected by mismodeling of the space geodetic techniques. The determination of the reference point is a very difficult task (Sarti et al. 2004). Other model components (e.g. atmospheric delay, loading deformations) influence the final coordinates as well.

[1]http://www.iers.org/IERS/EN/Organization/WorkingGroups/SiteSurvey/sitesurvey.html.

Different strategies to combine the space geodetic techniques exist. It is also possible to tie them at the space segment by using the eccentricities at the satellites (space ties) (Thaller et al. 2011). In our study we used neither local nor space ties. We utilize the pole coordinates as global ties (Seitz et al. 2012) using GNSS and SLR data. Together with proper datum realization, this approach enables us to estimate components of the LTs. The impact of the measured LTs on parameters of a TRF can also be investigated from the connection of the LTs and the Earth rotation parameters (ERP) but in the opposite direction. By using LTs to combine different space geodetic techniques, potential discrepancies can be shown in the estimated technique-wise ERPs. In such a study, using all the measured (and not error-free) LTs at co-located GPS and VLBI sites leads to an offset of about 0.013 ms in the UT1-UTC estimates (Thaller et al. 2007).

After demonstrating the data base of our study in Sect. 2.1 we explain our combination strategy (cf. Sect. 2) which allows us to validate LT components. In Sect. 3 we present our results and the paper will be summarized by conclusions in Sect. 4.

2 Strategy

2.1 Data

The input data of our study consist of GNSS and SLR normal equation systems (NEQs) spanning a 17-year period from 1994 to 2010. In the case of GNSS, the NEQs are generated on a daily basis containing observations of GPS and also of GLONASS since the 1st of January, 2002 of 334 globally distributed stations. The SLR NEQs are based on 7-day solutions including observations to LAGEOS-1 and LAGEOS-2 from 73 stations of a global network. Both GNSS and SLR NEQs were obtained from homogeneously reprocessed GPS, GLONASS and SLR observations (Fritsche et al. 2014). In order to determine and combine the NEQs the same reduction models and the same software (Bernese GNSS Software, Dach et al. 2007) were applied to the data ensuring a combined solution of highest consistency.

2.2 Single-Technique Solutions

In order to derive a reliable combined solution, single-technique solutions of GNSS and SLR have been produced first. This included a preliminary analysis of the GNSS and SLR station position time series to eliminate position outliers, to detect discontinuities and velocity changes and to select core stations for the definition of the geodetic datum of the network. For the detection of outliers and station events the program FODITS (Find Outliers and Discontinuities In

Table 1 Offsets and trends with their standard deviations, respectively, estimated from the differences Δx_P and Δy_P of the single-technique solutions of GNSS and SLR w.r.t. the "IERS 08 C04" (Bizouard and Gambis 2011) EOP series and of the GNSS-only minus the SLR-only solution as well at the reference epoch 01.01.2005

Offset [mas] Trend [mas/a]	GNSS-only	SLR-only	GNSS-SLR
Δx_P	−0.04±0.02	0.14±0.11	−0.19±0.09
	0.00±0.00	0.00±0.00	0.00±0.00
Δy_P	−0.06±0.02	0.11±0.11	−0.17±0.09
	0.00±0.00	0.00±0.00	0.00±0.00

Time Series, Ostini 2012) of the Bernese GNSS Software was used. FODITS adapts a functional model consisting of outliers, coordinate discontinuities, velocity changes, and periodic functions to the position time series and tests the significance of the parameters. Station coordinate discontinuities and velocity changes usually occur in the case of station movements due to natural reasons like earthquakes or due to hardware replacements, e.g. antenna or dome changes. Earthquakes can introduce jumps up to 3 m, e.g. at the station Concepción, Chile, due to the Mw 8.8 Chile Earthquake in 2010. In order to reduce the number of station events and to get a more stable long-term solution, only events introducing significant jumps (>1 cm) into the time series have been considered. Stations with observation time spans longer than 3 years and a station repeatability better than 3 mm and 1 cm in the case of GNSS and SLR, respectively, were selected as core stations. A homogeneous global distribution of the stations was also a selection criterion. Consequently, 76% of GNSS and 40% of SLR stations of the global network were selected as core stations unless they have been affected by earthquakes.

Since the pole coordinates have to be combined, an assessment of associated single-technique solutions was performed. This was done in order to examine systematic effects. Table 1 shows the offsets and trends estimated from the differences of the single-techniques solutions of GNSS-only and SLR-only and w.r.t. the "IERS 08 C04" (Bizouard and Gambis 2011) EOP series as a reference solution. The values w.r.t. their standard deviations indicate that there were neither systematic effects nor outliers. Therefore, a combination of the pole coordinates has been performed in the next step.

2.3 Inter-Technique Solution

The inter-technique solution was obtained from the direct combination of the GNSS and SLR data at the normal equation level. The strategy comprises the combination of the pole coordinates as common global parameters. Degree-1 surface load coefficients as an additional common parameter of GNSS and SLR have been combined as well. UT1-UTC/LOD was constrained to their a priori values at 0:00 UTC at each day identically for all solutions within this study. For the validation of LTs as presented in this paper we added a constraint that imposes the velocities at co-located sites to be the same for all markers and techniques. This assures that our estimates of LT components do not change with time. Further information concerning the combination strategy can be found in Glaser et al. (2015).

By applying this combination strategy the geodetic datum of the network was realized from the GNSS and SLR observations with a common origin. Similar to the technique-only solutions, the GNSS network scale was realized by the GNSS observations, and the SLR network scale was realized by the SLR observations. The agreement of the single-technique network scale w.r.t. the combined solution is at the level of 0.01 ppb and 0.00 ppb/a. In the case of the orientation, no-net rotation (NNR) conditions were imposed to the set of core stations. For GNSS the NNR condition is imposed w.r.t. the X, Y and Z axis and for SLR only w.r.t. the Z axis because the combination of the pole coordinates is equivalent to a common rotation around the X and Y axes for GNSS and SLR. The temporal change of the orientation is realized only by NNR w.r.t. the X, Y and Z axis for the GNSS, and no condition is applied to the SLR stations since the station velocities are forced to be equal at the co-located sites.

3 Results

The final solution allowed the computation of LT vectors using the adjusted coordinates at co-located stations. These "computed" LTs could then be compared with the measured LTs[2]. The differences of our computed and the measured LTs are presented in Table 2, and histograms of these differences are shown in Fig. 1. It has to be emphasized here that for the computed east component one degree of freedom remained (definition of zero meridian for GNSS and SLR) since the combination of the pole coordinates is equivalent to a combination of a common rotation around the X and Y axes of the GNSS and SLR station network. Therefore, the scatter of the differences in the east component is an indicator of the accuracy of the measured LTs whereas the mean shift in the east component cannot be conclusive as it is related to the limit of the applied datum definition. In contrast, with our combination approach the north and the height components can directly be compared and analysed for each co-located site. The differences between the estimated and the measured LTs are smallest in the north component, 96% of the 50 differences are below 1 cm (50% in the height

[2]http://itrf.ensg.ign.fr/local_surveys.php → "Local Ties used in ITRF2008" → "Local ties SINEX files".

Table 2 Differences of estimated and measured local ties (LTs) according to the ITRF2008 in the north, east and height component at the epoch of the measured LT (available from SINEX files of the LTs from ITRF2008 determination) for the 50 LTs

Station 1 IERS DOMES #	Station 2 IERS DOMES #	Epoch of the LT measurement	North [mm]	East [mm]	Height [mm]
7090 50107M001*	YAR1 50107M004*	2003 11 01 12 00 00	5.2	13.0	13.9
7090 50107M001*	YARR 50107M006*	2003 11 01 12 00 00	4.5	12.7	12.9
7105 40451M105*	GODE 40451M123*	2008 01 01 00 00 00	7.9	18.8	−15.0
7110 40497M001*	MONP 40497M004	1999 10 07 00 00 00	2.2	18.1	−8.1
7119 40445M004*	7210 40445M001*	2006 09 17 00 00 00	2.6	−0.8	−1.9
7119 40445M004*	MAUI 40445S008*	2006 09 17 00 00 00	4.6	15.7	−21.4
7124 92201M007*	THTI 92201M009*	2007 10 05 00 00 00	−3.5	12.0	−6.1
7210 40445M001*	MAUI 40445S008*	2006 09 17 00 00 00	1.9	16.5	−19.5
7231 21602S004	WUHN 21602M001*	2003 12 08 00 00 00	−16.6	4.2	−15.1
7249 21601S004	BJFS 21601M001*	2003 06 20 00 00 00	−1.6	10.0	−0.5
7308 21704S002	7328 21704M001	1999 10 12 00 00 00	8.9	−1.9	31.5
7308 21704S002	KGNI 21704S005*	1999 10 12 00 00 00	2.6	3.7	**87.8**
7328 21704M001	KGNI 21704S005*	1999 10 12 00 00 00	−6.3	5.6	**56.3**
7335 21701M002*	KSMV 21701S007	1999 10 19 00 00 00	5.1	21.8	**51.3**
7355 21612M002	URUM 21612M001*	1999 01 01 00 00 00	−13.8	−6.8	26.8
7403 42202M003	AREQ 42202M005	2007 05 03 00 00 00	2.4	30.9	−22.4
7405 41719M001	CONZ 41719M002*	2003 03 21 00 00 00	−8.9	18.0	−15.5
7501 30302M003*	HARB 30302M009*	2003 08 02 00 00 00	5.5	21.0	1.3
7501 30302M003*	HRAO 30302M004	2003 08 02 00 00 00	6.9	23.1	1.9
7806 10503S014	METS 10503S011*	2000 10 25 00 00 00	−8.7	13.5	−12.5
7806 10503S014	METZ 10503M005*	2000 10 25 00 00 00	−8.3	13.5	−8.2
7810 14001S001	7810 14001S007*	1996 04 04 00 00 00	−4.1	3.5	−8.7
7810 14001S001	ZIMM 14001M004*	1996 04 04 00 00 00	−0.8	21.8	−10.3
7810 14001S007*	ZIMM 14001M004*	1996 04 04 00 00 00	3.3	18.3	−1.6
7811 12205S001*	BOR1 12205M002*	1994 01 10 00 00 00	1.9	19.4	−22.9
7824 13402S007	SFER 13402M004*	2009 07 11 12 00 00	−6.9	23.6	−24.8
7825 50119S003*	7849 50119S001*	2001 07 28 12 00 00	−0.2	2.1	−3.7
7825 50119S003*	STR1 50119M002	2001 07 28 12 00 00	1.3	11.1	−7.5
7825 50119S003*	STR2 50119M001	2001 07 28 12 00 00	−3.5	7.9	−10.2
7835 10002S001*	7845 10002S002	1999 10 11 00 00 00	−2.0	4.8	−11.8
7835 10002S001*	GRAS 10002M006*	1999 10 11 00 00 00	−5.4	22.6	−11.5
7836 14106S009*	7841 14106S011	1994 08 09 00 00 00	−1.9	22.8	32.9
7836 14106S009*	POTS 14106M003*	1994 08 09 00 00 00	−1.9	23.4	0.4
7837 21605S001	SHAO 21605M002*	2003 11 28 00 00 00	−6.7	14.4	−7.1
7840 13212S001*	HERS 13212M007*	2008 06 25 00 00 00	−1.1	17.3	−12.1
7840 13212S001*	HERT 13212M010*	2008 06 25 00 00 00	3.0	17.1	−3.9
7841 14106S011	POTS 14106M003*	1994 08 09 00 00 00	−0.0	0.6	−32.5
7843 50103S007*	TIDB 50103M108	1995 09 20 00 00 00	2.2	2.5	−37.2
7845 10002S002	GRAS 10002M006*	1999 10 11 00 00 00	−3.3	17.8	0.3
7849 50119S001*	STR1 50119M002	2001 07 28 12 00 00	1.5	9.0	−3.8
7849 50119S001*	STR2 50119M001	2001 07 28 12 00 00	−3.3	5.8	−6.6
7941 12734S008	MATE 12734M008	2004 10 26 00 00 00	−2.2	23.7	−6.1
8834 14201S018*	WTZZ 14201M014	2002 09 23 00 00 00	−3.1	18.3	5.3
CAGL 12725M003*	CAGZ 12725M004*	1995 06 10 00 00 00	3.2	−0.4	−3.9
HARB 30302M009*	HRAO 30302M004	2003 08 02 00 00 00	1.4	2.1	0.5
HERS 13212M007*	HERT 13212M010*	2008 06 25 00 00 00	4.1	−0.2	8.1
LHAS 21613M001*	LHAZ 21613M002*	2000 02 25 00 00 00	−3.4	−2.3	14.4
METS 10503S011*	METZ 10503M005*	2000 10 25 00 00 00	0.4	0.0	4.2
STR1 50119M002	STR2 50119M001	2001 07 28 12 00 00	−4.8	−3.3	−2.7
YAR1 50107M004*	YARR 50107M006*	2003 11 01 12 00 00	−0.7	−0.3	−1.0

Core stations at the measuring epoch are marked with an asterisk. Attention should be drawn to the stations Koganei, Japan (7308 21704S002, 7328 21704M001, KGNI 21704S005), and Kashima, Japan (7335 21701M002, KSMV 21701S007) with differences larger than 5 cm in the height component (bold).

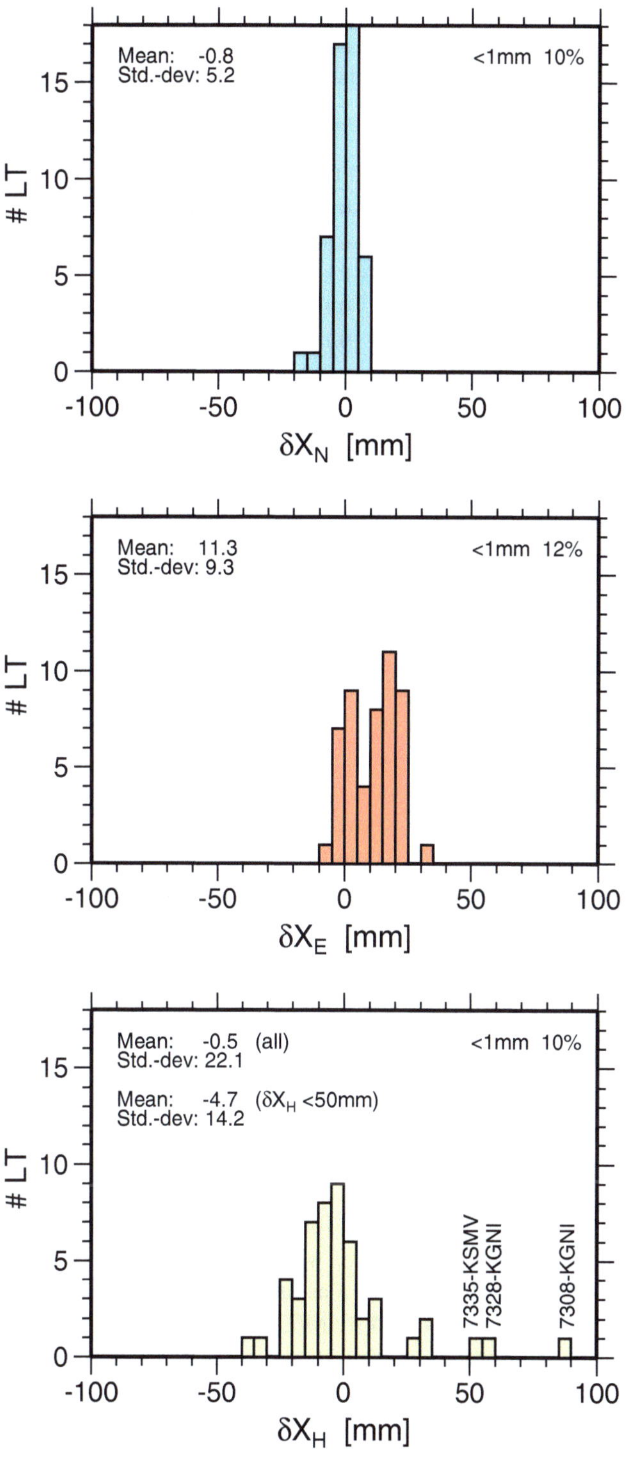

Fig. 1 Histograms of the differences between the estimated and the measured local ties in the north, east and height component, respectively

and 42% in the east component, respectively), cf. Fig. 1. The requirements for the accuracy of the measured LTs of 1 mm can be achieved for 5 LTs in the north and height and for 6 in the east component. The standard deviation of the differences can be as large as 22.1 mm in the height component. Here, 3 out of the 50 differences exceed 5 cm. A potential explanation could be the short observation time span at the SLR stations Koganei, Japan (7328 21704M001), and Kashima, Japan (7335 21701M002) with only 1 or 2 years of observations during 1994–2010.

It should be mentioned that there may exist a scale inconsistency between GNSS and SLR. In our combined solution the scale was realized inherently by the observations of each technique. The mean of the differences of the estimated and the measured LTs in the height component (cf. Fig. 1) of −4.7 mm (excluding 3 LTs with differences larger than 50 mm) indicates that the scale agreement of GNSS and SLR can be expected to be on the level of a few mm.

Since the estimation of LTs in this study results from a global network solution, a proper datum realization (Sect. 2.3) was carried out in order to achieve a reliable estimation of the LTs. However, the estimated LTs differ depending on whether a LT station is a core station or not. Therefore, Table 2 depicts an overview over all 50 differences between the estimated and the measured LTs. LT stations being core stations of the network are marked with an asterisk. Within the least-squares adjustment the residuals between estimated and measured LTs were estimated for all stations and minimized in the case of the core stations. Disregarding LTs between core stations, then 3 stations for the height and 2 for the north component still remain where the differences are below 1 mm. For the largest difference (>5 cm in the height component, cf. Table 2 in bold) the SLR stations (7308 21704S002; 7328 21704M001) and the GNSS station (KSMV 21701S007) were not set as a core stations, respectively. The station Wettzell, Germany, is well known for a high accuracy of the LT local survey. However, the differences are in the range of a few mm but the GNSS station WTZZ 14201M014 is not a core station. The co-located GNSS stations WTZA 14201M013 and WTZR 14201M010 with longer observation time spans of over 1 and 3 years compared to WTZZ 14201M014, respectively, are core stations in this solution.

Another method to evaluate the impact of measured LT on the TRF solution is the following: We apply the measured LTs between SLR and GNSS sites (altogether 50) at co-located sites to the SLR-only solution and compare the obtained coordinate set with the GNSS-only solution via a Helmert transformation. In addition, we excluded the sites with differences larger than 5 cm, which resulted in 47 remaining LTs. In an ideal case (absence of any measurement and model errors) the coordinates at co-located sites should be identical. In Table 3 we show the resulting transformation parameters. The standard deviation s_0 of an adjusted coordinate difference is 16 mm for all LTs, which can be considered as an estimate for the accuracy of the alignment of GNSS and SLR coordinates using LT information. However, we excluded all coordinate differences

Table 3 Transformation parameters of a 14-parameter Helmert transformation (translation T_X, T_Y, T_Z, rotation R_X, R_Y, R_Z, scale D and their temporal changes with their corresponding standard deviations) at the reference epoch $t_0 = 01.01.2005$ and the standard deviation s_0 of the transformation between the GNSS-only and the SLR+LT (SLR plus 50 measured LTs and SLR plus 47 measured LTs with $\delta X_H \leq 5$ cm, cf. Fig. 1) solution at co-located stations of GNSS and SLR

				SLR+LT			
GNSS				50 LT (all)		47 LT ($\delta X_H \leq 5$ cm)	
T_X	[mm]	$\dot{T}_X$	[mm/a]	2.97 ± 3.37	−0.34 ± 0.03	−0.43 ± 3.01	−0.37 ± 0.03
T_Y	[mm]	$\dot{T}_Y$	[mm/a]	−8.41 ± 3.38	0.40 ± 0.03	−6.67 ± 2.91	0.42 ± 0.03
T_Z	[mm]	$\dot{T}_Z$	[mm/a]	−1.88 ± 3.31	−0.12 ± 0.03	−0.81 ± 2.95	−0.11 ± 0.03
R_X	[mas]	$\dot{R}_X$	[mas/a]	0.128 ± 0.138	−0.012 ± 0.001	0.106 ± 0.123	−0.012 ± 0.001
R_Y	[mas]	$\dot{R}_Y$	[mas/a]	0.151 ± 0.115	−0.007 ± 0.001	0.110 ± 0.114	−0.007 ± 0.001
R_Z	[mas]	$\dot{R}_Z$	[mas/a]	−0.045 ± 0.120	0.003 ± 0.001	−0.038 ± 0.120	0.003 ± 0.001
D	[ppb]	$\dot{D}$	[ppb/a]	0.29 ± 0.54	0.03 ± 0.00	0.87 ± 0.46	0.03 ± 0.00
s_0	[mm]			16.46		14.58	

(after transformation) which were above a 95% confidence level. Looking at the obtained transformation parameters a significant translation in Y direction is evident. The other two translation parameters are smaller than their standard deviation. The estimated translation parameters (especially T_Y) show that inaccurate measured LTs introduced in ITRF determination will lead to a shifted GNSS station network on the amount of 0.8 cm. As far as the transformation is concerned for the case that stations with differences larger than 5 cm have been excluded, the main impact is reflected in the translation parameters and in the scale compared to the transformation with all 50 LTs. The translations w.r.t. all axes become smaller (86% in T_X, 21% in T_Y, and 57% in T_Z) and the scale gets 3-times larger. The smaller standard deviation of the scale indicates that the scale itself can be determined more accurate in this case. If we assume that the remaining 47 LTs are correct, the estimated value of 0.87 ppb shows the scale difference between GNSS and SLR, which compares well to the 5 mm mean discrepancy found for the LT height component (cf. Fig. 1).

4 Discussion and Conclusion

In this study a combination of GNSS and SLR was performed at normal equation level covering 17 years (1994–2010) of data. Instead of tying together the single-technique solutions via LTs, we combined the pole coordinates as joint parameters corresponding to a common rotation around the X and Y axes of the GNSS and SLR network. In order to cope with the rank deficiency remaining in the Z component, an NNR condition around the Z axis to the SLR core stations was imposed. Together with a proper datum realization and an additional constraint forcing that the velocities at co-located sites are identical, components of the LTs were estimated and evaluated. The evaluation of the east component is limited by the definition of the geodetic datum, whereas a validation of north and height components is fully feasible. The validation of the north and height component showed that the differences between the estimated and the measured LTs of the ITRF2008 determination are in the range of a few mm and reach almost 7 cm at the station located in Koganei, Japan. Nevertheless, 94% of the differences in the height component are smaller than 5 cm; 30% smaller than 0.5 cm. In the north component, 70% of the differences are below 0.5 cm. Using all the measured LTs, we show a translation of 0.8 cm in the direction of the Y axis of the GNSS-only network w.r.t. the SLR-only network with applied measured LTs. Excluding the stations Koganei and Kashima with differences in the height component being larger than 5 cm when performing the Helmert transformation, the translation parameters decrease by about 50%. This indicates that a network solution can benefit in terms of a more accurate origin when LTs with smallest discrepancies are used within the combination.

The LTs can also be compared to the coordinate differences of the space geodetic techniques. These discrepancies within the ITRF2008 determination[3] are in the range of some mm up to a few cm. The largest discrepancies between GNSS and SLR are also found in the height component at station Kashima, Japan, with 7 cm (KSMV 21701S007, 7335 21701M002) and are as large as 7 cm between KGNI 21704S005 and 7328 21704M001. However, the comparison of the discrepancies is based on a combined solution where the single-technique solutions were tied together using the LTs. An independent validation can also be performed by using space ties. As shown by Thaller et al. (2011), the differences are in the range of a few cm by using 1 year of data. The largest differences occur in the height component between KGNI 21704S005 and 7308 21704S002 (~10 cm), KGNI 21704S005 and 7328 21704M001 (~8 cm), and San Fernando, Spain (7824 13402S004 – SFER 13402M004) (~7 cm) (Thaller et al. 2013).

[3] http://itrf.ensg.ign.fr/ITRF_solutions/2008/ITRF2008.php → "ITRF2008 tie and space geodesy discrepancies".

Our results confirm the discrepancies between the measured LTs and the space geodetic measurements similar to those already demonstrated in previous studies (Altamimi et al. 2011; Thaller et al. 2011, 2013; Seitz et al. 2012). Since the uncertainty of the LTs seems to be larger than some centimeters in several cases, further investigations on this topic have to be carried out. Along with the IERS Working Group on Site Survey and Co-Location, the reasons for the discrepancies have to be investigated. Besides a standardization of the local survey and providing the full variance-covariance information for all LTs, the transformation from the local networks to the global network should be evaluated as well. In addition, some more co-located sites on the southern hemisphere could help to improve the global distribution of the co-located sites. Moreover, errors and mismodeling in space geodetic data analysis as another error source for discrepancies with LTs should be also examined. The comparison with the measured LTs shall help to look in detail whether there exist possibilities to improve the instrumental setup and calibration (e. g. for GNSS antennas and radomes in combination with specific local situations like horizontal mask or multipath effects; for SLR a careful calibration in order to avoid range biases which may map into the station coordinates). Certainly, the height component will be the quantity where such instrumental or mismodeling effects will have the largest impact.

Hence, the potential of space geodetic combination can be fully exploited. Only in this way, the requirements for a global terrestrial reference frame with an accuracy of 1 mm and a stability of 0.1 mm/a can be achieved in the framework of the Global Geodetic Observing System that aims on a better understanding of our changing planet.

Acknowledgements We would like to thank the German Research Foundation and the Swiss National Science Foundation for the financial support within the project "Geodätische und geodynamische Nutzung reprozessierter GPS-, GLONASS- und SLR-Daten". We are very grateful to three anonymous reviewers for their valuable comments which helped to improve the manuscript.

References

Altamimi Z, Collilieux X, Métivier L (2011) ITRF2008: an improved solution of the international terrestrial reference frame. J Geodesy 85(8):457–473. DOI 10.1007/s00190-011-0444-4

Bizouard C, Gambis D (2011) The combined solution C04 for Earth orientation parameters consistent with international terrestrial reference frame 2008. http://hpiers.obspm.fr/iers/eop/eopc04/C04.guide.pdf

Dach R, Hugentobler U, Fridez P, Meindl M (eds) (2007) Bernese GPS software version 5.0. Astronomical Institute, University of Bern, Bern

Fritsche M, Sośnica K, Rodríguez-Solano C, Steigenberger P, Wang K, Dietrich R, Dach R, Hugentobler U, Rothacher M (2014) Homogeneous reprocessing of GPS, GLONASS and SLR observations. J Geodesy 88(7):625–642. DOI 10.1007/s00190-014-0710-3

Glaser S, Fritsche M, Sośnica K, Dach R, Rodríguez-Solano CJ, Hugentobler U, Wang K, Rothacher M, Dietrich R (2015) A consistent combination of GNSS and SLR with minimum constraints. J Geodesy. DOI 10.1007/s00190-015-0842-0

Krügel M, Angermann D (2005) Analysis of local ties from multi-year solutions of different techniques. In: Proceedings of the IERS workshop on site co-locations, Verlag des Bundesamtes fr Karthographie und Geodäsie, Frankfurt am Main, IERS Technical Note No. 33

Ostini L (2012) Analysis and quality assessment of GNSS-derived parameter time series. Ph.D. thesis, Philosophisch–naturwissenschaftliche Fakultät der Universität Bern

Ray J, Altamimi Z (2005) Evaluation of co-location ties relating the VLBI and GPS reference frames. J Geodesy 79(4–5):189–195. DOI 10.1007/s00190-005-0456-z

Sarti P, Sillard P, Vittuari L (2004) Surveying co-located space-geodetic instruments for ITRF computation. J Geodesy 78(3):210–222. DOI 10.1007/s00190-004-0387-0

Seitz M, Angermann D, Bloßfeld M, Drewes H, Gerstl M (2012) The 2008 DGFI realization of the ITRS: DTRF2008. J Geodesy 86(12):1097–1123. DOI 10.1007/s00190-012-0567-2

Thaller D, Krügel M, Rothacher M, Tesmer V, Schmid R, Angermann D (2007) Combined Earth orientation parameters based on homogeneous and continuous VLBI and GPS data. J Geodesy 81(6–8):529–541. DOI 10.1007/s00190-006-0115-z

Thaller D, Dach R, Seitz M, Beutler G, Mareyen M, Richter B (2011) Combination of GNSS and SLR observations using satellite co-locations. J Geodesy 85(5):257–272. DOI 10.1007/s00190-010-0433-z

Thaller D, Roggenbuck O, Sośnica K, Mareyen M, Dach R, Jäggi A (2013) Validation of GNSS-SLR local ties by using space ties. In: IERS workshop on local surveys and co-locations, Paris

Part II

Geodetic Measurement Techniques

Combination of Space Geodetic Techniques on the Observation Level with c5++: Common Nuisance Parameters and Data Weighting

Thomas Hobiger and Toshimichi Otsubo

Abstract

The multi-technique space geodetic analysis software c5++ can be used to combine several techniques on the observation level. A variance component estimation (VCE) feature has been implemented recently and is now being tested concerning the impact on the combination of observational data from different space geodetic techniques. This goes along with studying which nuisance parameters can be estimated as common unknowns at co-located instruments and how offsets between the different techniques need to be considered in order to enable such an estimation strategy. It can be shown that the combination on the observation level is a sophisticated method to merge data from different space geodetic techniques with the goal of improving the quality of the target parameters. Moreover, it can be shown that formal errors of local ties between GPS and VLBI are at least 6 times too small on average.

Keywords

Combination • GGOS • GPS • Variance component estimation • VLBI

1 Introduction

Having co-located instruments at space geodetic observatories allows to take benefit from sharing infrastructure at the site and makes it possible to combine (Coulot et al. 2007) solutions from different techniques. Local ties relate between the reference points of the individual space geodetic techniques, but one can also take advantage from the close proximity of instruments and estimate nuisance parameters, in particular troposphere and clocks, in site-wise models rather than parameterizing these unknowns for each technique separately. In order to realize such an approach, a suitable model for the nuisance parameters as well as proper weighting of the observations from the individual techniques are necessary. These aspects and the application of such an analysis strategy to VLBI and GPS data will be described in the following sections.

T. Hobiger (✉)
Department of Earth and Space Sciences, Chalmers University of Technology, Onsala Space Observatory, SE-439 92 Onsala, Sweden
e-mail: thomas.hobiger@chalmers.se

T. Otsubo
Geoscience Laboratory, Hitotsubashi University, 2-1 Naka, Kunitachi, Tokyo 186-8601, Japan

2 Combination on the Observation Level with c5++

The analysis package "c5++" (Hobiger et al. 2010) supports combination of space geodetic data from VLBI, GPS, and SLR on the observation level, but also enables processing of single-technique solutions. c5++ uses the same geodetic and geophysical models (Petit and Luzum 2010) for all space geodetic techniques in order to obtain a set of utmost consistent target parameters when combining two or more techniques. The feasibility of such an approach has been demonstrated by Hobiger et al. (2014) and furthermore by Hobiger and Otsubo (2014). However, proper data weighting was not considered at that time. This drawback has been

T. van Dam (ed.), *REFAG 2014*, International Association of Geodesy Symposia 146,
DOI 10.1007/1345_2015_152

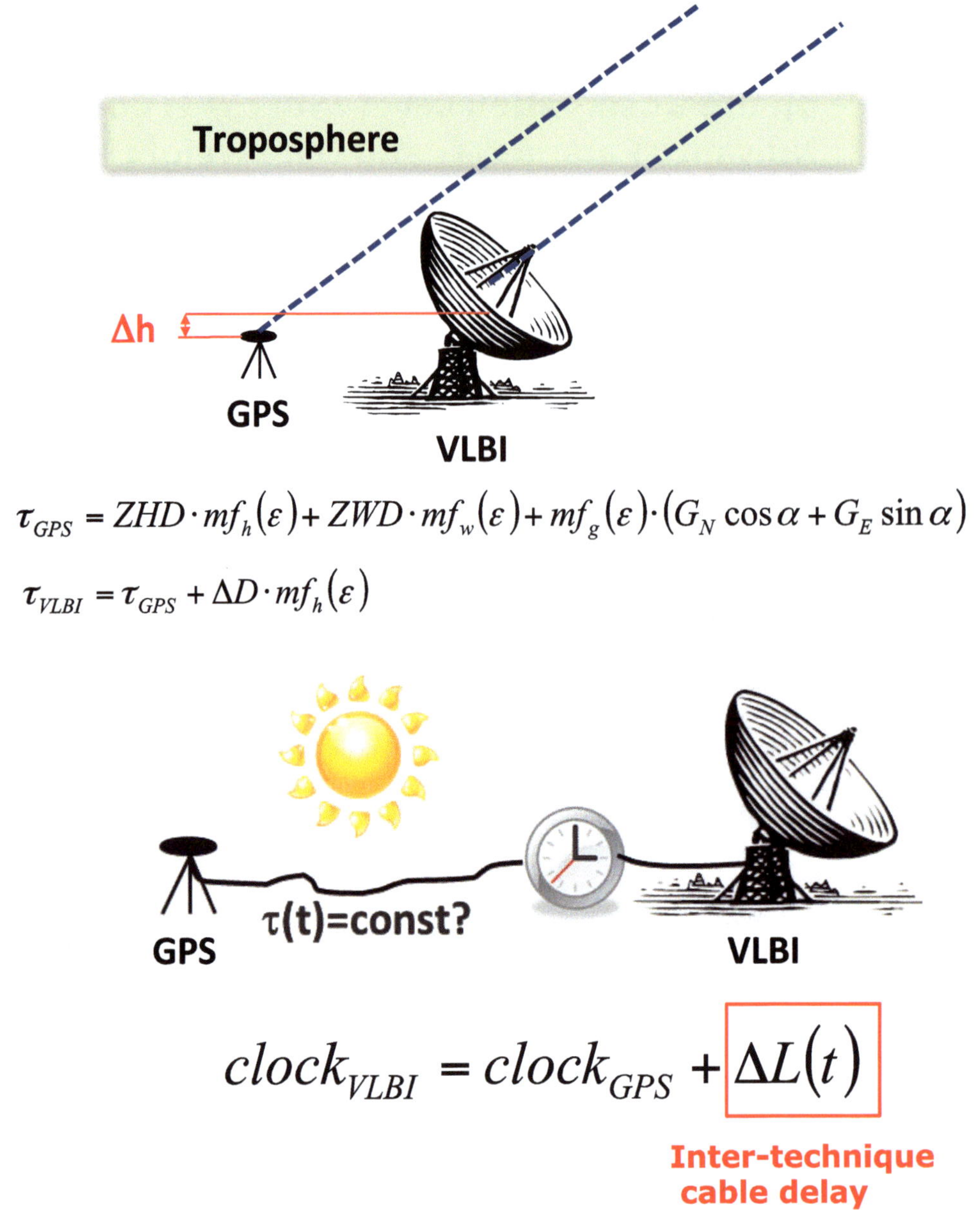

Fig. 1 Conceptual sketch showing how station-wise troposphere (*upper plot*) and clock (*lower plot*) models are realized. α, ε and m_f denote azimuth and elevation angles as well as mapping functions for the troposphere delays concerned

overcome in the meantime, by using a suitable stochastic model based on variance component estimation (cf. Sect. 2.2) which has been implemented within the parameter estimation process. Together with this approach one can also utilize certain parameters which can be combined at each site in order to take additional advantage from co-locating two or more space geodetic instruments.

2.1 Common Parameters

In the following, combination of space geodetic techniques will be restricted to GPS and VLBI as this enables to estimate common parameters related to clocks as well as station wise troposphere delay models. As for the latter one, Hobiger and Otsubo (2014) consider inter-technique troposphere delays (cf. Fig. 1) by estimating an additional constant zenith delay $\Delta D(\Delta h)$ at each site, i.e.

$$ZHD_{VLBI} = ZHD_{GPS} + \Delta D(\Delta h), \quad (1)$$

where ZHD_{VLBI} and ZHD_{GPS} denote the a-priori zenith hydrostatic delays at the VLBI and GPS sites. In doing so, site dependent common troposphere parameters, i.e. zenith wet delay (ZWD) and gradients (G_N, G_E), can be estimated as common parameters. Values of $\Delta D(\Delta h)$ are estimated as constant offsets for each 24 h session.

The difference between the GPS and VLBI clocks (cf. Fig. 1) is thought to be caused by temperature induced cable delay variations and internal electronic delays. It can either be estimated as a constant offset or parameterized by a piece-wise linear offset (PLO) model with a temporal resolution that allows to follow cable delay changes with sufficient temporal resolution (Hobiger and Otsubo 2014). It is obvious that the choice of this inter-technique delay potentially absorbs any benefit that could result from the estimation of a common clock between the two techniques. A too high temporal resolution bears the risk that a large fraction of valuable information from the second technique is not being reflected in the estimated clock. On the other side, an insufficient model representation of the mostly diurnal cable length changes potentially degrades the combined solution. According to Hobiger and Otsubo (2014) a temporal resolution of 3 h is suitable for most stations and thus is being used here in the following.

2.2 Variance Component Estimation

c5++ parameter estimation relies on a Gauss-Markov model (Koch 1997) but did not make use of variance component estimation (VCE) until very recently. With this new feature implemented, it is now possible to estimate variance components for different groups of observations. As GPS observations normally come without any stochastic information it is feasible to estimate variance components for code- and carrier-phase observations at each site. This approach is realized by estimating a component that refers to an observation in zenith distance and grows with the magnitude of the troposphere mapping function when applied to observations at arbitrary elevation angles. As for VLBI and local ties, the situation is slightly different as these types of observations are already provided with formal errors. Thus, only scaling factors, one for all VLBI observations and one for all local tie vectors, will be determined by means of VCE. A subtle grouping of the local tie measurements would likely be beneficial, but the relatively low number of observations bears the risk that the VCE method does not converge or leads to singularities. c5++ uses the simplified algorithm proposed by Förstner (1979) with which one can determine the variance factor σ_{ii}^2 of the i-th group of observations by

$$\sigma_{ii}^2 = \frac{v_i^T P_i v_i}{r_i}. \qquad (2)$$

Thereby, v_i denotes the residual vector of the prior adjustment and P_i is the inverse of the variance-covariance matrix. Together with the degree of freedom r_i of the i-th group, it is possible to iteratively determine the variance components while solving the least-squares adjustment problem. Since correlations among the observations are not considered by this algorithm one can update the variance component parameters after each iteration without the need of maintaining the large variance-covariance matrix as required for the original algorithm described in Helmert (1907).

3 Combination of GPS and VLBI on the Observation Level During CONT11

CONT11 was a campaign of continuous VLBI sessions, between September 15th and September 30th 2011, which were planned as daily sessions starting at 0 UT. Thus, data or products can be combined with other space geodetic techniques over the same time-span. In total 14 VLBI stations participated in CONT11 but only a fraction of the network stations shared a common frequency standard with the co-located GPS receiver (Rieck et al. 2012) whereas other stations did not continuously contribute to the network or had operational troubles during the 2 weeks. Thus, only six stations (Fig. 2) remain suitable for studying the concept of combination on the observation level as described before.

Single Technique Solutions Before performing a combined solution, single-technique GPS and VLBI solutions were obtained. Earth orientation parameters were not estimated and GPS data was resampled to have a temporal resolution of 5 min.

Combined Solution The solution which combines space geodetic observations on the observation level has been obtained with the same settings for the individual techniques which were applied to the single-technique solutions. However, local tie vectors were introduced as additional observations that relate between the reference points of GPS and VLBI stations. In doing so, no-net translation (NNT) and no-net rotation (NNR) conditions became obsolete as the local ties orientate the VLBI station network with respect to the PPP derived GPS site coordinates. Other than Hobiger and Otsubo (2014) who used the 30 s RINEX GPS observations, thinning out these data to a sampling rate of 5 min implicitly lowers the weight of GPS with respect to VLBI.

3.1 Results

In order to judge whether combination on the observation level leads to any improvement one needs to investigate how target parameters benefit from such an approach. Station coordinates are the most obvious group of parameters for

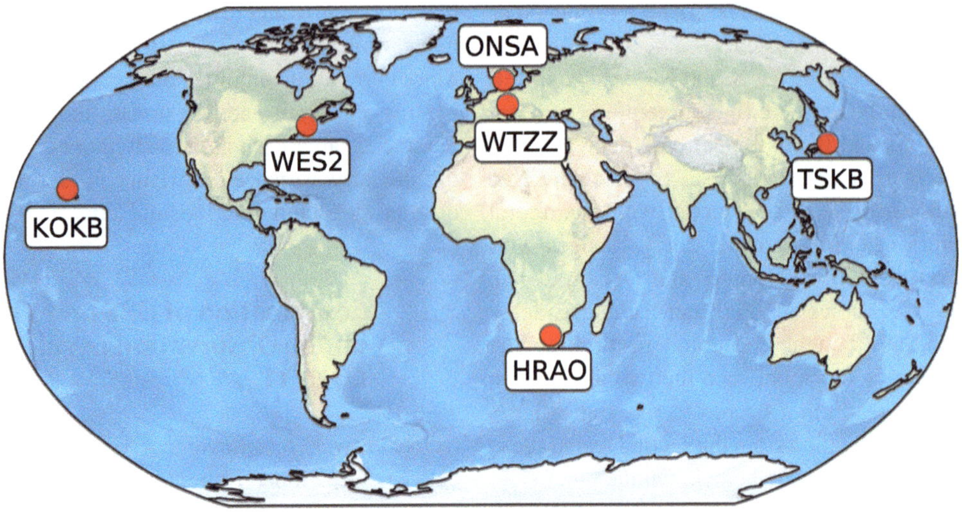

Fig. 2 Location of all stations which participated in the CONT11 campaign and were considered for combination on the observation level. Sites are abbreviated with their IGS name, i.e. Hartebeesthoek (HRAO), Kokee Park (KOKB), Onsala (ONSA), Tsukuba (TSKB), Westford (WES2) and Wettzell (WTZZ)

Table 1 Station position repeatabilities (in mm) from single-technique (S) and combined (C) solutions

		HRAO			KOKB			ONSA			TSKB			WES2			WTZZ		
[mm]		S	C	Δ	S	C	Δ	S	C	Δ	S	C	Δ	S	C	Δ	S	C	Δ
GPS	N	1.5	1.6	0.0	1.7	1.5	−0.1	1.0	1.0	0.0	2.4	2.4	0.0	1.0	1.1	0.1	1.0	1.1	0.1
	E	1.4	1.2	−0.2	3.1	2.6	−0.5	2.5	2.5	0.0	3.5	2.9	−0.7	2.0	1.7	−0.3	2.5	2.4	0.0
	U	4.0	4.1	0.1	4.8	4.7	0.0	3.7	3.4	−0.2	4.9	4.3	−0.7	4.1	5.0	0.9	4.8	4.4	−0.4
VLBI	N	5.6	5.6	0.0	3.4	5.4	1.9	2.5	2.9	0.4	7.4	5.1	−2.3	3.0	3.0	0.0	2.7	1.1	−1.6
	E	5.6	4.8	−0.8	7.0	6.3	−0.7	4.6	4.3	−0.3	12.1	8.2	−3.9	5.3	7.7	2.3	3.6	2.5	−1.1
	U	10.5	6.2	−4.4	9.5	11.8	2.3	6.8	6.6	−0.2	9.7	8.1	−1.6	8.8	8.1	−0.7	4.6	4.4	−0.2

Δ denotes the difference between these solutions where values <0 indicate an improvement when combining data on the observation level

such a test as they are usually of main interest in space geodetic analysis. Hobiger et al. (2014) and Hobiger and Otsubo (2014) have already shown that combination on the observation level leads to small but consistent improvement of station position repeatability when compared to single-technique solutions. This has to be confirmed here as well, given that two new features, VCE and thinning out of GPS data, were used for the analysis. Station position repeatabilities, measured as root mean squared (RMS) error of the coordinate time series, during the 15 day period are computed for single-technique and combined solutions and summarized in Table 1. In general, it can be noticed that the VLBI-only station position repeatability is worse than those of GPS-only solution. When data are combined on the observation level and local tie vectors are added as virtual observations, an improvement can be seen for most of the VLBI and GPS sites. Compared to the results presented by Hobiger and Otsubo (2014) VCE and a strongly reduced number of GPS observations, lead to more independence of the individual station position solutions although local-tie vectors are applied. As shown later, the estimated variance components tend to reduce the weight of the local ties and thus decouple the behavior of co-located instruments. Moreover, VCE helps to find proper weights for each observation type while preserving the gain from combination on the observation level.

3.2 Troposphere Ties

In order to judge whether the concept of site-dependent common troposphere models is feasible or not, one can check the magnitude and consistency of the obtained troposphere ties (cf. Sect. 2.1). The magnitude of the troposphere tie should match with the predicted value, which is basically related to the pressure (height) difference of the GPS and VLBI reference point. As this difference is stable over time, one would expect that the scattering of the daily troposphere tie estimates is also small. Table 2 lists the mean of the estimated troposphere ties at all sites and compares them with theoretical values from Teke et al. (2011). Except for stations TSKB and WES2, estimated troposphere ties agree well with the expected values derived from height differences and average atmosphere conditions. The estimates for WES2

Table 2 Mean troposphere ties (and their empirical standard deviation) between VLBI and GPS

IGS name	Estim. [mm]	Emp. [mm]
HRAO	−0.5 ± 2.6	−0.5
KOKB	−2.1 ± 2.3	−2.7
ONSA	−3.6 ± 2.1	−4.2
TSKB	−2.1 ± 2.7	−6.1
WES2	−7.0 ± 2.0	−0.6
WTZZ	−0.4 ± 1.0	−0.9

The right column lists the corresponding empirical values from Teke et al. (2011)

are consistent with the value derived by Thaller (2008) and might be related to the VLBI radome. Differences at TSKB could possibly be explained by the uncalibrated GPS radome.

3.3 Clock Ties

Other than troposphere ties, which are supposed to be constant over at least 24 h, clock ties do not fulfil this requirement. They are parameterized in a way which allows to account for at least diurnal signals in order to reflect temperature induced cable length variations. Figure 3 depicts these estimated inter-technique delays for station KOKB. A clear and repetitive diurnal pattern with a magnitude of about ±150 ps can be confirmed for that site. Clock ties at other sites reveal similar patterns, but differ slightly in phase and magnitude depending on local temperature variations and electrical lengths involved in either VLBI or GPS instruments. Although a three-hourly temporal resolution for the clock ties appears to be a good choice it could be possible that at other sites which are not considered in this study an even higher temporal resolution is necessary in order to account for such inter-technique delay changes.

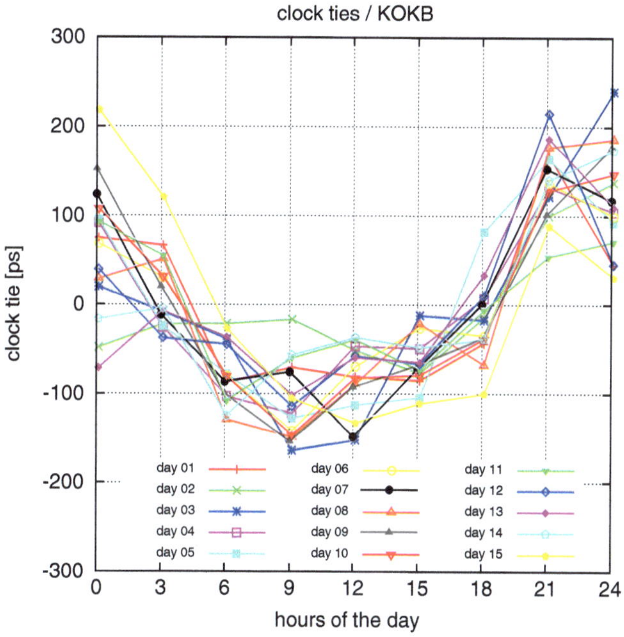

Fig. 3 Daily variations of inter-technique clock differences ("clock ties") at station KOKB. A mean value has been removed from each daily solution in order to increase readability

3.4 Interpretation of the Obtained Variance Components

Obtained variance components can be interpreted depending on how the components were estimated. First, absolute variance components are estimated for GPS code- and carrier-phase observations. These variance components reflect the estimated uncertainty of observations in zenith distance which is then multiplied with the corresponding mapping function in order to be scaled to an observation at a given elevation angle. Thus the obtained GPS variance components have to be interpreted as the estimated uncertainties in zenith direction.

On the other side, VLBI and local tie measurements have already formal uncertainties, either from the way how raw data were processed or from prior adjustment calculations. Thus, only scaling factors need to be estimated for these two kind of observations. Estimated variance components lower than 1.0 can thus be taken as an index that the formal errors that are provided with the observations are too pessimistic, whereas values larger than 1.0 point out that these formal errors are too small in order to represent the true stochastic nature of the observations. Table 3 summarizes the obtained variance components as mean values over the 15 daily analysis batches together with their corresponding standard deviation. As for VLBI, it can be stated that the formal errors provided from the correlators appear to be in good agreement with the stochastic model obtained from VCE. However, increasing or decreasing the number of VLBI or GPS sites will likely lead to a scaling factor which is not so close to 1.0 as in this study. On the other side, a mean value larger than 6.0 for the scaling factor of the local tie uncertainties makes clear that the formal errors provided within the SINEX files are too optimistic. In order to find out at which site the stochastic model needs to be adjusted it would be necessary to estimate variance components for each local tie at a given station. This is not feasible at the moment due to the strong unbalance of observations, but efforts are being made to overcome this drawback and provide more realistic formal errors for these crucial inter-technique 3D vectors.

Table 3 Mean values of the estimated variance components for different observation types

Technique		VC	Unit
VLBI		1.014 ± 0.109	–
TIE		6.481 ± 1.945	–
HRAO	Code	480.7 ± 6.3	mm
	Carrier	3.4 ± 0.6	mm
KOKB	Code	388.6 ± 7.2	mm
	Carrier	4.4 ± 0.5	mm
ONSA	Code	436.7 ± 9.7	mm
	Carrier	3.4 ± 0.5	mm
TSKB	Code	300.4 ± 9.2	mm
	Carrier	4.2 ± 0.9	mm
WES2	Code	377.8 ± 15.4	mm
	Carrier	4.2 ± 0.7	mm
WTZZ	Code	644.4 ± 9.1	mm
	Carrier	5.1 ± 0.3	mm

Scaling factors for the formal errors of VLBI and local ties observations were obtained for each daily batch. Variance components for GPS code- and carrier-phase observations were estimated in zenith direction and projected into slant direction with the corresponding troposphere mapping function

The obtained variance components for GPS observations reveal several interesting conclusions as well. First, estimated variance components appear to be very stable over the period of CONT11, showing values of several decimetres for code-phase observations and 3–5 mm for carrier phase observations. Although the ratio between code- and carrier-phase uncertainties matches well with the rule-of-thumb value of 100, it is clear that site and receiver specific effects have a huge impact, in particular for code-phase measurements. For example, code-phase observations at TSKB appear to have a formal error of about 300 mm in zenith, whereas the same type of observation at WTZZ is identified to have a variance component that is twice as large. Some of these differences might be explainable by different multi-path environments, but it is very indicative that there is a strong relation concerning which type of receiver is involved at a given site. HRAO and KOKB had identical receivers during CONT11 and ONSA and WES2 used receivers from the same manufacturer. On the other side, TSKB and WTZZ had equipped receivers from other vendors which have features for multi-path suppression and tracking of weak signals which can easily lead to different stochastic properties of the obtained code-phase delay observables. Carrier phase variance components appear to be consistent for all sites revealing formal errors of about 3–5 mm for observations in zenith direction. The strong significance, i.e. the low scattering, of the obtained GPS variance components could be used when improving and providing site-specific stochastic models for single-technique and combined GPS analysis.

4 Discussion and Outlook

Combination on the observation level is a powerful concept that enables us to gain maximum benefit from the strength of individual space geodetic techniques, whilst parameters can be estimated site-wise when instruments are co-located, share common infrastructure or are affected by the same environmental conditions. Many space geodetic techniques are currently not only improved concerning measurement precision, but also undergo system upgrades which allow to obtain more observations per session. As for GNSS, it is thus of high importance to consider biases between the different satellite systems in order to gain maximum benefit from the analysis of multi-GNSS data. Also VLBI, for which it is envisaged that the VLBI2010 system (Petrachenko et al. 2012) will replace the current S/X-band systems, it is important to understand and handle intra-system biases before combining this technique with data from other space geodetic instruments. In addition, the introduction of VLBI phase delay observables, will bring new challenges, but can also lead to a significant improvement in measurement precision, and thus, making it more competitive against GNSS and SLR.

Local ties are a crucial part for combing observations or results from different techniques. Thus, it is of high importance that the 3D vectors, which relate between the reference points of the individual techniques are well known, monitored and are made accessible for the analysts. Moreover, realistic stochastic information needs to be provided that does not only account for the inner accuracy of the local survey networks, but also accounts for other error sources that emerge when connecting electrical, optical or mechanical reference points of the individual techniques.

Similar to local ties, other means of connecting two or more instruments have been studied here. Troposphere ties appear to be most straightforward way of taking benefit of co-located instruments operating in the microwave domain. Hobiger and Otsubo (2014) show that site-wise troposphere models have a positive impact when being considered for the combination on the observation level. In addition, common time and frequency technology could be used as an optional way to combine space geodetic techniques. However, if one wants to take benefit from common frequency standards, it is recommended to either calibrate and monitor or estimate inter-technique cable delay changes. It was shown here that the estimation of clock ties is feasible, but one would expect that the knowledge of relative or even absolute timing offsets will likely improve such a concept. First efforts that could support such an approach have been made for VLBI as reported by Panek et al. (2013).

Variance component estimation has been applied in this study as well and turned out to be a very powerful tool in order to give proper weight to the individual techniques. In particular optimal weighting of local ties turned out to provide a good balance between maintaining a certain freedom for the individual techniques and tying them together by the introduction of these 3D vectors. Other than in the previous study (Hobiger and Otsubo 2014) VCE helps to find the "natural" weights when combining different observation types. This leads to slightly less coupling of co-located instruments. However, this reflects better the performance of each technique while still improving the site position repeatabilities.

Although only VLBI and GPS have been utilized in this study, one could extend the concept of combination on the observation level and include other space geodetic techniques as well. If orbit parameters are estimated as well, GNSS satellites which are also tracked by SLR would increase the number of implicit links between the techniques and lead to an even more consistent estimation of unbiased target parameters. Although c5++ would support such an approach after a few minor updates, improvements concerning the mathematical estimation approach and the handling of the large number of unknowns should be made.

Given that these constraints can be lifted, combination of space geodetic data on the observation level appears to be a promising and feasible strategy which could support the goals of GGOS and eventually help to realize the next generation of reference frames which are required for monitoring global change.

Acknowledgements The International GNSS Service (Dow et al. 2009) and the International VLBI Service for Geodesy and Astrometry (Schuh and Behrend 2012) are acknowledged for providing data. Design and implementation of c5++ were a huge effort, requiring the participation and support of many individuals.

References

Coulot D, Berio P, Biancale R, Loyer S, Soudarin S, Gontier AM (2007) Toward a direct combination of space-geodetic techniques at the measurement level: methodology and main issues. J Geophys Res 112: 2156–2202

Dow JM, Neilan RE, Rizos C (2009) The international GNSS service in a changing landscape of global navigation satellite systems. J Geodesy 83(3–4):191–198. doi:10.1007/s00190-008-0300-3

Förstner W (1979) Ein Verfahren zur Schätzung von Varianz- und Kovarianzkomponenten. Allgemeine Vermessungsnachrichten 11(12), 446–453

Helmert FR (1907) Die Ausgleichungsrechnung nach der Methode der kleinsten Quadrate, Zweite Auflage. Teubner, Leipzig

Hobiger T, Otsubo T (2014) Combination of GPS and VLBI on the observation level during CONT11—common parameters, ties and inter-technique biases. J Geodesy 88(11):1017–1028. doi: 10.1007/s00190-014-0740-x

Hobiger T, Otsubo T, Sekido M, Gotoh T, Kubooka T, Takiguchi H (2010) Fully automated VLBI analysis with c5++ for ultra-rapid determination of UT1. Earth Planets Space 45(2):75–79. doi:10.5047/eps.2010.11.008

Hobiger T, Otsubo T, Sekido M (2014) Combination of space geodetic observations from SLR and VLBI on the observation level with c5++. Adv Space Res 53(1):119–129. doi:10.1016/j.asr.2013.10.004

Koch KR (1997) Parameter estimation and hypothesis testing in linear models. Springer, Berlin

Panek P, Kodet J, Prochazka I (2013) Accuracy of two-way time transfer via a single coaxial cable. Metrologia 50(1):60–65. doi:10.1088/0026-1394/50/1/60

Petit G, Luzum B (2010) IERS conventions. IERS Technical Note 36, 179 pp

Petrachenko WT, Niell AE, Corey BE, Behrend D, Schuh H, Wresnik J (2012) VLBI2010: next generation VLBI system for geodesy and astrometry. In: Geodesy for planet Earth. International association of geodesy symposia. Springer, Berlin, pp 999–1005. doi:10.1007/978-3-642-20338-1_125

Rieck C, Haas R, Jarlemark P, Jaldehag K (2012) VLBI frequency transfer using CONT11. In: European frequency and time forum (EFTF), pp 163–165. doi:10.1109/EFTF.2012.6502358

Schuh H, Behrend H (2012) VLBI: a fascinating technique for geodesy and astrometry. J Geodyn 61:68–80. doi:10.1016/j.jog.2012.07.007

Teke K, Boehm J, Nilsson T, Schuh H, Steigenberger P, Dach R, Heinkelmann R, Willis P, Haas R, Garcia-Espada S, Hobiger R, Ichikawa R, Shimizu S (2011) Multi-technique comparison of troposphere zenith delays and gradients during CONT08. J Geodesy 85(7):395–413. doi:10.1007/s00190-010-0434-y

Thaller D (2008) Inter-technique combination based on homogeneous normal equation systems including station coordinates, Earth orientation and troposphere parameters. Scientific Technical Report STR 08/15, Deutsches GeoForschungsZentrum. ISSN:1610–0956

The Processing of Single Differenced GNSS Data with VLBI Software

Younghee Kwak, Johannes Böhm, Thomas Hobiger, and Lucia Plank

Abstract

Space geodetic techniques such as Very Long Baseline Interferometry (VLBI) and Global Navigation Satellite Systems (GNSS) are used for the determination of celestial and terrestrial reference frames and Earth orientation parameters. It is potentially valuable to combine the observations from the different techniques to fully exploit the strengths and unique characteristics of the techniques. Today, discrepancies of locally measured ties between reference points of two techniques and the space geodesy results are a potential issue in the determination of reference frames. To improve the link between the techniques, tests are under way to observe GNSS signals with VLBI radio telescopes directly, and to observe GNSS signals in GNSS antennas with subsequent processing in the VLBI system ("GNSS-VLBI Hybrid System") including VLBI correlation. In both cases, the GNSS data type is the difference in travel time from the satellite to two ground stations. However, it is still difficult to acquire those observations and thus we apply post-processed phase measurements from a precise point positioning (PPP) solution with the c5++ software to build those difference values which are then used in the Vienna VLBI Software (VieVS). We take seven GNSS sites, exclusively Global Positioning System (GPS) in this study, co-located with CONT11 VLBI sites to validate the models in VieVS for single differenced GNSS data, and estimate geodetic parameters. We find root mean square values of post-fit residuals for the VLBI-like observations of about 3.3 cm, compared to less than 2.0 cm from the GNSS PPP solution. At this stage, we do also find degradation in station coordinate repeatabilities (by a factor of 2 to 8), which is related to the systematic residuals.

Keywords

Combination at the observation level • GNSS • GNSS-VLBI hybrid system • VLBI

1 Introduction

The GNSS-VLBI (GV) hybrid system is an observation method to combine GNSS and VLBI at the observation level (Kwak et al. 2011). In the GV hybrid system, GNSS antennas are regarded as small VLBI antennas that receive GNSS L1 and L2 signals. In other words, VLBI observations are made to quasars and GNSS satellites at the same time and processed in the same way, making use of the big radio telescopes for the quasar signals and the comparatively small

Y. Kwak (✉) • J. Böhm
Vienna University of Technology, Gusshausstrasse 27-29, 1040 Vienna, Austria
e-mail: younghee.kwak@tuwien.ac.at

T. Hobiger
Department of Earth and Space Science, Chalmers University of Technology, Onsala Space Observatory, SE-439 92 Onsala, Sweden

L. Plank
University of Tasmania, Private Bag 37, Hobart 7001, Australia

T. van Dam (ed.), *REFAG 2014*, International Association of Geodesy Symposia 146,
DOI 10.1007/1345_2015_201

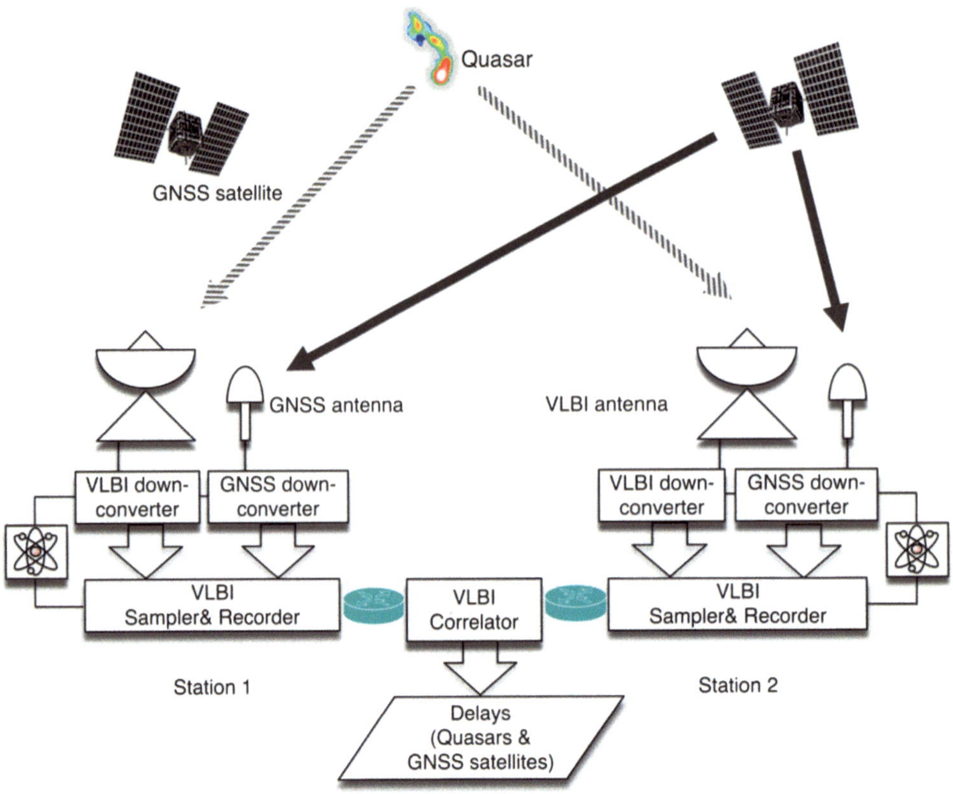

Fig. 1 Schematic diagram of the GNSS-VLBI hybrid system

and cheap GNSS antennas to receive the GNSS satellite signals (Fig. 1).

This system was successfully implemented at two test sites, Kashima and Koganei, and its performance was validated in 24 h experiments by Kwak et al. (2011). Nevertheless, the GV hybrid system is not fully exploited yet and needs further development of the instrument itself as well as analysis strategy. In this paper, we focus on the GNSS part of the GV hybrid system, generating VLBI-like observation data to GNSS satellites based on real GNSS observation data, exclusively Global Positioning System (GPS) in this study, and testing the Vienna VLBI Software (VieVS, Böhm et al. 2012) modules for processing those data based on a global network discussed in Sect. 2.

2 A Global Network

In the previous validation experiment of Kwak et al. (2011), GV hybrid observations were carried out on a single and short (109 km) baseline, which is insufficient to estimate global parameters such as satellite orbits, source coordinates, and Earth orientation parameters (EOP). In a global network, sites must be stable and homogeneously distributed and – for practical reasons – it is effective to use existing GNSS antennas co-located with VLBI sites for the GV hybrid system. The International VLBI Service for Geodesy and Astrometry (IVS) CONT campaign (a series of continuous IVS VLBI sessions over 15 days, http://ivscc.gsfc.nasa.gov/program/cont11/) network has a reasonably balanced geographical distribution of stations between the northern and southern hemispheres. Moreover, at most of the CONT sites, GNSS data are simultaneously acquired through International GNSS Service (IGS) stations. In this paper, we generate the GNSS part data of the GV hybrid observations based on co-located real GNSS observations during the CONT11 campaign which was held in 2011 (15 September 2011 00:00 UT through 29 September 2011 24:00 UT). For the following analyses, we use exclusively seven sites where an identical clock is applied for VLBI and GNSS (Fig. 2). In Table 1, we give an overview of the specific characteristics of VLBI observations to quasars and to GNSS satellites.

3 VieVS for VLBI Observations to GNSS Satellites

In VieVS, tools for the processing of VLBI satellite delays have been implemented (Plank et al. 2014) and validated by the analysis of Selene same beam data (Plank et al. 2012),

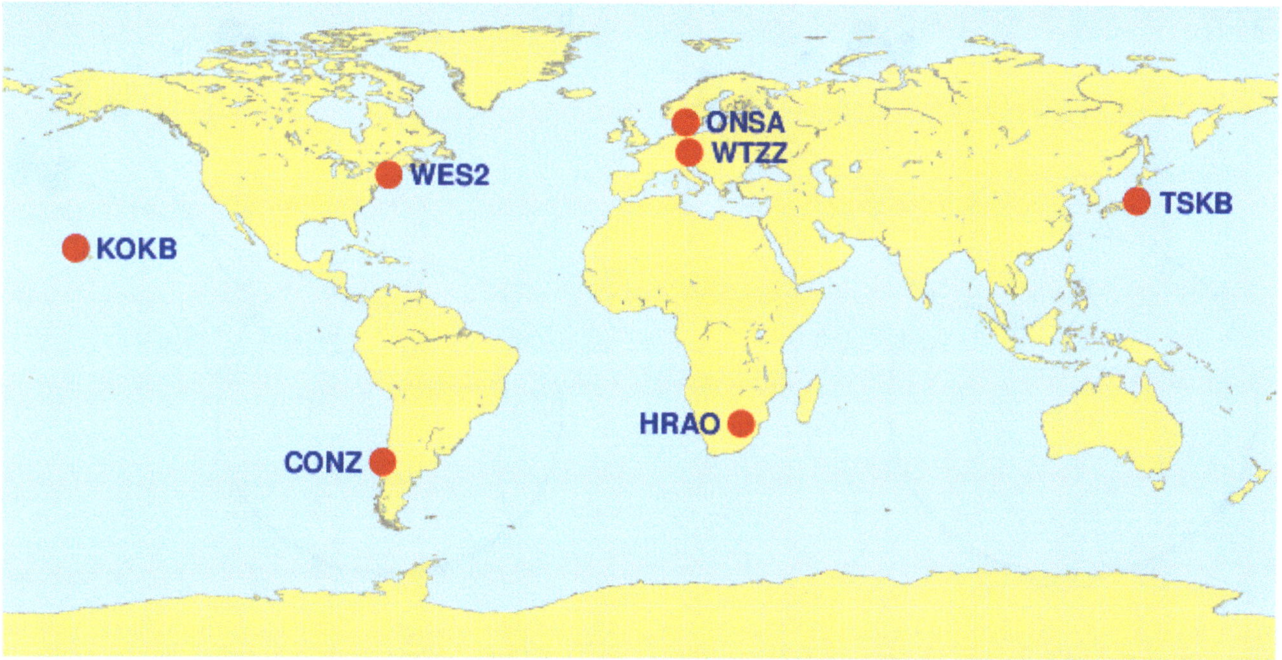

Fig. 2 Global network: co-located CONT11 sites using the same clock for both VLBI and GNSS. The sites are abbreviated with their IGS name, i.e. Hartebeesthoek (HRAO), Kokee Park (KOKB), Onsala (ONSA), TIGO Concepcion (CONZ), Tsukuba (TSKB), Westford (WES2) and Wettzell (WTZZ)

Table 1 Differences between VLBI observations to quasars and to GNSS satellites (partly brought from Plank et al. 2014)

	To quasars	To GNSS satellites
Antenna	Radio telescope	Standard GNSS antenna
Backend	VLBI system (e.g. K5 or Mark5)	VLBI system
Correlation	VLBI software correlator (e.g. K5 software correlator or DiFX)	VLBI software correlator
Scan	1 source per scan	All satellites in the sky (usually 6–9)
Spectrum	White noise	Limited frequencies (e.g. L1 and L2 band)
Signal model	Plane wave front	Curved wave front
	Infinite distance	Finite distance (about 20,000 km)

Chang'E-1 VLBI tracking data, and VLBI observations to GLONASS satellites (Haas et al. 2014). It was further shown by simulations (Plank et al. 2013) that VLBI observations to GNSS satellites might be used for the determination of the frame tie between the VLBI and the GNSS system. However, the telescopes can only point at one direction per scan and need to slew the antennas to continuously track the satellite. The necessary amendments for scheduling, the actual antenna steering, and analysis have been added to VieVS (Hellerschmied et al. 2015). In the case of GV hybrid observations, non-directional GNSS antennas will receive every signal from each satellite in the sky at the same time. Therefore, VieVS is being modified to handle multi directional scans.

For the purpose of carrying out a realistic test, we generate VLBI-like GNSS observations, i.e. single differenced GNSS data, based on real GNSS observations. For more details, see Sect. 4. Currently, VieVS is able to read and process those observations.

4 Data and Analysis

In either case, satellite tracking or GV hybrid system, the data type for GNSS satellites is the difference in travel time from the satellite to two ground stations. The only difference is that the GV hybrid system employs GNSS antennas while the satellite tracking makes use of VLBI radio telescopes. Those data are acquired from the VLBI correlator directly and we assume that we receive the signals which are emitted at the same time (t_0) from a source. Therefore, receiving times (t_1 and t_2) are different at each site plus the baseline keeps moving during the time difference (retarded baseline effect, Fig. 3a). Usually the reference epoch is the receiving time of one station (t_1), and thus the position of the other station is corrected to determine the baseline at t_1.

In this work, because we do not have any real observations at the moment, we generate the VLBI-like data based on normal GNSS measurements.

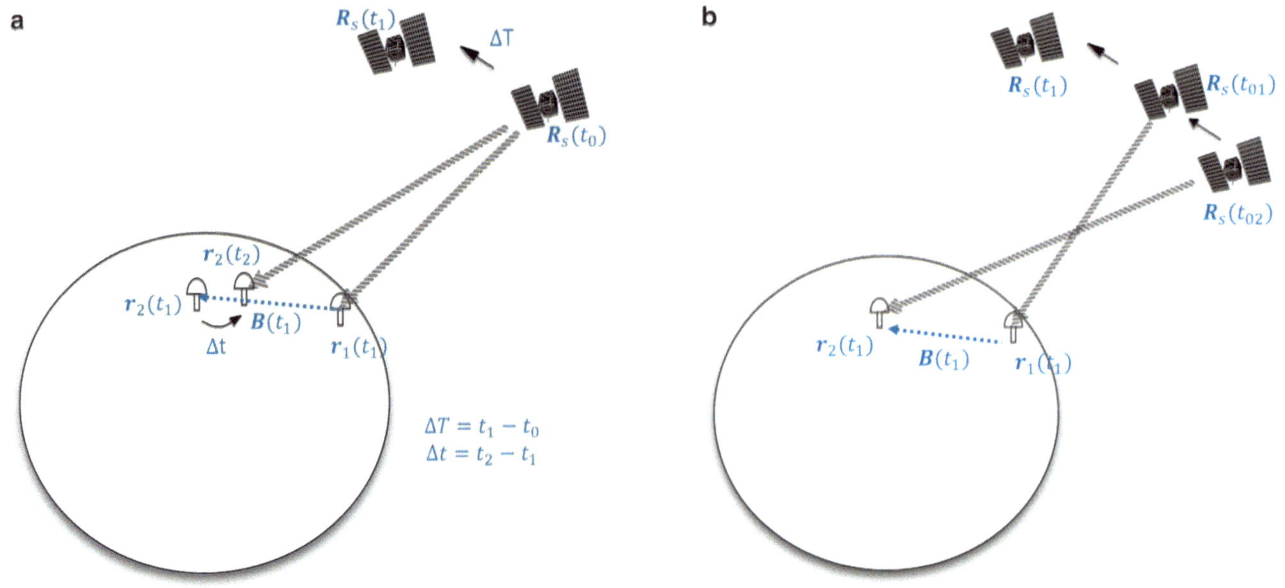

Fig. 3 Positions of the satellites and stations during VLBI observations to a GNSS satellite. (**a**) When two stations receive the identical signal which was emitted at time t_0 from the satellite, reception epochs are t_1 and t_2 for stations 1 and 2, respectively (GV hybrid observation). (**b**) When the phase measurements at the same reception times are provided, the corresponding emission times are t_{01} and t_{02} for stations 1 and 2, respectively, so that the satellite is located at different positions at the different emission times (this work)

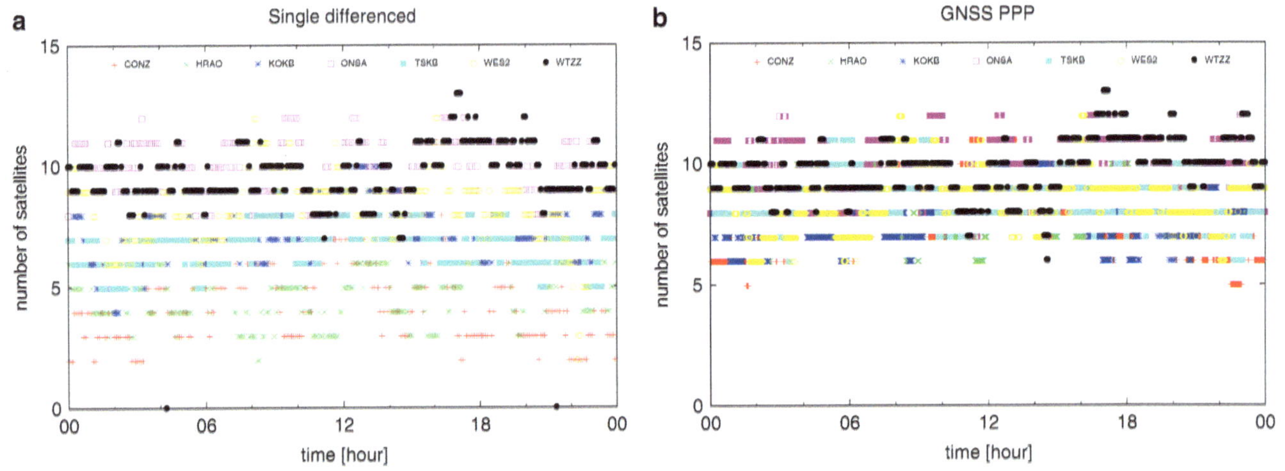

Fig. 4 Number of satellites observed at the seven sites per observation epoch during the first 24-h session. Each symbol depicts a particular site. (More explanation about the legend are added in Fig. 2)

1. First, we process GNSS data using precision point positioning (PPP) technique with the c5++ software.
2. The ionospheric delays are cancelled out using linear combination of double frequencies and the estimated phase ambiguities, phase center variations (PCV) and phase wind-up effects are subtracted from the original phase measurements.
3. We take differences of the corrected phase measurements between two stations which are receiving the same satellite signals like as single difference of the GNSS processing technique.
4. Those difference values (hereafter "single differenced GNSS data") are the input data in VieVS.

The GNSS PPP solutions use 2,599,951 observations in total with a 30-s interval for 7 stations during CONT11. On the other hand, the data sampling interval of difference values is 5 min, thus the total number of data is 199,689 throughout 15 days. Two baselines (CONZ-TSKB and HRAO-KOKB) do not have any observations at all since the lengths of baselines are too long to observe common satellites. The number of observations is clearly below that of the GNSS PPP solutions even though the sampling interval is the same, because the satellite has to be visible from at least two sites. Figure 4 shows number of satellites observed at the sites per observation epoch during the first 24-h session. The stations which have relatively short baselines, e.g. ONSA and

WTZZ, in single differenced GNSS data can observe as many satellites as GNSS PPP solutions while southern hemisphere stations, i.e. CONZ and HRAO, can see mostly less than 5 satellites per epoch.

The single differenced GNSS data include geometric delays (distance differences between a satellite and two stations), atmospheric delays, clock errors of stations, and measurement errors. To process the data properly, we do not need to take account for the retarded baseline effect here but should consider the different positions of the satellite at different emission times (t_{01} and t_{02}, Fig. 3b). In VieVS, we slightly modified the original model of VLBI observation to GNSS satellites to adapt for those features.

Table 2 Models and a priori used to process single differenced GNSS data in this work

Satellite position	IGS Final Orbit
Station position	GNSS PPP solution
Solid Earth tide	IERS 2010 conventions (Petit and Luzum 2010)
Ocean loading	FES2004 (Lyard et al. 2006)
Earth orientation parameters	IGS Final
Troposphere delay	Zenith hydrostatic delays from GPT (Böhm et al. 2007)
	VMF (Böhm et al. 2006)
	No a priori for troposphere gradient
Ionosphere	Corrected by using ionospheric linear combination in the PPP processing

Table 3 Parameterization in this analysis

	Parameters	Interval
Station clocks	Piecewise linear offsets	5 min
	Clock rate and quadratic term	1 day
Zenith wet delay	Piecewise linear offsets	2 h
Gradients	East and north components	6 h
Earth orientation parameters	Piecewise linear x-pole, y-pole, dUT1, no nutation terms	1 day
Station coordinates	NNR/NNT constraints for all stations to PPP solution	1 day

For the estimation, we introduced the IGS Final Orbit for GNSS satellite positions and keep them fixed. The IGS Final was taken as a priori EOP (Table 2). We also corrected for tidal effects (solid Earth tides, tidal ocean loading, tidal atmosphere loading, pole tide and ocean pole tide) at each station. We estimated clock parameters, zenith wet delays, troposphere gradients, polar motion, dUT1, and station coordinates (Table 3). Since all stations are connected to atomic clocks, we estimated piecewise linear offsets for clocks. Figure 5a shows the post-fit residual plot of the first day in CONT11 using VieVS, while Fig. 5b shows the post-fit residual plot of the GNSS PPP solution for the first day as derived from c5++. The root mean square value of the post-fit residuals of the VieVS solution is 3.3 cm, and that of the GNSS PPP solution is less than 2.0 cm. The analyses of the other sessions yield similar statistical values. The residuals of the VieVS solution still contain systematic variations, however, the origin is not fully clear at the moment and needs further investigations.

Station position repeatabilities of GNSS PPP and single differenced solutions over the 15 days are plotted for each station in Fig. 6. Average values of all stations are 5.1 mm, 6.6 mm and 11.6 mm for north, east and up components, respectively, of single differenced solutions, compared to 1.1 mm, 2.4 mm, and 4.0 mm for the GNSS PPP solutions. The degradation in coordinate repeatabilities is likely due to unmodelled systematic effects rather than to the number of observations. However, the reason for the systematic effect still has to be found.

Zenith wet delays of GNSS PPP and single differenced solutions during 15 days are plotted for each station in Fig. 7.

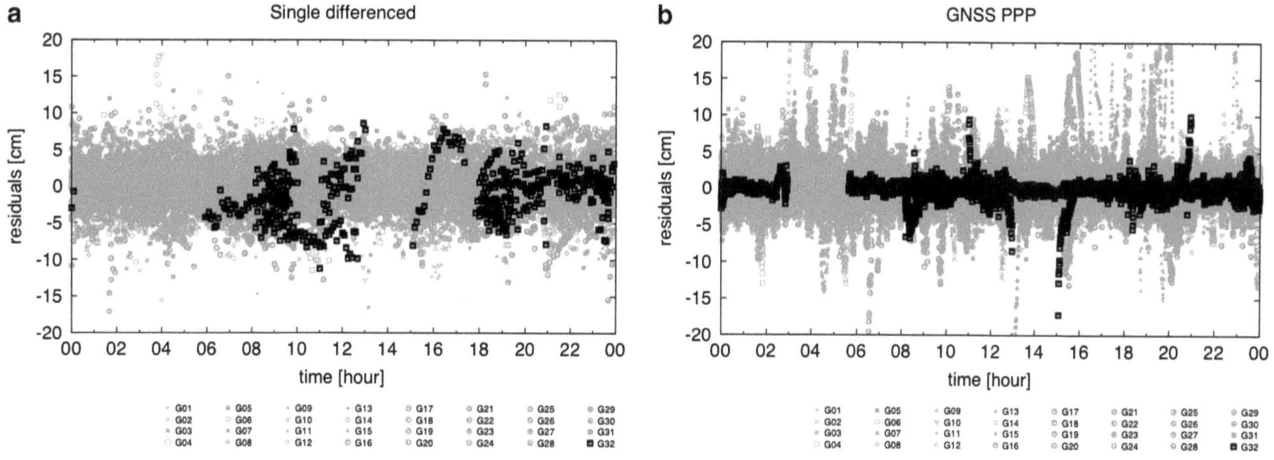

Fig. 5 Post-fit residuals of single differenced GNSS data processed by VieVS (**a**) and GNSS PPP solutions from c5++ (**b**) for all stations and satellites during the first 24-h session. Each symbol depicts a particular satellite. Here, we mark only one satellite (G32) in *black* to see systematic variation clearly while other satellites are shown in *gray*

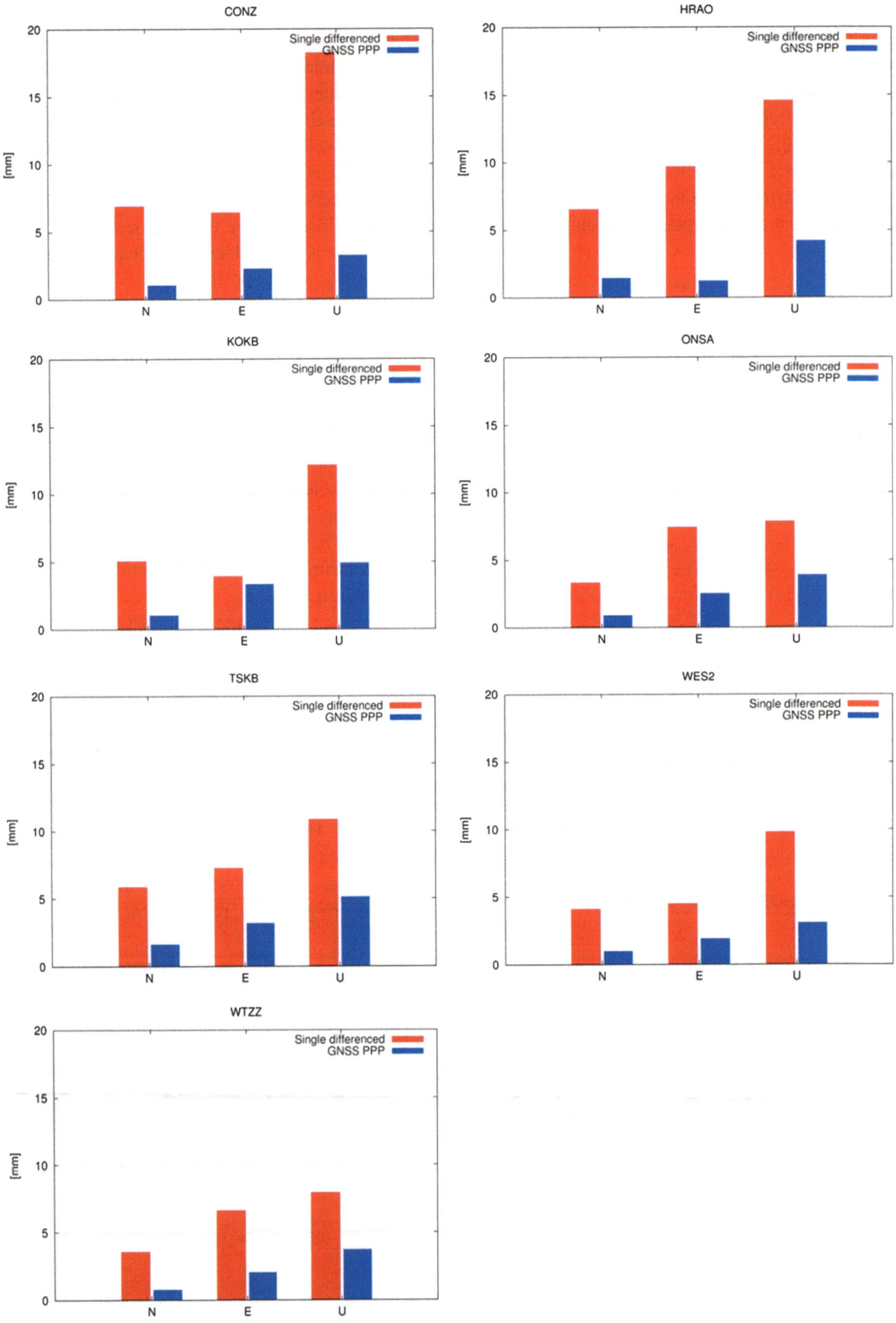

Fig. 6 Station position repeatabilities of GNSS PPP (*blue*) and single differenced (*red*) solutions for each station during 15 days

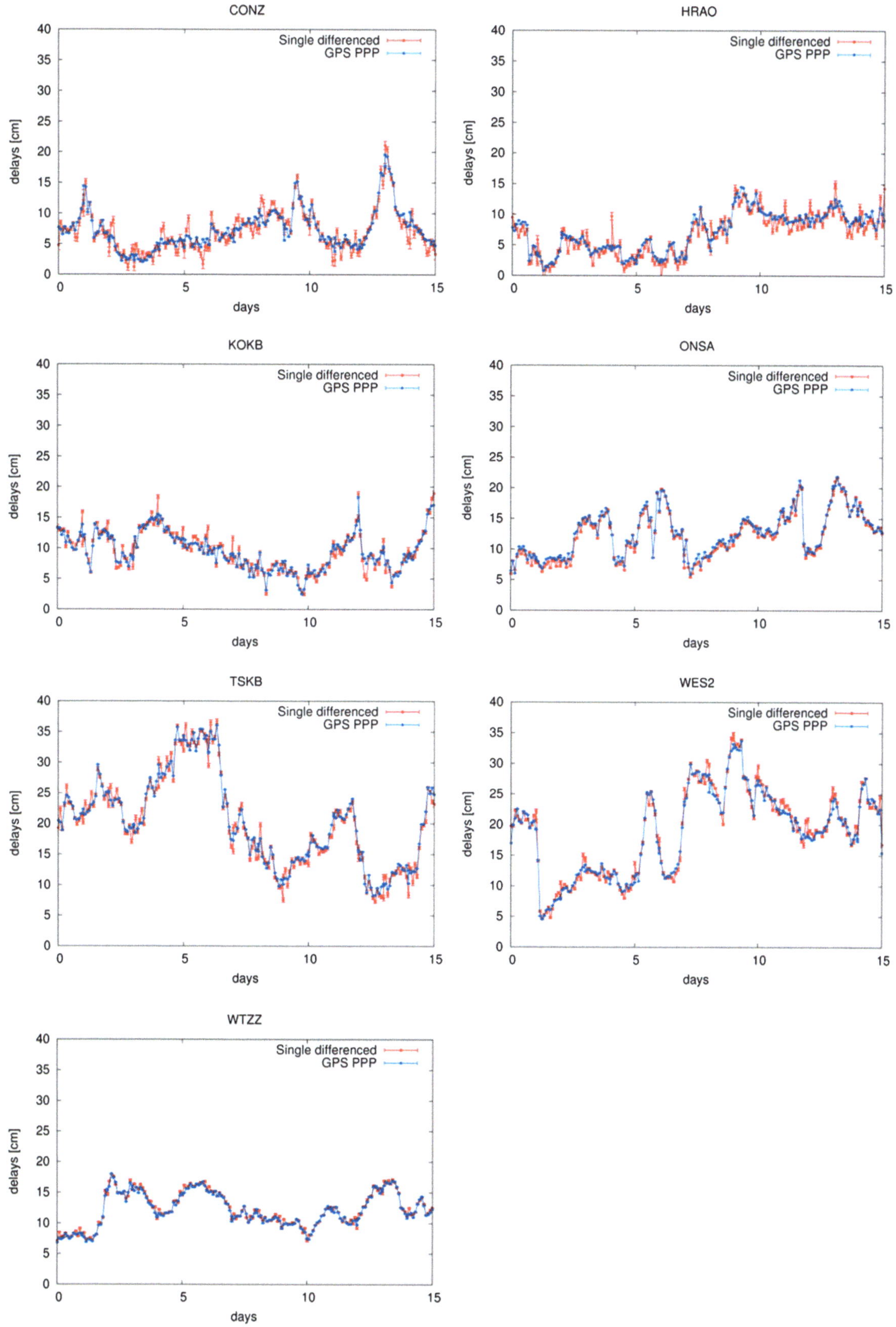

Fig. 7 Zenith wet delays which are estimated in GNSS PPP (*blue*) and single differenced (*red*) solutions for each station during 15 days

Differences between the series are in the range between 4.2 and 14.9 mm. Those disagreements do also mirror the station position repeatabilities of single differenced solutions.

During processing of the single differenced GNSS data, we gave identical weights to all observations. However, as we can see from Fig. 5b, there are big errors at low elevations which should be accounted for in the stochastic model. Furthermore, with single differenced data we have to consider different elevations at both sites at the same time. In the case of real observations, we can also utilize measurement errors from the correlator directly as weights.

5 Conclusions

The purpose of this work is to develop and test VieVS modules for the GNSS part of GV hybrid observations which will be obtained by GNSS antennas, VLBI backends and VLBI correlators. In the meantime, the modules will be also capable of analyzing observations with VLBI radio telescopes to GNSS satellites. Since we do not have any real observation data currently, we created VLBI-like GNSS observations, i.e. single differenced data, based on real GNSS observations. The data were successfully processed in VieVS and we found centimeter level accuracy of the models involved. Compared to GNSS PPP solutions, the residuals are still worse, which needs to be further investigated.

This approach, VLBI correlation of GNSS signals, also has a potential to be used to get a better understanding of the satellite transmissions since the sum of the phase delay around a triplet of VLBI baselines does not sum to zero due to source structure while the phase differences of GNSS measurements computed for baselines would sum to zero.

In the future, we will combine VLBI and GNSS data and estimate site-wise common parameters (Hobiger and Otsubo 2014) together, i.e. troposphere parameters and clock parameters in addition to station coordinates and Earth orientation parameters. Then, we will assess the impact of the GV hybrid combination solution on the geodetic products.

Acknowledgements We thank the three anonymous reviewers and the associate editor for their helpful comments and suggestions. This work has been supported by project Hybrid GPS-VLBI (M1592) which is funded by the Austrian Science Fund (FWF) and Fellowship FS1000100037 of the Australian Research Council.

References

Böhm J, Werl B, Schuh H (2006) Troposphere mapping functions for GPS and very long baseline interferometry from European centre for medium-range weather forecasts operational analysis data. J Geophys Res 111:B02,406. doi:10.1029/2005JB003629

Böhm J, Heinkelmann R, Schuh H (2007) Short note: a global model of pressure and temperature for geodetic applications. J Geod 81(10):679–683. doi:10.1007/s00190-007-0135-3

Böhm J, Böhm S, Nilsson T, Pany A, Plank L, Spicakova H, Teke K, Schuh H (2012) The new Vienna VLBI software VieVS. In: Kenyon S, Pacino MC, Marti U (eds) Geodesy for planet earth, international association of geodesy symposia, vol 136. Springer, Berlin/Heidelberg, pp 1007–1011. doi:10.1007/978-3-642-20338-1-126

Haas R, Neidhardt A, Kodet J, Plötz C, Schreiber U, Kronschnabl G, Pogrebenko S, Duev D, Casey S, Marti-Vidal I, Yang J, Plank L (2014) The Wettzell-Onsala G130128 experiment VLBI-observations of a GLONASS satellite. In: D DB, Baver K, Armstrong K (eds) IVS 2014 general meeting proceedings. Science Press, Beijing

Hellerschmied A, Böhm J, Plank L, Neidhardt A, Kodet J, Haas R (2015) Scheduling VLBI observations to satellites with VieVS. In: Proceedings of IAG Commission 1 Symposium 2014. doi:10.1007/1345_2015_183

Hobiger T, Otsubo T (2014) Combination of GPS and VLBI on the observation level during CONT11 – common parameters, ties and inter-technique biases. J Geod 88(11):1017–1028. doi:10.1007/s00190-014-0740-x

Kwak Y, Kondo T, Gotoh T, Amagai J, Takiguchi H, Sekido M, Ichikawa R, Sasao T, Cho J, Kim T (2011) Validation experiment of the GPS-VLBI hybrid system. In: Alef W, Bernhart S, Nothnagel A (eds) Proceedings of the 20th EVGA meeting, pp 154–157

Lyard F, Lefevre F, Letellier T, Francis O (2006) Modelling the global ocean tides: modern insights from FES2004. Ocean Dyn 56:394–415. doi:10.1007/s10236-006-0086-x

Petit G, Luzum B (2010) IERS Conventions (2010). IERS technical note 36. Verlag des Bundesamtes fü Kartographie und Geodäsie

Plank L, Böhm J, Madzak M, Tierno Ros C, Schuh H (2012) Processing SELENE differential VLBI data. In: D DB, Baver K (eds) IVS 2012 General meeting proceedings, NASA/CP-2012-217504, pp 291–295

Plank L, Böhm J, Krasna H, Schuh H (2013) VLBI satellite tracking for precise coordinate determination: a simulation study. In: Zubko N, Poutanen M (eds) Proceedings of the 21st meeting of the European VLBI group for geodesy and astronomy, Reports of the Finnish Geodetic Institute, pp 105–109

Plank L, Böhm J, Schuh H (2014) Precise station positions from VLBI observations to satellites: a simulation study. J Geod 88:659–673. doi:10.1007/s00190-014-0712-1

IVS Contribution to ITRF2014

S. Bachmann, L. Messerschmitt, and D. Thaller

Abstract

The contribution of the International VLBI Service for Geodesy and Astrometry (IVS) to the ITRF2014 generation is presented. The ITRF2014 is an inter-technique combined product of the IERS ITRS Center, combining contributions from the four space geodetic techniques -DORIS, GNSS, SLR, and VLBI. The VLBI contribution is provided by the IVS Combination Center at the German Federal Agency for Cartography and Geodesy (BKG), and is comprised of station coordinates and Earth Orientation Parameters derived from data collected between 1979 and 2014. The status and results of the IVS contribution, as well as a comparison with other space geodetic techniques, are shown.

Keywords

EOP • Intra-technique combination • ITRF • TRF • VLBI

1 Introduction

The International VLBI Service for Geodesy and Astrometry (IVS) is organized under the umbrella of the International Association of Geodesy (IAG) and the International Astronomical Union (IAU), and contributes to the International Earth Rotation and Reference Systems Service (IERS). IVS is tasked with organizing VLBI observing sessions and analyzing VLBI data in order to provide products to the international community. The IVS Combination Center is located at the German Federal Agency for Cartography and Geodesy (BKG). The task of the Combination Center is to produce combined IVS products, including Earth Orientation Parameters (EOP), station coordinates, and Terrestrial Reference Frames (TRF) based on VLBI observations. The routine solutions consist of twice-weekly observations of 24 h sessions R1 and R4. These stations are derived from a combination of SINEX files supplied by multiple IVS Analysis Centers (AC). The files are combined based upon the normal equations of each individual contribution. The current tasks and components of the IVS are summarized in the IVS annual reports.[1] The International Terrestrial Reference Frame (ITRF) is an inter-technique combined product that incorporates the four space geodetic techniques (DORIS, GNSS, SLR, and VLBI) into one frame. The latest realization of the ITRF is ITRF2008 (Altamimi 2011), and in early 2014, the IERS ITRS Center sent all ITRF-affiliated technical services a Call for Participation for the next ITRF. The data to provide for the original Call for Participation included observations until the end of 2013. In November 2014, the services were requested to provide an additional year of data for 2014. The data presented at REFAG (and in this article) are based on the data included in the original Call

IAG Commission 1 Symposium 2014: Reference Frames for Applications in Geosciences (REFAG2014), 13-17 October, 2014 Kirchberg, Luxembourg.

S. Bachmann (✉) • L. Messerschmitt • D. Thaller
Federal Agency for Cartography and Geodesy, Frankfurt am Main, Germany
e-mail: Sabine.bachmann@bkg.bund.de

[1]http://ivscc.gsfc.nasa.gov/

T. van Dam (ed.), *REFAG 2014*, International Association of Geodesy Symposia 146,
DOI 10.1007/1345_2015_136

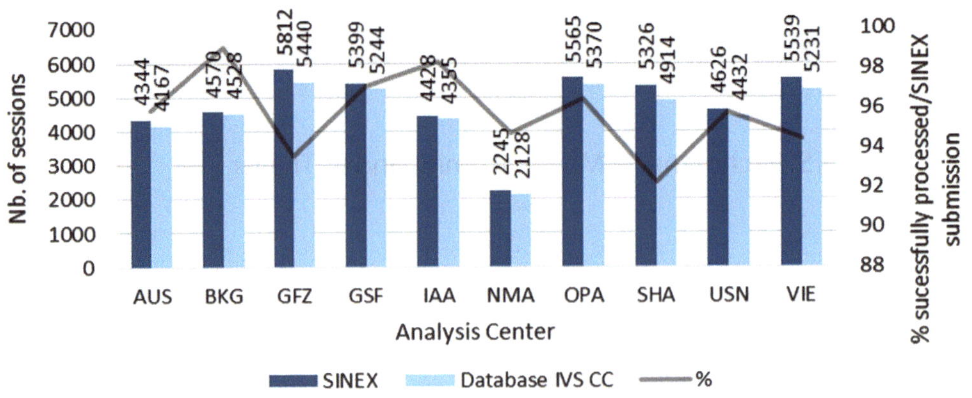

Fig. 1 Availability of SINEX files and successfully combined sessions (Database IVS CC)

for Participation without the extension for 2014. The strength of VLBI data for the ITRF2014 lies in the contribution of the scale and the long time span of observation data.

2 Data and Combination Strategy

The VLBI contribution to ITRF is similar to the routine products in that it is a combined solution based on normal equations with an identical approach as the routine combination, but with more ACs and more data. This means that in addition to analyzing the R1 and R4 sessions, all 24 h VLBI sessions are utilized. Figure 1 gives an overview of the availability of SINEX files and successfully combined sessions. The left bar of each pair (dark blue color) shows the number of input SINEX files for each AC, while the right bar of each pair (light blue color) shows the number of successfully combined sessions. A session is considered successfully combined when it has the correct formal input format, the normal equations can be solved, and no signs of outliers in station coordinates or EOP are identified. The different numbers of submitted sessions between the ACs are a result of the different time spans of the analyzed sessions. Only a few number of ACs submitted sessions from the end of the 1970s onward, but most were operational by the mid-1980s or early 1990s. Overall, about 6,000 24 h sessions have been processed. The grey line in Fig. 1 shows the percentage of successfully processed sessions with respect to submitted SINEX files. The percentage value varies between 92 and 99% with a mean percentage of successfully analyzed sessions of 95.8%.

For the purposes of ITRF inter-technique combination, the combined SINEX files are used as input data for the IVS contribution (Fig. 2). Since ITRF2005, the following main developments have been observed:

- Increasing number of IVS ACs between 2008 and 2014:
  ITRF2008: 7 ACs using 4 different s/w packages
  ITRF2014: 10 ACs using 5 different s/w packages
- Increasing number of observed sessions:
  ITRF2008: between 3,433 and 4,592 depending on the AC
  ITRF2014: between 4,344 and 5,812 depending on the AC
- Extended time span:
  ITRF2008: 1979.0–2009.0
  ITRF2014: 1979.0–2015.0
- Implementation of IERS2010 Conventions (subdaily dUT model and precession/nutation model, Petit and Luzum 2010).

Table 1 gives an overview of the current IVS ACs. Twelve ACs are institutions from a large number of countries, spread all over the world and using different software packages. Five of the ACs are contributing to the routine combination as well as to the ITRF2014. Among the remaining ACs, some only participate in the ITRF2014 project, others only contribute to the rapid routine combination using R1 and R4 sessions, and a few are currently under review in order to become an operational AC within the IVS.

Figure 2 depicts the combination strategy at the BKG IVS Combination Center, which is described in greater detail in Böckmann et al. (2010a). The foundation of this combination strategy is a repository of session-wise SINEX files containing normal equations available at the IVS data center.[2] VLBI contributes a full set of EOP – pole coordinates in X- and Y-Pole plus their rates, UT1-UTC, LOD and the nutation angles dX and dY – and geocentric station coordinates X, Y, Z. An epoch transformation to 12 h UT is performed in order to align the different epochs in the various SINEX files with the other space-geodetic techniques. A transformation to equal a priori station coordinates is also performed. No-net

[2] ftp://ivs.bkg.bund.de/pub/vlbi/ITRF2013/daily_sinex/

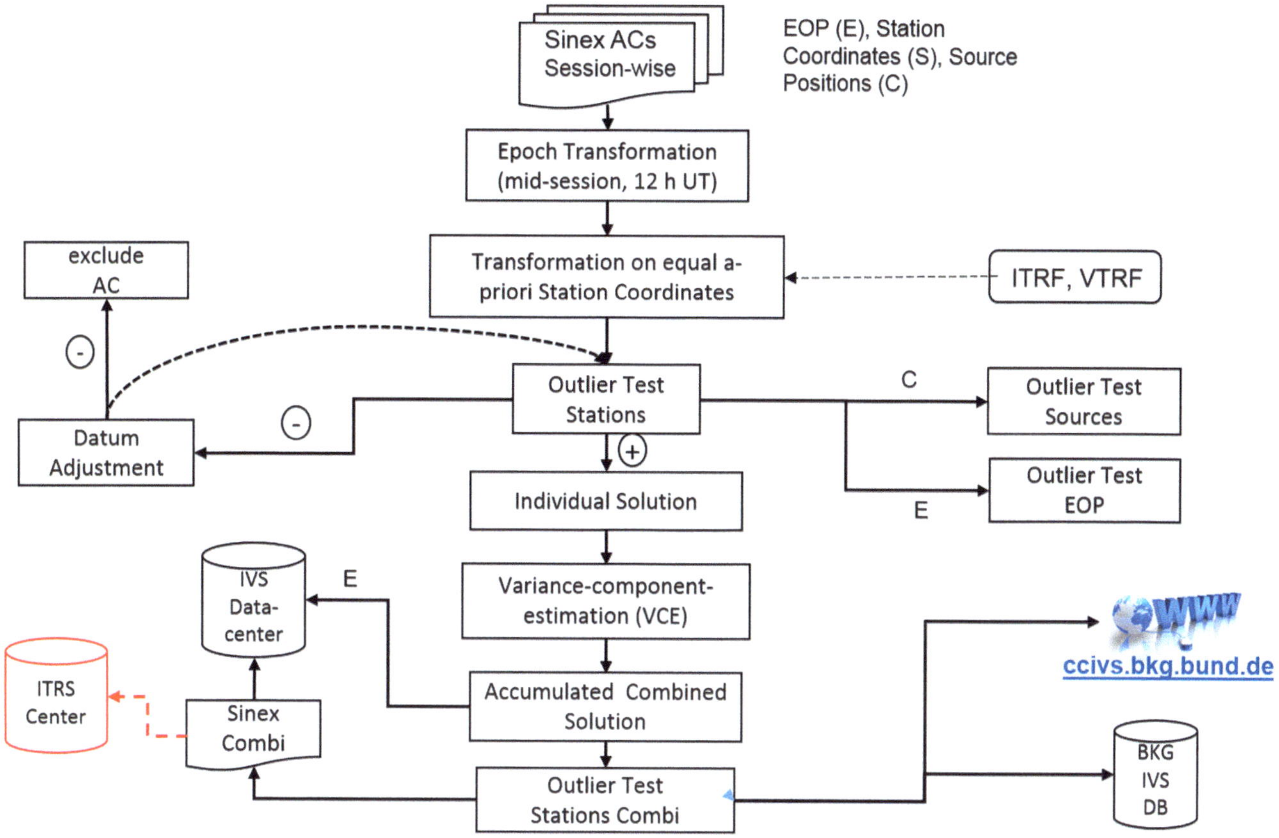

Fig. 2 Combination strategy applied at the IVS Combination Center at BKG

Table 1 Current IVS analysis centers

AC	Name	Software	Operational AC	ITRF2014 contr.
AUS	Geoscience Australia, Australia	OCCAM(LSC)	No	Yes
BKG	Federal Agency for Cartography and Geodesy, Germany	Calc/(nu)Solve	Yes	Yes
CGS	Centro di Geodesia Spaziale, Italy	Calc/(nu)Solve	Under review	No
DGFI	German Geodetic Research Institution	OCCAM(LSM)	Yes	No
GFZ	German Research Center for Geosciences	VieVS	Under review	Yes
GSFC	Goddard Space Flight Center, USA	Calc/(nu)Solve	Yes	Yes
IAA	Institute of Applied Astrometry, Russia	Quasar	Yes	Yes
NMA	Norwegian Mapping Authority, Norway	Geosat	No	Yes
OPAR	Observatory of Paris, France	Calc/(nu)Solve	Yes	Yes
SHAO	Shanghai Observatory, China	Calc/(nu)Solve	No	Yes
USNO	US Naval Observatory, USA	Calc/(nu)Solve	Yes	Yes
VIE	Vienna University of Technology, Austria	VieVS	No	Yes

rotation (NNR) and no-net translation (NNT) constraints on datum stations are added prior to the inversion of the normal equations. Stations with a sufficient long observation history (>2.5 years) are used as datum stations. Earthquake affected stations are excluded from datum definition until a (new) stable position can be estimated (~1.5–2 years after the Earthquake). The resulting solution is used to apply outlier tests for station coordinates and EOP based on a Least Median of Squares (LMS) method, which is described in Bachmann and Lösler (2012). If a station is identified as outlier, the choice of the datum definition can be corrected by choosing different datum stations. The AC corresponding to the outlier will then be excluded from the combination, if the outlier test still finds statistically deviant data. The final weighting factor of each individual contribution to the combined solution is determined by a variance-component estimation (VCE) based on individual solutions. The combined solution is generated via an aggregation of individ-

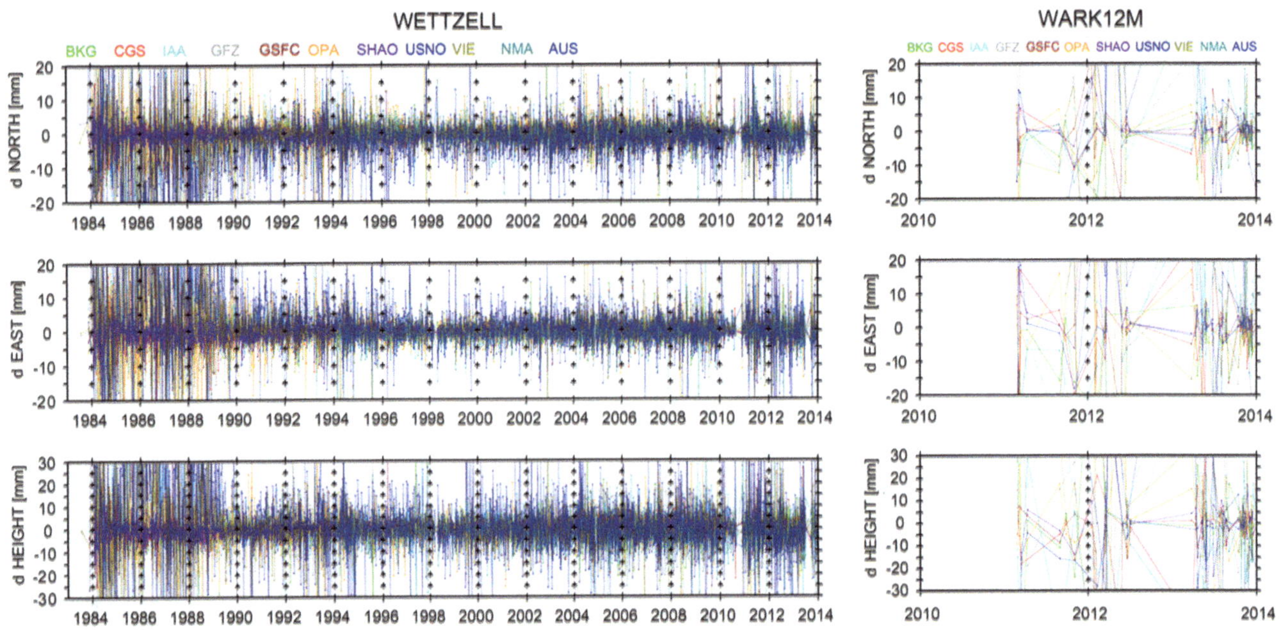

Fig. 3 Time series of station WETTZELL (*left*) and station WARK12M (*right*)

ual contributions that are each adjusted by their respective weighting factors. Combined SINEX files for the ITRF2014 contribution are then submitted to the IERS ITRS Center for the inter-technique combination. SINEX files resulting from the intra-technique combined solution will be submitted to the IVS Datacenter. Figure 1 in Seitz et al. (2012) provides an overview of the inter-technique combination process and the integration of the IVS contribution into the generation of a multi-technique TRF.

EOPs and station coordinates will also be published on the IVS Combination Center's database and homepage[3] as soon as the ITRF2014 has been officially released by the IERS ITRS Center.

3 Preliminary Results

Since the inter-technique combination for ITRF2014 is not yet finished, the results presented hereafter are still preliminary and changes in the input data are still expected. The outcomes and challenges of the IVS ITRF2014 contribution are presented by means of station coordinates, VLBI Terrestrial Reference Frame (VTRF) and EOP time series.

4 Station Coordinates and VLBI TRF

The basis of data for the different stations is quite diverse. Figure 3 shows two station time series. The left side of Fig. 3 shows station WETTZELL with nearly 30 years of regular and frequent VLBI observations, whereas the right side of Fig. 3 shows station WARK12M, which began VLBI observations in 2011. The station plot for WETTZELL shows a low scatter due to a long observation history, a geometric advantageous location and the participation in sessions with globally well distributed stations. The data for WARK12M is more scattered due to its location and its participation in mainly regional sessions. The primary challenge of the IVS combinations is to cope with the data diversity of the different stations and AC contributions, because the station coordinates contribute directly to the VTRF. Figure 4 shows the overall weighted root mean squares (WRMS) of the north, east and height components of the combined and individual contributions over all stations with more than 30 observed sessions. The WRMS values rely on equal sessions for all ACs. Since AC NMA data only contain sessions starting in 1994, only these sessions are considered

[3]http://ccivs.bkg.bund.de/

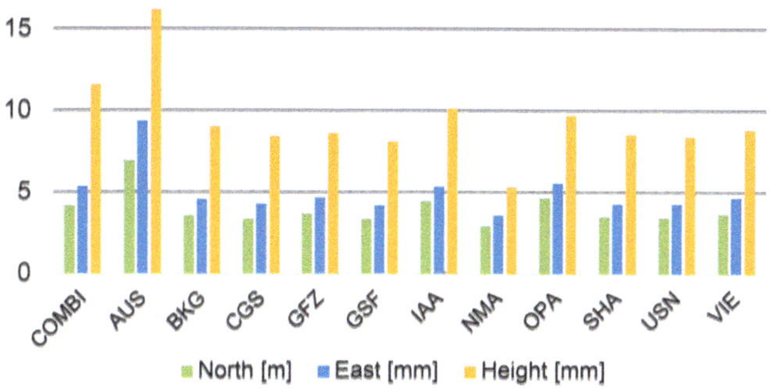

Fig. 4 WRMS of north, east and height component for the combined solution and all ACs over all stations using equal sessions

Table 2 WRMS quantiles in north, east and height for stations with more than 30 observed sessions for the combined solution

Quantile (%)	N (mm)	E (mm)	H (mm)
10	2.96	2.84	8.50
25	3.49	3.51	10.28
50	4.43	5.24	13.7
75	5.86	6.98	16.52
90	7.04	9.46	20.89
95	8.10	10.21	26.74
Min.	2.77	2.14	7.49
Max.	8.49	13.75	32.15

for the WRMS calculation. The WRMS is about 4–5 mm for the north and east components and 9–11 mm for height component. Table 2 shows the WRMS quantiles for the same set of stations. The median WRMS is 4–5 mm for north and east and 13.7 mm for height, while none of the stations reaches a WRMS of less than 1 mm. 95% of the stations have a WRMS of better than 8.1 mm for north, 10.2 mm for earth and 26.7 mm for the height component. The determination of the height component by VLBI is less accurate than the north and east components mainly due to the correlation with the troposphere. The hypotheses of the combination strategy is that the WRMS of a combined solution is better than the WRMS of each individual contributions. It has thus to be investigated with the final data sets if AC AUS leads to an elevated WRMS of the combined solutions as Fig. 4 and Table 2 may assume.

Station coordinate information is used to establish station velocities and to generate a VLBI based terrestrial reference frame (VTRF). Table 3 shows the Helmert transformation parameters of the VTRF generated by ITRF2014 VLBI contributions and the preceding ITRF2008. As expected, the rotation angles as well as the translations are very small due to the no-net rotation and no-net translation conditions applied on the datum stations in order to keep the two frames consistent. The standard deviation of these parameter sets shows that they are of negligible significance, except for the scale.

Table 3 Helmert transformation between VTRF and ITRF2008 (epoch 2005.0)

	Position	σ_{Pos} +/.
t_x (mm)	0.1	0.6
t_y (mm)	−0.00	0.6
t_z (mm)	−0.8	0.5
r_x (mas)	−0.035	0.02
ry (mas)	−0.014	0.02
r_z (mas)	−0.006	0.02
Scale (ppb)	0.38	0.08

Finally, leaving the scale as the only parameter directly accessible by VLBI and, together with SLR, setting the scale of the ITRF2014. The value of the calculated scale (0.38 ppb, Table 3) is comparable to the one detected by Böckmann et al (2010b), i.e. 0.4 ppb when comparing the ITRF2005 to the VLBI contribution.

5 Earth Orientation Parameters

Figure 5 shows the WRMS of the differences between the individual contributions with respect to the combined solutions (left side) and with respect to C04 EOP series[4] (right side). The degree of precision of each contribution is reflected by the WRMS, which is in a range of 0.05 to 0.15 mas for the different ACs. An exception to the aforementioned range is the AC AUS, where the higher WRMS can be explained by a different implementation of the IERS Conventions. The comparison with the C04 series shows that the combined solution has a lower WRMS value than the individual contributions. These statistical values are generated using only sessions that have been analyzed by all ACs, i.e. only sessions since 1994 have been used for the comparisons.

6 Conclusion and Outlook

At the time of the REFAG meeting, i.e., October 2014, the data set for the IVS contribution to ITRF2014 is almost complete. Final reprocessing efforts have been submitted to the IVS data center, and ACs have completed all necessary minor changes and adjustments. The IVS has received an unprecedented number of contributions for this contribution to the ITRF – a tremendously positive situation that brings

[4]http://www.iers.org/IERS/EN/DataProducts/EarthOrientationData/eop.html

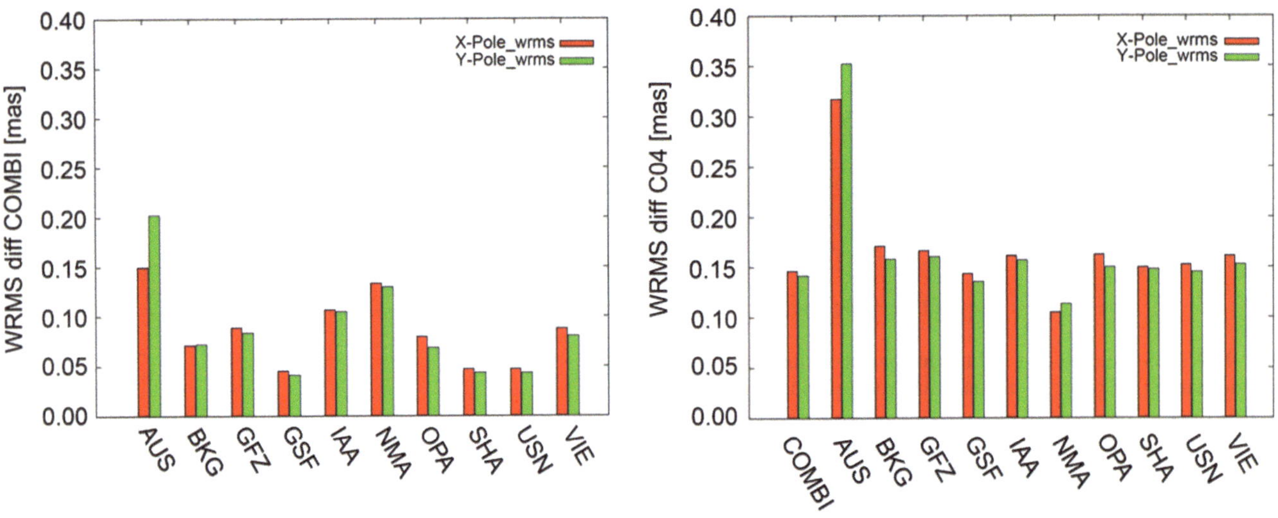

Fig. 5 WRMS of X-Pole and Y-Pole w.r.t. combined solution (*left side*) and w.r.t. C04 series (*right side*) using equal sessions for all ACs

additional challenges when combining individual contributions and ensuring overall data homogeneity. The results for station coordinate time series are promising. The strength of VLBI and its contribution to the ITRF2014 is its ability to determine the scale and to set the scale for the ITRF2014 together with SLR. The EOP results show good agreement between the individual contributions as well as when compared to the IERS 08 C04 series.

At the end of November 2014, finally the decision has been made by the IERS ITRS Center to extend the time span of data contributions for one additional year for all space geodetic techniques. Therefore, the results presented here have to be considered as preliminary.

References

Altamimi Z, Collilieux X, Métivier L (2011) ITRF2008: an improved solution of the international terrestrial reference frame. J Geod 85:457–473

Bachmann S, Lösler M (2012) IVS combination center at BKG – robust outlier detection and weighting strategies. In: Behrend D, Baver KD (eds) IVS 2012 general meeting proceedings of NASA/CP-2012-217504, pp 266–270

Baver KD, Behrend D, Armstrong KL (eds) (2014) International VLBI Service for Geodesy and Astrometry 2013 Annual Report, NASA/TP-2014-217522, http://ivscc.gsfc.nasa.gov/publications/ar2013/index.html

Bizouard C, Gambis D (2010) The combined solution C04 for Earth Orientation Parameters consistent with International Terrestrial Reference Frame 2008 IERS Earth Orientation Product Centre, http://hpiers.obspm.fr/iers/eop/eopc04/C04.guide.pdf

Böckmann S, Artz T, Nothnagel A, Tesmer V (2010a) International VLBI Service for Geodesy and Astrometry: Earth orientation parameter combination methodology and quality of the combined products. J Geophys Res 115. doi:10.1029/2009JB006465

Böckmann S, Artz T, Nothnagel A (2010b) VLBI terrestrial reference frame contributions to ITRF2008. J Geod 84:201–219. doi:10.1007/s00190-009-0357-7

Petit G, Luzum B (2010) IERS Conventions (2010) IERS Technical Note No. 36, International Earth Rotation and Reference Systems Service (IERS), http://www.iers.org/IERS/EN/Publications/TechnicalNotes/tn36.html

Seitz M, Angermann D, Bloßfeld M, Drewes H, Gerstl M (2012) The 2008 DGFI realization of the ITRS: DTRF2008. J Geod 86(12):1097–1123. doi:10.1007/s00190-012-0567-2

Antenna Axis Offsets and Their Impact on VLBI Derived Reference Frames

Tobias Nilsson, Julian A. Mora-Diaz, Virginia Raposo-Pulido, Robert Heinkelmann, Maria Karbon, Li Liu, Cuixian Lu, Benedikt Soja, Minghui Xu, and Harald Schuh

Abstract

We have estimated the antenna axis offsets of VLBI (Very Long Baseline Interferometry) telescopes using VLBI data from 1995 to 2013. These were compared to values estimated in terrestrial surveys. We generally found an agreement of a few millimeters, however, occasionally the differences were larger than 1 cm. Furthermore, we investigated the impact of the axis offset on the terrestrial and the celestial reference frames derived by VLBI. This was done by comparing the reference frames estimated simultaneously with the axis offset with those estimated when the axis offsets were fixed to their a priori values. For a telescope with azimuth-elevation mount we found that an error in the axis offset of 1 cm causes an error in the station height of about 1.3 cm.

Keywords

Axis offsets • Reference frames • VLBI

1 Introduction

A VLBI (Very Long Baseline Interferometry) radio telescope is a huge structure (diameter 5–100 m), and parts of it are moving around when the telescope is pointed in different directions. Nevertheless, just as for other space geodetic techniques such as GNSS (Global Navigation Satellite Systems), all measurements need to be referred to a unique fixed reference point. The natural selection of a reference point for a VLBI telescope is the intersection of the two rotation axes of the telescope. This point is the only point invariant to any rotation around the two axes, i.e. its position does not depend on what direction (azimuth and elevation angles) the telescope is pointing in. However, there are antennas where the two axes do not intersect, either due to the antenna construction with having a mechanical axis offset of up to a few meters or due to misalignment of the two axes. Thus in reality, the reference point is defined as the point on the primary axis which is closest to the secondary axis. The offset between this point and the secondary axis is called the axis offset and needs to be considered in the VLBI data analysis. The location of the reference point and the axis offset is illustrated in Fig. 1 for a telescope with an azimuth-elevation mount. For more details and descriptions of other antenna mounts see Nothnagel (2009).

There are two ways of determining the antenna axis offset of a VLBI telescope. One way is to measure it in a local terrestrial survey (see e.g. Harvey 1991; Combrinck and Merry 1997; Lösler et al. 2013). Typically this is done by measuring the position of one or more markers placed on the antenna dish as the antenna is pointed in different directions. When the antenna is rotated around one of its axes, the reference markers will make an almost circular motion around this axis. Thus it is possible to determine the positions of the two axes of rotation and hence the axis offset, from these measurements. The other possibility is to estimate the axis offsets in the VLBI data analysis. It is however often difficult to separate the axis offsets from other estimated

T. Nilsson (✉) • J.A. Mora-Diaz • V. Raposo-Pulido • R. Heinkelmann • M. Karbon • L. Liu • C. Lu • B. Soja • M. Xu • H. Schuh
Department 1: Geodesy and Remote Sensing, Helmholtz Centre Potsdam, GFZ German Research Centre for Geosciences, 14473 Potsdam, Germany
e-mail: nilsson@gfz-potsdam.de

T. van Dam (ed.), *REFAG 2014*, International Association of Geodesy Symposia 146,
DOI 10.1007/1345_2015_126

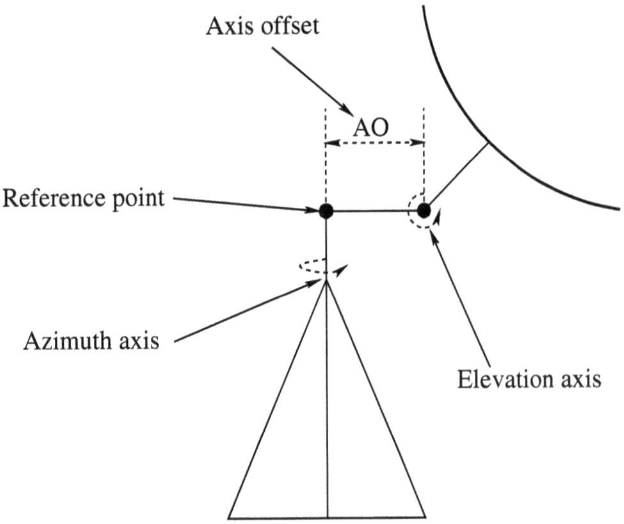

Fig. 1 Sketch of an azimuth-elevation mounted VLBI telescope. The location of the two axes, the reference point, and the axis offset are shown

parameters in the data analysis, thus the precision obtained from a single 24 h VLBI session is normally relatively low (Kurdubov and Skurikhina 2010). Hence, in order to obtain a reasonable precision the axis offsets are normally estimated in global solutions, where data from many VLBI sessions are combined. This is most reasonable because every observation taken by an antenna can contribute to this parameter, assuming that the axis offset is constant over time. Nevertheless, it is generally considered that the values obtained from VLBI data analysis are not as accurate as those obtained in a high precision terrestrial survey. Unfortunately, axis offsets from terrestrial surveys are only available for a few VLBI stations (see Table 1); thus for the other stations values estimated in a global VLBI solution have to be used. Within the IVS (International VLBI Service for Geodesy and Astrometry, Schuh and Behrend 2012) the former Analysis Coordinator, Axel Nothnagel, has compiled a list of recommended axis offsets (as well as thermal deformation parameters, etc.) for all geodetic VLBI antennas (http://vlbi.geod.uni-bonn.de/Analysis/Thermal/antenna-info.txt, Nothnagel 2009). In this list the axis offsets are from terrestrial surveys in the cases when such measurements are available, otherwise from a global VLBI solution made by Dan MacMillan (NVI Inc./NASA Goddard Space Flight Center) in 2013.

The aim of this study is to investigate how accurate the axis offsets can be determined in a global VLBI solution, as well as to study how errors in the axis offsets propagate to errors in other parameters estimated in the VLBI data analysis, particularly the station and the radio source coordinates. This is done by making two global VLBI solutions with data from the period 1995–2013, one with and one without estimating the axis offsets. The accuracy of the axis offsets estimated in the former solution are evaluated by comparing it to the recommended IVS values, especially those obtained from terrestrial surveys. Furthermore, we compare terrestrial and celestial reference frames (TRF and CRF) obtained from the two solutions in order to evaluate how these frames are affected by the estimation of axis offsets.

2 Data Analysis

We calculated the global VLBI solutions using VLBI data from the period 1995–2013. We used all geometrically stable VLBI sessions from the GFZ VLBI contribution to ITRF2013 (Heinkelmann et al. 2014), containing at least four stations. In total, there were 2,506 sessions, containing observations by 72 stations of 3,384 sources. The data were first analyzed with the GFZ version of the Vienna VLBI Software (Böhm et al. 2012), VieVS@GFZ, to set up the normal equations for each session. Then the VieVS module VIE_GLOB (Krásná et al. 2014a) was used to calculate the global solutions. In VIE_GLOB, we have recently implemented the possibility to estimate axis offsets as global parameters in addition to the global parameters available in standard VieVS. In the first global solution we estimated axis offsets for all stations, while in the second the axis offsets were fixed to the IVS recommended values. In both solutions station coordinates, station velocities, and radio source coordinates were estimated as global parameters (18 stations, observed in only a few sessions or which showed non-linear behavior, as well as the special handling sources, were session-wise reduced). The terrestrial datum was defined by applying no-net-translation (NNT) and no-net-rotation (NNR) condition relative to the ITRF2008 (Altamimi et al. 2011) coordinates of good quality stations. The datum of the CRF was realized by applying NNR condition to the ICRF2 (Fey et al. 2009) defining sources.

3 Results

3.1 Axis Offsets

In Table 1 the axis offsets estimated in the first global solution are listed for the stations which participated in more than ten sessions. For comparison, the IVS recommended values are listed. We can see that our estimates generally agree with the IVS values within 1 cm. The agreement is very good for most stations which participated in a large number of sessions and where the IVS values are from a VLBI solution. For example, for ALGOPARK, FORTLEZA, MATERA, TSUKUB32, and WESTFORD the differences are below 1 mm. This is probably because, to a very large extent, the sessions used in this work were also used for obtaining the IVS values. For stations which participated in

Table 1 Estimated axis offsets for stations which participated in at least ten sessions analyzed in this work

			IVS values		Our results			Coordinate difference		
Station	Mount	# Sess.	Axis offset (cm)	Sigma (cm)	Axis offset (cm)	Sigma (cm)	Difference (cm)	East (cm)	North (cm)	Up (cm)
AIRA	AZEL	42	−0.81	0.13	−0.20	0.26	−0.61	0.03	−0.06	−0.97
ALGOPARK	AZEL	554	0.33	0.04	0.34	0.06	−0.01	−0.01	0.02	−0.05
BADARY	AZEL	214	**0.25**	–	−0.34	0.09	0.59	−0.05	−0.02	0.88
BR-VLBA	AZEL	157	**213.46**	0.02	213.07	0.04	0.39	0.01	−0.01	0.54
CHICHI10	AZEL	32	−0.47	0.14	0.13	0.35	−0.60	0.05	−0.03	−0.78
CRIMEA	AZEL	97	**−0.18**	0.02	−0.27	0.19	0.09	−0.05	−0.01	0.20
DSS15	AZEL	32	−0.37	0.16	−0.49	0.22	0.12	0.00	−0.02	0.23
DSS45	AZEL	53	−0.11	0.18	0.33	0.25	−0.44	0.13	−0.07	−0.62
DSS65	AZEL	52	−0.26	0.09	0.08	0.16	−0.34	−0.04	−0.00	−0.46
DSS65A	AZEL	20	−0.26	0.09	−0.62	0.25	0.36	−0.01	0.01	0.63
EFLSBERG	AZEL	25	**1.03**	0.10	1.12	0.19	−0.09	−0.02	−0.00	−0.09
FD-VLBA	AZEL	161	**212.98**	0.29	213.20	0.05	−0.22	0.00	0.00	−0.24
FORTLEZA	AZEL	1156	0.35	0.05	0.42	0.06	−0.07	−0.06	0.05	−0.02
GGAO7108	AZEL	36	−0.26	0.25	2.41	0.39	−2.67	0.05	0.07	−5.24
GILCREEK	XYNS	887	728.46	0.01	729.20	0.02	−0.74	0.02	−0.01	0.44
HART15M	AZEL	25	**149.50**	–	149.53	0.31	−0.03	–	–	–
HARTRAO	EQUA	654	**669.53**	0.19	670.06	0.05	−0.53	−0.12	0.10	0.04
HN-VLBA	AZEL	156	213.01	0.03	212.99	0.05	0.02	−0.01	0.01	0.06
HOBART12	AZEL	145	1.29	0.10	1.24	0.12	0.05	0.06	0.44	0.37
HOBART26	XYEW	476	**819.13**	0.15	820.18	0.05	−1.05	0.10	0.08	0.65
KASHIM34	AZEL	70	−0.19	0.10	0.26	0.13	−0.45	0.02	−0.04	−0.83
KASHIMA	AZEL	45	−0.11	0.11	−0.07	0.20	−0.04	0.04	−0.04	−0.41
KATH12M	AZEL	103	0.43	0.13	0.47	0.13	−0.04	0.06	0.11	0.25
KOKEE	AZEL	1660	51.80	0.03	52.16	0.03	−0.36	0.08	−0.05	−0.37
KP-VLBA	AZEL	162	213.13	0.03	213.13	0.04	0.00	0.01	−0.00	0.05
LA-VLBA	AZEL	161	213.21	0.02	213.23	0.04	−0.02	0.00	0.00	0.02
MATERA	AZEL	662	−0.20	0.03	−0.16	0.04	−0.04	−0.04	−0.00	0.03
MEDICINA	AZEL	366	**183.01**	–	182.65	0.04	0.36	−0.03	−0.00	0.55
METSAHOV	AZEL	47	−0.24	0.20	−0.78	0.28	0.54	−0.03	−0.02	0.73
MIAMI20	AZEL	20	0.46	0.52	1.13	0.55	−0.67	–	–	–
MK-VLBA	AZEL	159	213.41	0.04	213.57	0.06	−0.16	0.07	−0.04	−0.15
NL-VLBA	AZEL	157	213.08	0.03	213.15	0.04	−0.07	−0.00	0.01	−0.07
NOTO	AZEL	105	183.16	0.08	183.53	0.11	−0.37	−0.05	−0.01	−0.48
NRAO20	AZEL	316	50.91	0.04	51.09	0.06	−0.18	−0.01	0.02	−0.26
NRAO85 3	EQUA	56	670.65	0.04	671.09	0.11	−0.44	–	–	–
NYALES20	AZEL	1282	**52.42**	0.02	52.14	0.02	0.28	−0.01	−0.02	0.46
OHIGGINS	AZEL	94	**0.00**	–	−1.59	0.62	1.59	−0.22	0.12	1.67
ONSALA60	AZEL	448	**−0.60**	0.04	−0.97	0.04	0.37	−0.02	−0.01	0.53
OV-VLBA	AZEL	160	213.10	0.03	213.20	0.04	−0.10	0.01	−0.01	−0.11
PARKES	AZEL	24	**0.06**	0.10	−5.27	0.58	5.33	0.02	−0.21	17.39
PIETOWN	AZEL	162	213.77	0.02	213.75	0.05	0.02	–	–	–
SANTIA12	XYNS	37	0.21	0.14	0.01	0.22	0.20	–	–	–
SC-VLBA	AZEL	156	213.46	0.05	213.34	0.06	0.12	−0.02	0.03	0.26
SESHAN25	AZEL	205	**0.02**	–	−0.23	0.08	0.24	−0.03	−0.05	0.08
SVETLOE	AZEL	398	**−0.75**	0.50	−0.69	0.06	−0.06	−0.03	−0.01	−0.01
SYOWA	AZEL	51	2.33	0.45	2.25	0.54	0.08	−0.01	−0.09	−0.11
TIGOCONC	AZEL	1081	**0.00**	0.03	−0.72	0.10	0.72	−0.07	0.11	0.90
TSUKUB32	AZEL	548	0.49	0.03	0.49	0.03	0.00	0.02	−0.04	−0.24

(continued)

Table 1 (continued)

			IVS values		Our results			Coordinate difference		
			Axis offset	Sigma	Axis offset	Sigma	Difference	East	North	Up
Station	Mount	# Sess.	(cm)	(cm)	(cm)	(cm)	(cm)	(cm)	(cm)	(cm)
URUMQI	AZEL	91	−0.58	0.09	0.02	0.12	−0.60	−0.09	−0.02	−0.93
WARK12M	AZEL	32	**0.10**	0.02	−0.10	0.28	0.20	0.00	0.40	0.37
WESTFORD	AZEL	930	31.82	0.02	31.80	0.03	0.02	−0.01	0.02	−0.00
WETTZELL	AZEL	1927	**−0.01**	0.00	0.12	0.02	−0.13	−0.03	−0.00	−0.10
YARRA12M	AZEL	92	**0.00**	–	−0.10	0.15	0.10	0.19	0.29	0.45
YEBES	AZEL	29	0.03	0.28	−0.16	0.35	0.19	−0.02	0.01	0.38
YEBES40M	AZEL	101	200.04	0.05	200.40	0.08	−0.36	−0.04	−0.00	−0.45
YLOW7296	AZEL	48	0.56	0.25	0.27	0.35	0.29	0.00	−0.01	0.35
ZELENCHK	AZEL	326	**−1.15**	–	−0.83	0.08	−0.32	−0.05	−0.02	−0.50

Shown are also the mount type, formal errors of the axis offset estimates, as well as the IVS recommended values with formal errors (when available). The IVS values in bold are from local surveys at the stations, the others have been estimated in a VLBI solution. In addition, the differences in station positions between the solution with the axis offsets fixed to the IVS values and the solution with estimation of axis offsets are shown (no values are given for station coordinates which were session-wise reduced)

a smaller number of sessions, the differences are generally larger, as expected since the estimates are less precise due to the smaller number of observations. The larger difference for GILCREEK (7.4 mm, sigma 0.2 mm) could be explained by the non-linear post-seismic motion of GILCREEK after the Denali Earthquake in 2002. As discussed later in Sect. 3.2, the estimated in station coordinates and axis offsets are highly correlated.

When comparing with the axis offsets measured in terrestrial surveys the differences are normally a few millimeters. For a few stations the differences are, however, larger than 1 cm. Differences larger than 1 cm are found at PARKES (5.21 cm, sigma 0.59 cm), OHIGGINS (1.59 cm, sigma 0.62 cm), and HOBART26 (1.05 cm, sigma 0.16 cm). These are all stations in the southern hemisphere where there are relatively few VLBI stations. Thus this effect could be related to bad observation geometry at these stations, caused by the fact that these stations are mostly needed to observe together with stations in the northern hemisphere.

As a test of the quality of the solution, we compared the axis offset estimates with those from a solution containing only half of the VLBI sessions. In general, the agreement was good as the differences were of the same magnitude or smaller than the formal errors. One exception was MEDICINA, where a difference of 0.25 cm was found, what is five times larger than the formal errors (0.06 cm and 0.04 cm for the solution with all and half of the sessions, respectively). Actually, the value estimated using only half of the sessions (182.90 cm, sigma 0.06 cm) was closer to the value from the local survey (183.01 cm) than the value from all sessions (182.65 cm, sigma 0.04 cm). One reason could be that there are some unidentified systematic errors in the observations of station in one or a few sessions, which affected the results.

3.2 Terrestrial Reference Frame

We compared the station coordinates estimated from the first (with axis offsets estimation) and second (axis offsets fixed to IVS values) global solutions. The differences are plotted in Fig. 2, and also listed in Table 1. We can see that the main differences are in the vertical components where the differences often reach several millimeters, while the differences in the horizontal components are mostly smaller than 1 mm. This is expected since the effect of the axis offset on the observed delay is proportional to $\cos\epsilon$ for telescopes with azimuth-elevation mount (the most common mount, see Table 1), where ϵ is the elevation angle. Thus we would expect a high correlation between the axis offset estimates and other parameters with an elevation angle dependent impact on the observed delay, e.g. the vertical coordinates ($\propto \sin\epsilon$) and the zenith wet delay ($\propto 1/\sin\epsilon$). In general we can note that the largest coordinate differences are seen at stations with large differences between the estimated axis offsets and the IVS values exist, what could be expected. For stations with an azimuth-elevation mount, the vertical coordinate and axis offset differences are clearly correlated. If the axis offset is increased by 1 cm, the estimated vertical coordinate will increase by about 1.3 cm (see Table 1). This relation seems to hold for all antennas independent of their locations. There is, however, no such clear correlation between the axis offsets and the horizontal coordinates. Furthermore, the effect of the axis offsets on the station coordinates is different for other antenna mounts. However, we are not able to give any general relation for these antennas due to the small amount of antennas with mounts different to azimuth-elevation in this study.

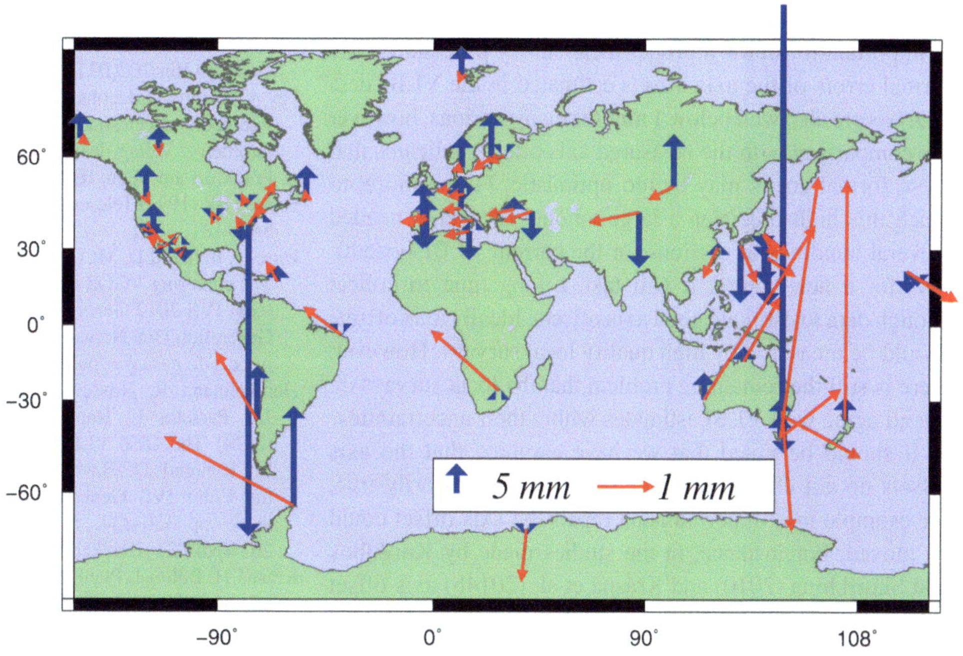

Fig. 2 Difference in station positions between a global VLBI solution where the axis offset were estimated and a second one where they were fixed to the IVS recommended values

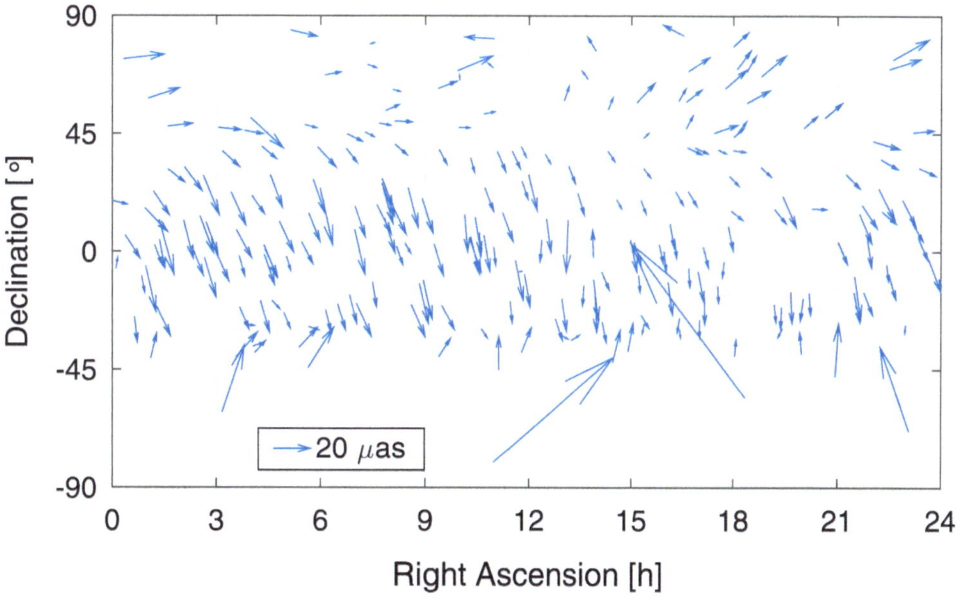

Fig. 3 Difference in radio source coordinates between a global VLBI solution where the axis offset were estimated and a second one where they were fixed to the IVS recommended values. Shown are sources observed in more than 50 sessions

3.3 Celestial Reference Frame

We also investigated the impact of the axis offsets on the radio source coordinates. The difference in coordinates between the two solutions of the radio sources observed in at least 50 sessions are shown in Fig. 3. For the majority of the sources the changes are less than 20 μas. This is small compared to the formal errors of the source coordinate estimates, which all are more than a factor of five larger. There seems to be a systematic change in the coordinates towards smaller declination, especially for sources between −45° and +45° declination. To explain the cause of this effect, further investigations are needed.

4 Conclusions

As we have seen, the axis offsets can have a significant impact on the station coordinates estimated by VLBI. Thus it is important for all IVS analysis centers to consistently apply as accurate axis offsets as possible in their VLBI data analyses. This will be especially important for the future VGOS (VLBI Geodetic Observing System, Hase et al. 2012). If the goal of 1 mm station position precision is to be reached, the axis offsets should be known at the sub-mm level. As seen in Table 1, the formal errors of some axis offsets estimated in terrestrial surveys achieve this requirement, however some not. To obtain an accurate measurement of the axis offset, it is important to follow a proper local survey procedure. The formal errors of the axis offsets estimated in the VLBI data analysis are also well below 1 mm for many stations, however the comparison with the measured axis offsets indicates that these formal errors may be too optimistic. Furthermore, to reach this high precision a large amount of data is needed (several hundreds of sessions for the current VLBI system), thus for a new station it will take a long time to collect enough data to estimate good axis offsets. Ideally axis offsets should be measured by high quality local surveys. However, there is still the remaining problem that the local surveys do not all agree with VLBI estimates within their uncertainties.

It should be noted that we have assumed that the axis offsets do not change in time. This is not necessarily true, for example in a major antenna repair the axis offset could be moved. For instance, in the studies made by Kurdubov and Skurikhina (2010) and Krásná et al. (2014b) axis offset changes of 1 cm at SVETLOE and 8 mm at HARTRAO stations, respectively, were detected. This could be one reason for the 0.53 cm (sigma 0.20 cm) difference between our estimates and the IVS values for HARTRAO. In the future we will extend this study by also considering possible changes in the axis offsets due to antenna repairs. We will also study the correlations between the axis offsets, station coordinates, and other parameters like the tropospheric delays in more details.

Acknowledgements We are grateful to the IVS providing the VLBI data used in this study. We would also like to thank the two anonymous reviewers for their valuable comments on the manuscript.

References

Altamimi Z, Collilieux X, Métivier L (2011) ITRF2008: an improved solution of the international terrestrial reference frame. J Geod 85(8):457–473. doi:10.1007/s00190-011-0444-4

Böhm J, Böhm S, Nilsson T, Pany A, Plank L, Spicakova H, Teke K, Schuh H (2012) The new Vienna VLBI software. In: Kenyon S, Pacino MC, Marti U (eds) IAG Scientific Assembly 2009, no. 136 in International Association of Geodesy Symposia. Springer, Buenos Aires, pp 1007–1011. doi:10.1007/978-3-642-20338-1_126

Combrinck WL, Merry CL (1997) Very long baseline interferometry antenna axis offset and intersection determination using GPS. J Geophys Res 102(B11):24741–24743. doi:10.1029/97JB02081

Fey A, Gordon D, Jacobs CS (eds) (2009) The second realization of the international celestial reference frame by very long baseline interferometry. Verlag des Bundesamts für Kartographie und Geodäsie, Frankfurt am Main, IERS Technical Note 35

Harvey BR (1991) Telescope axis surveys. Aust J Geod Photogram Surv 54:1–18

Hase H, Behrend D, Ma C, Petrachenko B, Schuh H, Whitney A (2012) The emerging VGOS network of the IVS. In: Behrend D, Baver KD (eds) IVS 2012 General Meeting Proceedings Launching the Next-Generation IVS Network, NASA/CP-2012-217504, pp 8–12. http://ivscc.gsfc.nasa.gov/publications/gm2012/hase.pdf

Heinkelmann R, Nilsson T, Karbon M, Liu L, Lu C, Mora-Diaz JA, Parselia E, Raposo-Pulido V, Soja B, Xu M, Schuh H (2014) The GFZ VLBI solution—characteristics and first results. In: Behrend D, Baver KD, Armstrong K (eds) Proceedings of the Eight IVS General Meeting: VGOS: The New VLBI Network, pp 330–334. ftp://ivscc.gsfc.nasa.gov/pub/general-meeting/2014/pdf/071_Heinkelmann_etal.pdf

Krásná H, Böhm J, Plank L, Nilsson T, Schuh H (2014a) Atmospheric effects on VLBI-derived terrestrial and celestial reference frames. In: Rizos C, Willis P (eds) Earth on the edge: science for a sustainable planet. International Association of Geodesy Symposia, vol 139. Springer, Heidelberg. doi:10.1007/978-3-642-37222-3_26

Krásná H, Nickola M, Böhm J (2014b) Axis offset estimation of VLBI telescopes. In: Behrend D, Baver KD, Armstrong K (eds) Proceedings of the Eight IVS General Meeting: VGOS: The New VLBI Network, pp 339-343. ftp://ivscc.gsfc.nasa.gov/pub/general-meeting/2014/pdf/073_Krasna_etal.pdf

Kurdubov S, Skurikhina E (2010) Antenna axis offset estimation from VLBI. In: Behrend D, Baver KD (eds) Proceedings 6th IVS General Meeting. Hobart, Tasmania, pp 247–250. http://ivscc.gsfc.nasa.gov/publications/gm2010

Lösler M, Haas R, Eschelbach C (2013) Automated and continual determination of radio telescope reference points with sub-mm accuracy: results from a campaign at the Onsala Space Observatory. J Geod 87(8):791–804. doi:10.1007/s00190-013-0647-y

Nothnagel A (2009) Conventions on thermal expansion modelling of radio telescopes for geodetic and astrometric VLBI. J Geod 83:787–792. doi:10.1007/s00190-008-0284-z

Schuh H, Behrend D (2012) VLBI: a fascinating technique for geodesy and astrometry. J Geod 61:68–80. doi:10.1016/j.jog.2012.07.007

Scheduling VLBI Observations to Satellites with VieVS

Andreas Hellerschmied, Johannes Böhm, Alexander Neidhardt, Jan Kodet, Rüdiger Haas, and Lucia Plank

Abstract

Observations of satellites with Very Long Baseline Interferometry (VLBI) radio telescopes provide a variety of new possibilities such as the integration of different geodetic techniques, which is one of the main goals of GGOS, the Global Geodetic Observing System of the IAG. Promising applications can be found, among others, in the field of inter-technique frame ties. With the standard geodetic VLBI scheduling software not being prepared to use satellites as radio sources so far, such observations were complicated due to the need to carefully prepare the required interchange files. The newly developed Satellite Scheduling Module for the Vienna VLBI Software (VieVS) offers a solution to this. It allows the user to prepare VLBI schedule files in a standardized format, providing the possibility to carry out actual satellite observations with standard geodetic antennas, e.g. of the IVS network. First successful observations of GLONASS satellites, based on schedules created with the new VieVS module, took place on the baseline Wettzell-Onsala in January 2014.

Keywords

Satellite observations • Scheduling • VieVS • VLBI

1 Introduction

Observations of satellite signals with the VLBI technique is a promising topic and has been discussed in the geodetic VLBI community in recent years. A number of test observations to GNSS satellites, e.g. Tornatore et al. (2014), and case studies (Plank et al. 2014) have been carried out in recent years. The last successful observation of this series of test experiments was carried out in January 2013 on the baseline Wettzell-Onsala (Haas et al. 2014). Nevertheless, VLBI satellite observations to obtain delay measurements as common in the geodetic VLBI are still far away from being applied operationally. The challenges start at the level of observation planning. Although the astronomical scheduling software SCHED (Walker 2015) is able to schedule observations to satellites and deep space probes, mostly for phase-referencing techniques, standard geodetic scheduling software do not provide this ability so far. Consequently, the planning of satellite VLBI experiments is quite labor-intensive at present, due to the large amount of manual interaction required to prepare suitable observation schedules and the related interchange files.

In order to promote the development of an operational process chain for satellite observations, similar to the standard observation procedure for the geodetic VLBI, the Vienna VLBI Software (VieVS; Böhm et al. 2012)

A. Hellerschmied (✉) • J. Böhm
Department for Geodesy and Geoinformation, Vienna University of Technology, Gußhausstraße 27-29, 1040 Vienna, Austria
e-mail: andreas.hellerschmied@geo.tuwien.ac.at

A. Neidhardt • J. Kodet
Technische Universität München, Geodetic Observatory Wettzell, Germany

R. Haas
Chalmers University of Technology, Onsala Space Observatory, Sweden

L. Plank
University of Tasmania, Hobart, TAS, Australia

T. van Dam (ed.), *REFAG 2014*, International Association of Geodesy Symposia 146,
DOI 10.1007/1345_2015_183

was upgraded with a dedicated scheduling module. The new software provides a flexible and easy-to-handle tool to generate realistic VLBI schedules in a standardized data format, feasible to conduct actual satellite observations in an automated manner.

2 Scheduling of VLBI Satellite Observations

2.1 Introduction to Scheduling

To carry out VLBI observations a proper time schedule is required. It has to be prepared prior to a VLBI session. In general a VLBI schedule defines, which antennas should observe a selected source simultaneously at a particular time with common visibility of the target.

The scheduling task is a complex optimization problem and is usually done by dedicated VLBI scheduling software packages, such as the program SKED (Gipson 2012), maintained at the NASA Goddard Space Flight Center, or VIE_SCHED (Sun et al. 2014), the standard scheduling module for quasar observations of VieVS. Although each of them are designed to schedule conventional VLBI observations to quasars, the current official releases do not routinely support satellites as signal sources. Scheduling programs are used to generate dedicated schedule files, comprising all information required to carry out an actual VLBI session. Additionally to the complete observation time plan, all setup parameters of the station equipment are defined in there. One commonly used standard file format is VEX (Whitney et al. 2002). Prior to a VLBI session, the schedule files are distributed to all participating stations. At each site the locally required information is extracted and written to local control files, which are used to control the antenna, the receiver back-end, and the recorder equipment automatically during the observations.

2.2 Framework Conditions for Satellite Scheduling

In standard geodetic VLBI, extragalactic radio sources (mostly quasars) are observed routinely. Due to their virtually infinite distance from Earth, their proper motion can be neglected. Hence, they can be treated as invariant points on the celestial sphere with constant coordinates that are usually represented in the schedules in terms of right ascension (Ra) and declination (Dec) in the ICRF frame. To take into account the Earth's rotation, the topocentric pointing angles of observing antennas have to be adjusted continuously. These corrections are usually applied automatically by the antenna control system.

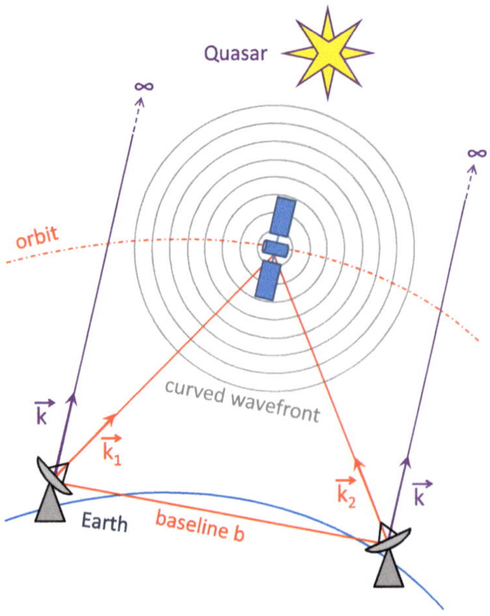

Fig. 1 Satellite observation geometry. Due to the finite distance between source and antenna, the incident wavefront of a satellite signal cannot be considered as plane. It is curved and the view directions differ between antennas at different locations ($\mathbf{k_1} \neq \mathbf{k_2}$). For observations to quasars, which can be assumed to be at an infinite distance, the directions from different sites ($\mathbf{k}$) can be assumed as parallel

If satellites are observed instead, the situation becomes more complex. Considering that satellites orbit the Earth rather fast, the complexity of scheduling increases as the sources are no longer stationary targets. The factor time becomes more critical, because of source positions and, therefore, topocentric view directions from the observing VLBI stations, change quickly in time. For this reason, satellites have to be tracked actively during observations. The applied tracking procedure already has to be considered at the level of scheduling, to include the required tracking data.

Unlike in standard VLBI, the source distance cannot be considered as infinite, which results in a significantly different observation geometry, as illustrated in Fig. 1: In case of quasar observations, the view direction vectors ($\mathbf{k}$) for two locations on the Earth surface are parallel. Due to the relatively short range between antennas and satellites[1], view direction vectors, represented by $\mathbf{k_1}$ and $\mathbf{k_2}$, cannot be assumed to be parallel. Therefore, a *cross-eyed observation configuration* has to be applied (Duev et al. 2012), where the coordinates of the same source defined in terms of topocentric Ra and Dec differ between sites.

Generally, VLBI observations are limited to satellites which actively emit radio signals that can be acquired with the available station equipment. Satellite signals differ from natural signals of quasars in terms of bandwidth,

[1]For example approximately 20,000 km in case of GNSS satellites.

frequency and flux density. Quasars emit very weak microwave radiation on a broad frequency range, which are routinely observed in the S and X band. In contrast to this, satellites broadcast signals on various carrier frequencies with comparatively narrow bandwidths, but with a much higher power level. It is a prerequisite for successful VLBI satellite observations to account for those signal properties during data acquisition. Hence, VLBI receiver systems have to be adopted accordingly, e.g. with L-band receivers for GNSS signals. However, varying signal characteristics also have to be considered for the generation of schedule files, because the recording parameters, e.g. channel frequency and bandwidth, have to be precisely adjusted to the signal characteristics.

3 VieVS Satellite Scheduling Module

The new satellite scheduling tool is included as a module to VieVS, which provides a versatile software package for VLBI analysis and scheduling. The integration into the Graphical User Interface (GUI) of VieVS allows a user-friendly handling.

Concerning the application field of the new satellite scheduling module, it has to be considered that satellite observation with VLBI is still an experimental approach. Further research and development is necessary to achieve operational applicability. To be able to carry out dedicated experiments, a tool is needed to generate suitable observation schedules in a simple and quick way.

The current scheduling programs for standard geodetic VLBI provide different optimization strategies for the selection of sources. The determination of the optimal integration time per scan (on-source-time) is done automatically. In geodetic VLBI this is quite important, because the source distribution (influencing the sky coverage) and the on-source-time (influencing the number of observations) have fundamental impacts on the quality of the final geodetic products. Satellites are not operationally observed so far. Therefore, an automatic source selection, which e.g. implements an approach to optimize the sky coverage, is currently not supported. For the calculation of appropriate on-source-times, which are necessary to achieve a certain signal-to-noise ratio in the acquired data, further information would be required. Among others, additional data on the flux density of the transmitted satellite signals, as well as the level of efficiency of particular receiver chains would be needed. However, the determination of these parameters for satellite observations is still a subject of research.

For those reasons, the most convenient scheduling approach seems to be an interactive procedure, where the scheduler is able to determine the observation sequence, as well as the on-source-times for each scan manually. In this case, the purpose of the scheduling program is to support and to verify the user's decisions. The program has to determine periods in which preselected satellites are available for observations from a defined station network, taking into account various observation conditions (see Sect. 3.1). The great advantage of this approach is that the user maintains full control over the schedule, which is particularly important for research and development experiments.

3.1 Conditions for a Valid Satellite Scan

One of the main tasks of the scheduling module is to determine the available observation periods for particular satellites and for a defined observation configuration. Therefore, the following main conditions are considered:

- **Common visibility:** The observed source has to be visible simultaneously from at least a sub-net of two stations representing one baseline, i.e. it has to be above the cut-off elevation at those station. The common visibility depends on several parameters, e.g. the baseline length and the orbit height[2].
- **Antenna slew rates:** In order to ensure, that the antennas are able to follow the satellites in the sky, the slew rate limits of the antenna axes must not be exceeded. The speed of a satellite as seen from a site on the Earth's surface depends on the orbit height (lower orbiting satellites move faster) and the local elevation angle (the required tracking speed increases with larger elevation angles). Therefore, the most challenging observations are those to satellites on a low orbit close to the zenith. Furthermore, the antenna mount type[3] plays a role in this context, because each mount type has certain areas (*keyholes*), where coordinate singularities occur (Salzberg 1967). Although an antenna may be able to point at a particular keyhole position, one of its axes would need an infinite rotation speed to be able to track continuously through this position.
- **Sun distance:** To avoid signal corruption or even hardware damage due to the strong radiation emitted by the sun, a defined minimal angular distance between the observed signal source and the Sun has to be kept during observations.
- **Cable wrap:** The cable wrap has to be tracked to determine slew times between consecutive scans and to check the slewing motion limits.

If these conditions are met simultaneously, the satellite is observable for that particular time and can be added to the schedule.

[2]Nevertheless, VieVS is able to schedule single station observations too.
[3]Mostly azimuth-elevation mounts are used for geodetic VLBI.

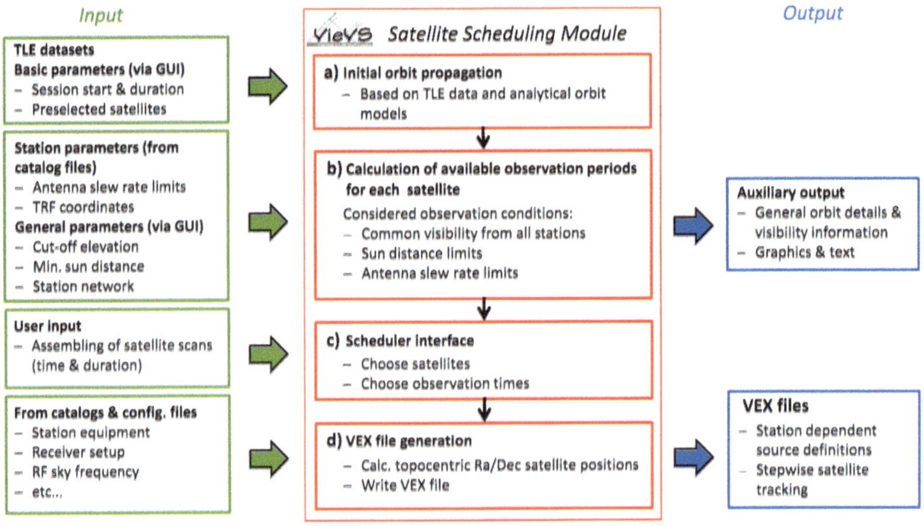

Fig. 2 Flowchart of the VieVS satellite scheduling module. The required input data and parameters are depicted on the *left side* (*green*). The actual program flow is illustrated in the *middle* (*red*) and the program output on the *right side* (*blue*)

3.2 Satellite Orbit Prediction

Satellite positions and velocities are needed during the observation planning to check if the observation conditions, described in Sect. 3.1, are fulfilled and to determine the possible observation periods. Furthermore, satellite positions are required to derive appropriate tracking data, which has to be included in the schedule files (see Sect. 3.4). It should be possible to do the scheduling several days before the actual observations are carried out. Therefore, the orbit data have to provide the possibility to *predict* satellite orbits into future with sufficient accuracy.

In the VieVS satellite scheduling module all orbit calculations are based on Two-Line Element (TLE) data, an orbit data format specified by the North American Aerospace Defense Command (NORAD). TLE datasets are mean Keplerian orbit element sets, which allow the calculation of satellite positions for particular times with designated analytical models (Hoots and Roehrich 1980). TLE data provide a great flexibility in the selection of satellite targets, because they are freely available on a daily basis for thousands of space objects via dedicated web services.

3.3 Program Workflow

The general observation setup can be defined in the VieVS GUI. The user is able to select the VLBI station network, preselect satellites for observations, set date and time for the planned session, and several further scheduling parameters, e.g. a general cut-off elevation angle. The most recent TLE data can be obtained from a web service via mouse click and is used to maintain a local TLE library, which is the basis for all orbit prediction tasks.

On program start the required input data are loaded: *SKED catalog files* (Vandenberg 1997) provide all necessary VLBI station parameters, such as TRF coordinates, information about the receiver equipment and antenna properties, e.g. slew rate limits. *VieVS-specific catalog files* have been set up to predefine various receiver settings intended for satellite observations specifically. *Configuration files* are used for different purposes, e.g. to define all VLBI receiver setup parameters for a complete station network by combining the basic information, which is provided by the catalog files. This information is later on incorporated into the generated schedule files.

The program workflow is illustrated in the flowchart in Fig. 2. The left side (green) illustrates the required input data, as well as user-defined parameters for each processing step, shown in the middle (red). The program output is drawn on the right side (blue).

(a) First, the allocated TLE data are used to calculate time series of positions and velocities for the selected satellites and for the chosen time. This is the basis for all subsequent computations.
(b) In the following step the available observation periods are calculated for each satellite, considering the conditions for a valid observation, discussed in Sect. 3.1. This information is then presented to the user in terms of auxiliary output in the form of graphics and descriptive text. Examples for the auxiliary graphical output are shown in Fig. 3 (elevation plot) and Fig. 4 (sky plot). The *elevation plot* illustrates the topocentric satellite elevations as a function of time for each station. This information is particularly important, because the fundamental condition, whether a satellite is observable, depends on it being positioned above the defined cut-off elevation. Violations of the observation conditions are also indicated in the plot. The lower sub-plot summarizes the available observation periods per satellite. Corresponding *sky plots* additionally illustrate the azimuthal directions.

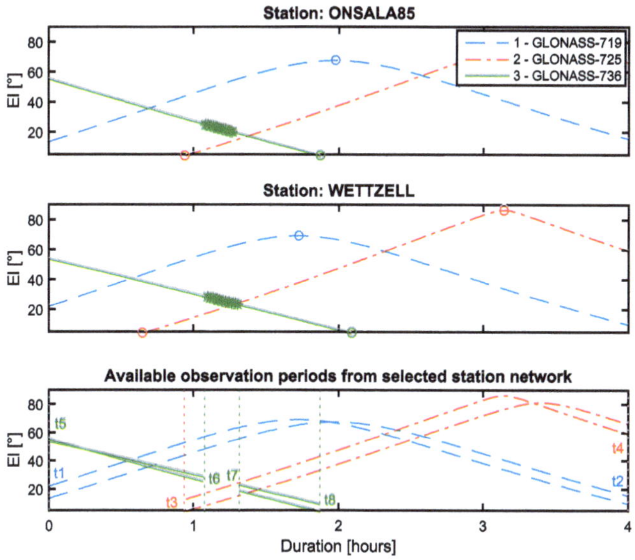

Fig. 3 Auxiliary output: elevation plot for stations Wettzell and Onsala85, showing the tracks of three GLONASS satellites from 6:00 to 10:00 UTC on August 30, 2014. The two *upper plots* depict the satellite elevation for both stations as a function of time elapsed since session start. The cut-off elevation is 5°. The highlighted section of the track of GLONASS-736 tags a violation of the Sun distance which is set to 4° (see Sect. 3.1). The plot on the very bottom summarizes the available observation periods for the selected observation setup, marked with *vertical dotted lines* and time tags (t_x) at the beginning and at the end

(c) By means of the provided auxiliary information, the user has to compile the desired schedule by applying the provided interface. The complete observation schedule is set up by assembling an arbitrary number of individual scans. For each scan the observed satellite and the on-source-time has to be defined. The source switch times, depending on various parameters such as slew rates and previous cable wrap positions, can be determined by the program.

(d) At the end, realistic VEX-formatted schedule files are generated, comprising all required information, to control actual VLBI satellite observations.

3.4 Implementation of Satellite Tracking

To be able to track satellites with VLBI antennas, information about the orbits have to be included in the schedule files. Apart from a few exceptions, such as the experimental satellite tracking module SatTrack in Onsala (Moya Espinosa and Haas 2007) and the satellite tracking procedure implemented in Wettzell (Hellerschmied et al. 2014), the standard antennas of the IVS network do not routinely support specific satellite tracking modes so far. Furthermore, the current VEX format (version 1.5b1) does not provide a dedicated possibility to incorporate orbit information in a suitable way, e.g. directly in terms of TLE data.

In order to be able to track satellites with the currently available technical means, a *stepwise tracking approach* is implemented in the schedules. Therefore, the satellite orbit is fragmented and approximated by a time series of discrete positions. Each position is defined as distinct source in the schedule, equal to the definition of a quasar. To implement a cross-eyed observation, as discussed in Sect. 2.2, the satellite positions are defined in terms of topocentric Ra and Dec individually for each station in separate site-specific VEX files. During observations the antenna is repositioned stepwise in the defined interval, in order to keep the satellite within the antenna beam and to prevent a signal loss during data acquisition.

4 Satellite Observation Experiments

First test observations, based on schedules created with the new VieVS Satellite Scheduling Module, were carried out in early 2014 on the baseline Wettzell-Onsala. Two sessions were conducted on January 16 (G140116a and G140116b), respectively on January 21 (G140121a and G140121b), as described by Hellerschmied et al. (2014). Each experiment had a duration of 1 h and scheduled several consecutive observations to GLONASS satellites in the L1 band. In a preliminary analysis of the acquired data, strong amplitudes and continuous phases were found for all satellites, indicating that the scheduling with VieVS worked properly at both sites.

5 Summary and Outlook

The new Satellite Scheduling Module of VieVS is a flexible tool for the scheduling of VLBI observations to satellites and therefore, fills a gap in the presently available software. It supports observation planning for numerous satellites and station networks. Although a combination of quasar and satellite observations in one schedule is not possible so far, it is planned to add this capability in the near future.

The generated schedule files in the current VEX format contain all required information to run the scheduled observations. The stepwise satellite tracking approach, implemented in these files, has the advantage that it is viable with all standard antennas of the IVS network, without the requirement of specific satellite tracking capabilities. Satellite positions are defined in the same manner as quasars in terms of celestial coordinates, which are directly processable by the antenna controls. More possibilities for an adequate

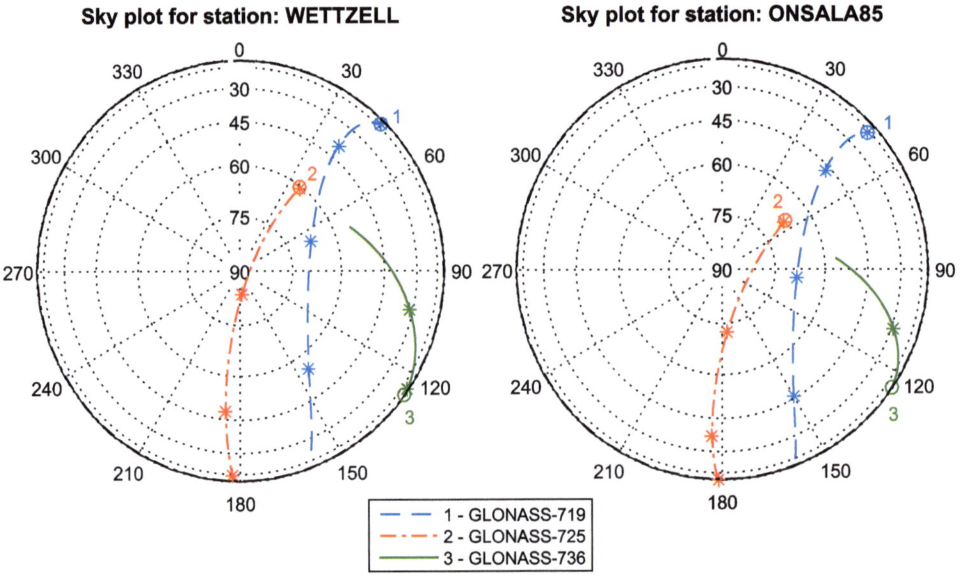

Fig. 4 Auxiliary output: sky plots for stations Wettzell (*left*) and Onsala85 (*right*), showing the tracks of three GLONASS satellites from 6:00 to 10:00 UTC on August 30, 2014

definition of satellites as radio sources in the VEX files (e.g. by TLE data) and, hence, for the implementation of satellite tracking, will be provided by the revised VEX format (VEX 2.0), which was already announced for 2015. The proper functionality of the new VieVS module was shown in four successful experiments in January 2014, where several GLONASS satellites were observed on the baseline Wettzell-Onsala.

The development of the Satellite Scheduling Module for VieVS is an important step towards an operational application of VLBI observations to satellites and is particularly important for further research activities in this field.

References

Böhm J, Böhm S, Nilsson T, Pany A, Plank L, Spicakova H, Teke K, Schuh H (2012) The New Vienna VLBI Software VieVS. In: Kenyon S, Pacino MC, Marti U (eds) Geodesy for planet earth. International Association of Geodesy Symposia, vol 136. Springer, Berlin/Heidelberg, pp 1007–1011

Duev DA, Molera Calvés G, Pogrebenko SV, Gurvits LI, Cimó G, Bocanegra Bahamon T (2012) Spacecraft VLBI and Doppler tracking: algorithms and implementation. Astron Astrophys 541:A43+

Gipson J (2012) Sked – VLBI scheduling software. NASA Goddard Space Flight Center. ftp://gemini.gsfc.nasa.gov/pub/sked/sked_Manual_v2012May09.pdf

Haas R, Neidhardt A, Kodet J, Plötz C, Schreiber U, Kronschnabl G, Pogrebenko S, Duev D, Casey S, Marti-Vidal I, Yang J, Plank L (2014) The Wettzell-Onsala G130128 experiment – VLBI-observations of a GLONASS satellite. In: Behrend D, Baver K, Armstrong K (eds) IVS 2014 General Meeting Proceedings – "VGOS: The New VLBI Network". Science Press, pp 451–455

Hellerschmied A, Plank L, Neidhardt A, Haas R, Böhm J, Plötz C, Kodet J (2014) Observing satellites with VLBI radio telescopes - practical realization at Wettzell. In: Behrend D, Baver K, Armstrong K (eds) IVS 2014 General Meeting Proceedings – "VGOS: The New VLBI Network". Science Press, pp 441–445

Hoots F, Roehrich R (1980) Spacetrack Report number 3: models for propagation of NORAD element sets. Tech. Rep., US Airforce Aerospace Defense Command, Colorado Springs

Moya Espinosa M, Haas R (2007) SATTRACK a satellite tracking module for the VLBI field system. In: Böhm J, Pany A, Schuh H (eds) Proceedings 18th European VLBI for Geodesy and Astrometry Working Meeting, pp 53–58

Plank L, Böhm J, Schuh H (2014) Precise station positions from VLBI observations to satellites: a simulation study. J Geod 88(7):659–673

Salzberg I (1967) Mathematical relationships of the MFOD Antenna axes. Tech. Rep. 67N39334, NASA Goddard Space Flight Center, Greenbelt, MD, United States, report/Patent Number: NASA-TM-X-55956, X-553-67-213

Sun J, Böhm J, Nilsson T, Krásná H, Böhm S, Schuh H (2014) New VLBI2010 scheduling strategies and implications on the terrestrial reference frames. J Geod 88(5):449–461

Tornatore V, Haas R, Casey S, Pogrebenko S, Molera Calvés G (2014) Direct VLBI Observations of Global Navigation Satellite System Signals. In: Rizos C, Willis P (eds) Earth on the edge: science for a sustainable planet, Proceedings IAG General Assembly, 2011. International Association of Geodesy Symposia, vol 139. Springer, Berlin/Heidelberg, pp 247–252

Vandenberg N (1997) sked's Catalogs – Program Reference Manual. NASA Goddard Space Flight Center. VLBI Software Documentation, Scheduling Program. ftp://gemini.gsfc.nasa.gov/pub/sked/docs/skcat.pdf

Walker RC (2015) The SCHED User Manual, Version 11.4 of March 15, 2015. http://www.aoc.nrao.edu/software/sched/

Whitney A, Lonsdale C, Himwich E, Vandenberg N, van Langevelde H, Mujunen A, Walker C (2002) VEX File Definition/Example, Rev. 1.5b1. http://www.vlbi.org/vex/docs/vex

Refined Tropospheric Delay Models for CONT11

D. Landskron, A. Hofmeister, and J. Böhm

Abstract

For the period of CONT11, a continuous VLBI campaign which lasted from September 15 through September 29, 2011, new *a* coefficients for the Vienna Mapping Functions 1 (VMF1; Böhm et al., J Geophys Res 111:B02406, 2006) and an extended calculation strategy for horizontal gradients were applied. The *a* coefficients of the mapping functions are usually calculated by ray-tracing through numerical weather models (NWM) from the European Centre for Medium-Range Weather Forecasts (ECMWF) with a spatial resolution of $0.25° \times 0.25°$ every 6 h. By enhancing the spatial resolution to $0.125° \times 0.125°$ and using a new ray-tracing program called RADIATE, the actual refractivities are described in more detail. Moreover, existing equations for the calculation of the azimuth-dependent part of the slant delay by means of horizontal gradients were extended in order to achieve better accordance with the data from ray-tracing, which improves the resulting baseline length repeatabilities.

Keywords

GNSS • Horizontal gradients • Ray-tracing • Tropospheric delay • VLBI • VMF1

1 Introduction

Modeling tropospheric delays is one of the major error sources in the analysis of Global Navigation Satellite Systems (GNSS) or Very Long Baseline Interferometry (VLBI) observations. Many approaches like numerous mapping functions and several ways of handling the azimuthal asymmetry have been developed throughout the last decades; however, we are still far away from accuracies like 1 mm in position and 0.1 mm in velocity as requested by the Global Geodetic Observing System (http://ggos.org). For example, Böhm and Schuh (2001) tested spherical harmonics as supplement to mapping functions and gradients, Hobiger et al. (2008) applied ray-tracing for precise point positioning in GPS, Gegout et al. (2011) developed so-called "adaptive mapping functions" to model azimuthal asymmetries, Zus et al. (2012) computed slant total delays in a numerical weather model, Nafisi et al. (2012) developed a ray-tracing program based on numerical weather models and Eriksson et al. (2014) investigated the impact of using ray-traced tropospheric delays on geodetic VLBI results.

In this paper, we compare ray-traced delays derived from operational analysis data of the European Centre for Medium-Range Weather Forecasts (ECMWF) with a spatial resolution of $0.125° \times 0.125°$ every 6 h against models usually applied for the reduction of tropospheric delays, such as the Vienna Mapping Functions (VMF1) and horizontal tropospheric gradients. In addition to the standard parameters, we also add specific VMF1 coefficients and gradients derived from the same ECMWF fields which are used for ray-tracing. Finally we assess the impact of the different approaches on geodetic parameters by analyzing the CONT11 dataset with the Vienna VLBI Software (VieVS; Böhm et al. 2012) in terms of baseline length repeatability

D. Landskron (✉) • A. Hofmeister • J. Böhm
Department of Geodesy and Geoinformation, Technische Universität Wien, Gußhausstraße 27-29, 1040 Vienna, Austria
e-mail: daniel.landskron@geo.tuwien.ac.at

T. van Dam (ed.), *REFAG 2014*, International Association of Geodesy Symposia 146,
DOI 10.1007/1345_2015_56

(BLR). For short time periods like CONT11, the BLR is simply the standard deviation of a set of measured baseline lengths.

2 Calculation of New Mapping Function Coefficients

The Vienna Mapping Functions 1 (VMF1) for a certain elevation are based on the continued fraction form by Herring (1992):

$$\mathrm{mf}(e) = \frac{1 + \cfrac{a}{1 + \cfrac{b}{1 + c}}}{\sin(e) + \cfrac{a}{\sin(e) + \cfrac{b}{\sin(e) + c}}} \tag{1}$$

where mf is the mapping function and e denotes the elevation angle. The coefficients b and c are determined by empirical functions as suggested by Böhm et al. (2006), whereas the coefficient a depends on the current weather situation. The VMF1 use ray-traced delays through a numerical weather model (NWM) with $0.25° \times 0.25°$ spatial resolution as data input. The main outputs of ray-tracing are slant delays and zenith delays, with their ratio being the slant factor. Subsequently, the a coefficients can be determined from those delays and the values of VMF1 are provided online and in real-time on the GGOS Atmosphere server (http://ggosatm.hg.tuwien.ac.at/DELAY/). By means of ray-tracing (using the new ray-tracing program RADIATE Hofmeister and Böhm 2014) through the currently available $0.125° \times 0.125°$ NWM data, new values for the a coefficients could be obtained in a least squares approach.

Applying the re-calculated mapping function coefficients in VLBI analysis with VieVS slightly improves the resulting mean BLR from 1.09 to 1.08 cm. For 54% of the baselines (42 of 78), the BLR then is lower. Moreover it is notable, that when using the new coefficients, almost all repeatabilities of baselines containing station HOBART12 get by far better, while almost all repeatabilities of baselines containing station KOKEE get by far worse. This scenario also appears in investigations of BLR using different gradients (see Sect. 3). The likeliest explanation for this is the geographic location of the VLBI station KOKEE. It is situated almost 1,200 m above sea level on the island Kauai in the archipelago of Hawaii. The rapid altitude changes of the island together with the comparably low temporal resolution of the NWM (6 h) may not be suitable for properly representing the resulting changes in air refractivity, which contribute to the measured VLBI delays. The cause for the significantly better BLR of station HOBART12 on the island of Tasmania is still subject of investigation. However, when removing the stations HOBART12 and KOKEE from the solution, the general pattern of the results stays the same. That is, application of the re-calculated mapping function coefficients cause a general improvement of BLR, albeit only in the sub-mm range. Figure 1 shows the differences in BLR for all baselines.

Furthermore, the impact of choosing NWM with a spatial resolution of $1° \times 1°$ was tested as well. However, as anticipated, the mean BLR gets slightly worse compared to the coefficients from the NWM with $0.125° \times 0.125°$ spatial resolution (from 1.09 cm to 1.10 cm); For 59% of the baselines (46 of 78), the BLR gets worse. Moreover, baselines containing station HOBART12 get considerably worse, while those containing KOKEE get slightly better.

3 Horizontal Gradients

It should be noted here that all BLR shown in this section used the standard VMF1 coefficients, not the re-calculated ones from the previous section.

Applying mapping functions does not account for azimuthal asymmetries, which occur systematically on a large spatial scale due to the fact that the atmosphere is thicker at the equator and thinner at the poles and randomly on a small spatial scale, because the refractivity of the air varies with the azimuth because of changing weather conditions. Because the influence of azimuthal asymmetry increases with decreasing elevation angles, particularly all low-elevation observations require correction of this effect.

Applying horizontal gradients by using gradient formulas accounts for the bulk of this problem. At present, the gradient model by Chen and Herring (1997) is the most used and widely accepted way of modeling azimuthal asymmetries. It reads

$$\Delta L(a,e) = \underbrace{\Delta L_0(e)}_{\text{isotropic part}} + \underbrace{mf_g(e)[G_n \cos(a) + G_e \sin(a)]}_{\text{anisotropic part}} \tag{2}$$

with

$$mf_g(e) = \frac{1}{\sin(e)\tan(e) + 0.0032}$$

where a is the azimuth, e the elevation, L_0 denotes the delay without horizontal gradients (product of the zenith delay and the mapping function) and G_n and G_e are the horizontal north and east gradients, respectively. Values for the horizontal gradients G_n and G_e depend on the time of

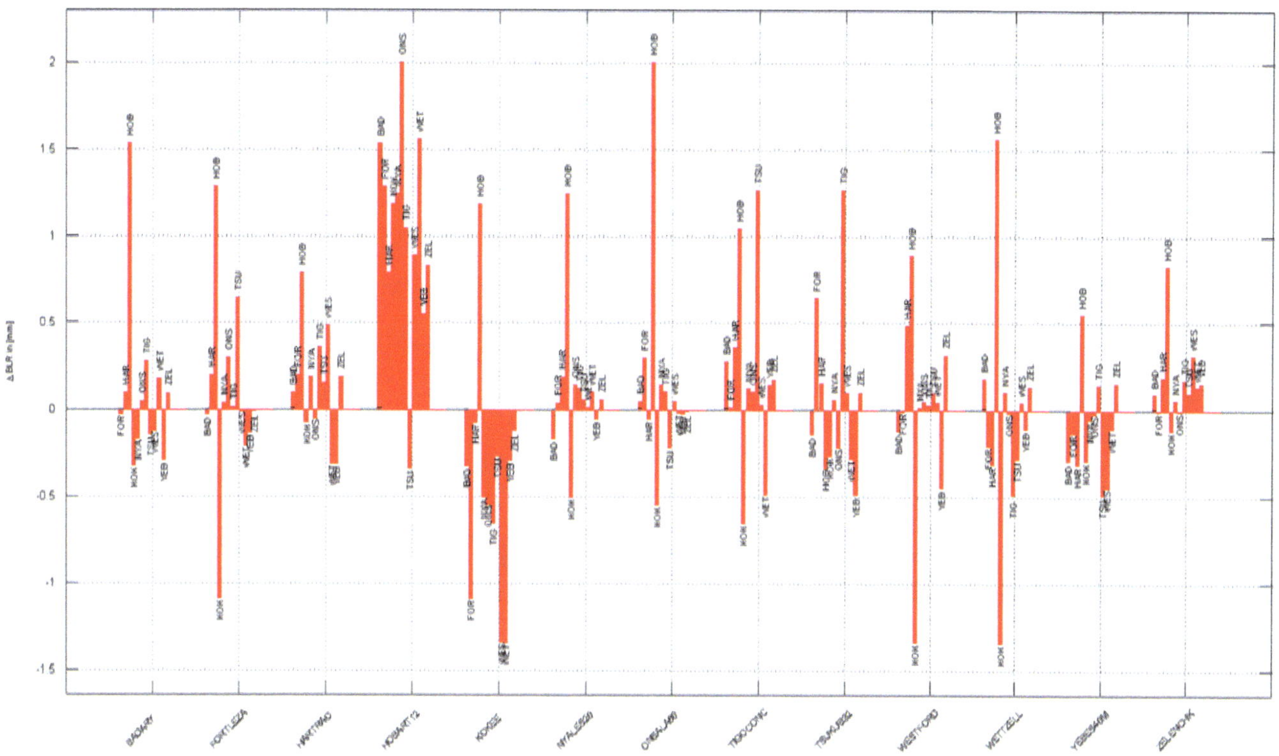

Fig. 1 The differences in baseline length repeatabilities ΔBLR between using the standard VMF1 coefficients and the re-calculated ones. All bars above the x-axis indicate that the repeatabilities of these baselines are lower for the re-calculated coefficients. Almost all bars above the x-axis with highest improvement correspond to baselines containing HOBART12, those below the x-axis with highest decline to KOKEE

the measurement and the location of the site. For example, gradients derived from horizontal gradients along the site vertical (Böhm and Schuh 2007) can be downloaded from the GGOS Atmosphere server (http://ggosatm.hg.tuwien.ac.at/DELAY/). In this paper, we suggest adding higher order terms in azimuth to the expression in Eq. (2).

$$\Delta L(a,e) = \Delta L_0(e) + mf_g(e)[G_n\cos(a) + G_e\sin(a) + G_{n_2}\cos(2a) + G_{e_2}\sin(2a)] \quad (3)$$

and

$$\Delta L(a,e) = \Delta L_0(e) + mf_g(e)[G_n\cos(a) + G_e\sin(a) + G_{n_2}\cos(2a) + G_{e_2}\sin(2a) + G_{n_3}\cos(3a) + G_{e_3}\sin(3a)] \quad (4)$$

The horizontal gradients G_n and G_e and additional variables G_{n_2}, G_{e_2}, G_{n_3} and G_{e_3} were calculated in a least squares adjustment using ray-traced tropospheric delays computed by the ray-tracing program RADIATE for each of the 14 stations and 60 epochs of CONT11. Therefore, 112 ray-traced delays at 16 evenly distributed azimuths (0:22.5:337.5) and 7 elevations (3°, 5°, 7°, 10°, 15°, 30°, 70°) were used per station. The basic idea of the expansion of the gradient formula is that the oscillation of the systematic residual delays due to the shape of the Earth's atmosphere can be described more precisely, as is evident in Fig. 2. This figure shows the residuals between ray-traced slant delays and delays calculated by the three gradient formulas for station WESTFORD.

Calculating and averaging these ray-traced delay residuals for the testing period of CONT11 for all 14 CONT11 stations shows that they decrease by 69% when using Eq. (2), by 78% when using Eq. (3) and by 80% when using Eq. (4). This already shows a clear benefit of the revised gradient formulas.

Horizontal gradients as derived from ray-traced delays using NWM can be used as a priori gradients which serve as input to the VLBI analysis with VieVS. In addition, there is also the possibility of estimating the standard gradients [Eq. (2)] during the analysis itself, which is common practice. This is handled by a least squares adjustment using

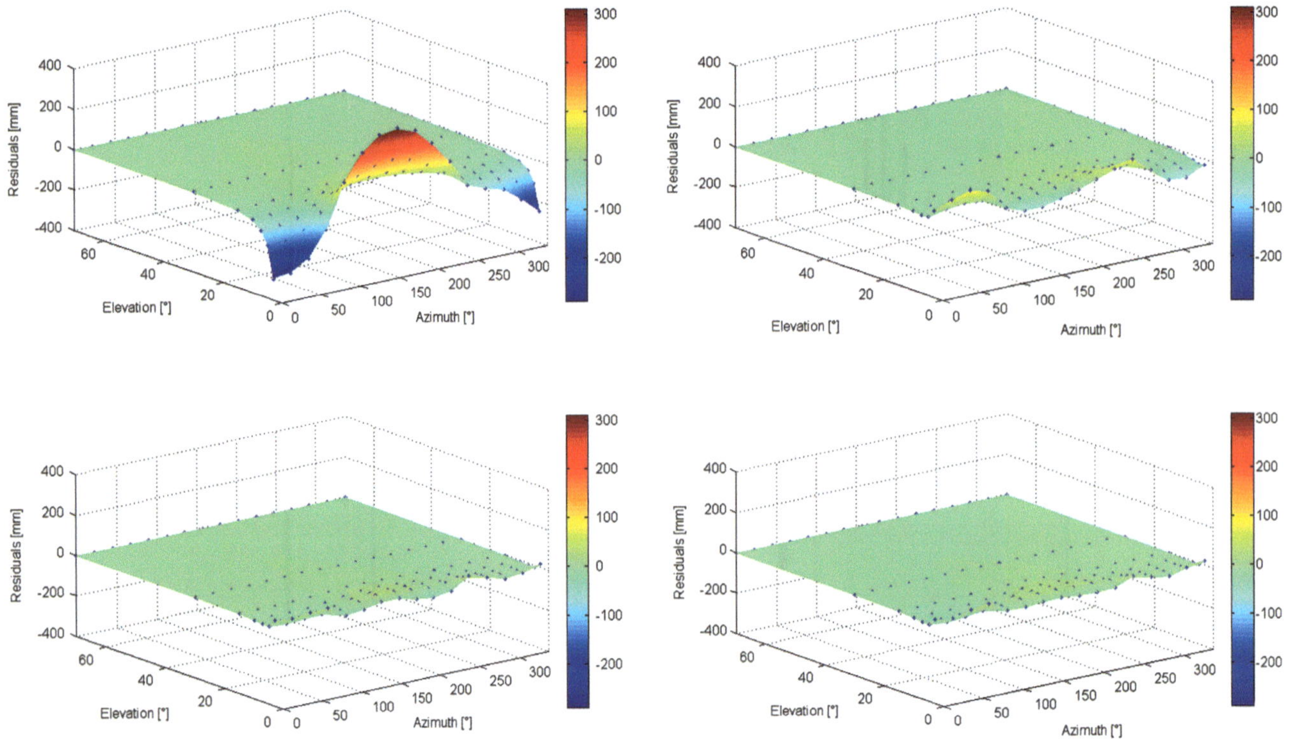

Fig. 2 *Top left*: Residuals of the total slant delays for station WESTFORD at September 26, 2011, 18:00 GMT after subtraction of a mean over the 16 constantly distributed azimuths. The high residuals in north (a = 0°) and south (a = 180°) direction are mostly due to the high latitude location of the site, since the extension of the Earth's atmosphere is higher at the equator and lower at the poles. *Top right*: Residuals after applying gradients by using Eq. (2); this lowers the residuals significantly. *Bottom left*: Residuals after applying gradients by using Eq. (3); again, the residuals are lowered considerably. *Bottom right*: Residuals after applying gradients by using Eq. (4); residuals hardly change compared to Eq. (3)

Table 1 Mean BLR in [cm] for cases (a) standard gradients estimated during VLBI analysis using gradient formula (2) and (b) no gradients estimated during VLBI analysis

A priori gradients	(a)	(b)
None	1.07	1.20
Using gradient formula (2)	1.05	1.10
Using gradient formula (3)	1.03	1.09
Using gradient formula (4)	1.04	1.09

For reasons of interpolation, the time span was shortened to September 16 through September 28, 2011 here

the standard gradient formula by Chen and Herring (1997). Table 1 shows the resulting mean BLR values for the different scenarios.

Table 1 reveals quite distinctively that if gradients calculated from ray-traced delays using NWM data with the revised gradient formulas (3) and (4) are used as a priori values, the resulting baselines have a slightly lower (better) mean BLR. Therefore, it is suggested that one use Eq. (3) for calculation of a priori gradients if the information is available. For 65% of the baselines (51 of 78), the BLR is lower compared to the standard gradient formula (2) (in case gradients are additionally estimated in VLBI analysis). The additional gradient variables G_{n_2} and G_{e_2} can be provided in real-time in the same way as the gradients G_n and G_e; thus, the download would not require additional work for the user.

As already mentioned before, north and east gradients are usually estimated in VieVS in a least squares adjustment using the normal gradient formula (2). It was also tested in this study whether the use of the two revised gradient formulas instead would bring a further improvement in BLR. That is, the additional gradient variables G_{n_2}, G_{e_2}, G_{n_3} and G_{e_3} were also estimated in the VLBI analysis. However, the resulting mean BLR thus increases considerably (in case of using no a priori gradients: 1.17 cm when using formula (3) and 1.22 cm when using formula (4), respectively). The reason for it is most likely that too many parameters are to be estimated which are highly correlated. In other words, there are not enough observations to properly de-correlate those parameters. As a result, gradients calculated by the extended gradient formulas presented in this paper can only be used as a priori gradients.

Table 2 Mean BLR in [cm] for cases (a) gradients estimated during VLBI analysis using gradient formula (2) and (b) no gradients estimated during VLBI analysis

A priori gradients	(a)	(b)
None	1.07	1.20
APG	1.07	1.20
DAO	1.07	1.20

For reasons of interpolation, the time span was shortened to September 16 through September 28, 2011 here

Lastly, the effect of using the empirical a priori gradients APG (Böhm et al. 2011) and DAO (Data Assimilation Office, GSFC, USA) was tested. As visible in Table 2, their usage causes no detectable improvement in BLR.

4 Conclusions

Tested for the time period of VLBI campaign CONT11, a re-calculation of the VMF1 coefficients using ray-traced delays from a denser numerical weather model (NWM) with a resolution of 0.125° × 0.125° yields only a small improvement in mean BLR compared to the existing VMF1 coefficients (1.08 cm compared to 1.09 cm). This derivation of new coefficients is based on the new ray-tracing program RADIATE which will also be used to calculate ray-traced delays for the complete history of VLBI observations. Furthermore, revised gradient formulas based on that by Chen and Herring (1997) were studied. They yield slight improvements in BLR when used for the calculation of a priori gradients derived from ray-traced NWM delays. Hence, gradient formula (3) is suggested for use, if available. The BLR is reduced from 1.07 cm to 1.03 cm in the case that the standard gradients are additionally estimated in a least squares adjustment within the VLBI analysis and from 1.20 cm to 1.09 cm in the case of no subsequent estimation, respectively. Thus, another goal is to provide those gradients not only for the complete history of VLBI observations but also to reflect about strategies to provide gradients on global grids for the application in the analysis of GNSS observations.

Acknowledgements The authors would like to thank the Austrian Science Fund (FWF) for financial support of the project RADIATE VLBI (Ray-traced Delays in the Atmosphere for geodetic VLBI) (P25320) which is the basis of this work.

References

Böhm J, Schuh H (2001) Spherical harmonics as a supplement to global tropospheric mapping functions and horizontal gradients. In: Proceedings of the 15th working meeting on European VLBI for geodesy and astrometry, pp 143–148

Böhm J, Schuh H (2007) Troposphere gradients from the ECMWF in VLBI analysis. J Geodesy 81(6–8):403–408

Böhm J, Werl B, Schuh H (2006) Troposphere mapping functions for GPS and VLBI from ECMWF operational analysis data. J Geophys Res 111:B02406

Böhm J, Spicakova H, Urquhart L, Steigenberger P, Schuh H (2011) Impact of a priori gradients on VLBI-derived terrestrial reference frames. In: Alef W, Bernhart S, Nothnagel A (eds) Proceedings of the 20th meeting of the European VLBI group for geodesy and astrometry, 29–30 March 2011, pp 128–132

Böhm J, Böhm S, Nilsson T, Pany A, Plank L, Spicakova H, Teke K, Schuh H (2012) The new Vienna VLBI software VieVS. In: Kenyon S, Pacino MC, Marti U (eds) Proceedings of IAG scientific assembly 2009, international association of geodesy symposia series, vol 136, pp 1007–1011

Chen G, Herring TA (1997) Effects of atmospheric azimuthal asymmetry on the analysis of space geodetic data. J Geophys Res 102(B9):20489–20502

Eriksson D, MacMillan DS, Gipson JM (2014) Tropospheric delay raytracing applied in VLBI analysis. J Geophys Res Solid Earth 119:9156–9170

Gegout P, Biancale R, Soudarin L (2011) Adaptive mapping functions to the azimuthal anisotropy of the neutral atmosphere. J Geodesy 85:661–677

Herring TA (1992) Modeling atmospheric delays in the analysis of space geodetic data. In: DeMunck JC, Spoelstra T.A.Th. (eds) Publications on geodesy. Proceedings of refraction of transatmospheric signals in geodesy, vol 36, pp 157–164, Netherlands Geodetic Commission Publications in Geodesy

Hobiger T, Ichikawa R, Takasu T, Koyama Y, Kondo T (2008) Ray-traced troposphere slant delays for precise point positioning. Earth Planets Space 60:e1–e4

Hofmeister A, Böhm J (2014) Ray-traced delays in the atmosphere for geodetic VLBI. In: Behrend D, Baver KD, Armstrong K (eds) IVS 2014 general meeting proceedings

Nafisi V, Madzak M, Böhm J, Ardalan AA, Schuh H (2012) Ray-traced tropospheric delays in VLBI analysis. Radio Sci 47:RS2020

Zus F, Bender M, Deng Z, Dick G, Heise S, Shang-Guan M, Wickert J (2012) A methodology to compute GPS slant total delays in a numerical weather model. Radio Sci 47:RS2018

Loading-Induced Deformation Due to Atmosphere, Ocean and Hydrology: Model Comparisons and the Impact on Global SLR, VLBI and GNSS Solutions

O. Roggenbuck, D. Thaller, G. Engelhardt, S. Franke, R. Dach, and P. Steigenberger

Abstract

Precise measurements of the Earth's shape, gravity field, and rotation provide critical data for many geoscientific disciplines. In order to obtain reliable data, an accurate, stable, and global reference frame is required. The International Terrestrial Reference Frame, where station positions are modeled linearly, is commonly used throughout the geoscientific community for this purpose. Mass redistribution in the geophysical fluids, namely atmosphere, oceanic, and continental hydrology, cause time dependent variations in station coordinates, and other parameters such as the geocenter coordinates. Tidally-induced loading is described in the IERS Conventions. Models for non-tidal loading are available through the Global Geophysical Fluid Center. An overview of these models is given and comparisons were carried out. Within these comparisons, the best agreement was observed between the two atmospheric models, whereas the biggest discrepancies were found between the hydrology models. The processing setup for the Very Long Baseline Interferometry (VLBI), Satellite Laser Ranging (SLR) and Global Navigation Satellite Systems (GNSS) solutions is described, with loading displacement values introduced at the observation level. By using of non-tidal loading models the RMS of the height is reduced for 93% of the GNSS stations, with a max. reduction of 50%. The model impact on the station height in Wettzell derived by GNSS, SLR and VLBI shows a good agreement. In SLR results the Blue-Sky effect is visible. Applying the loading models reduced the seasonal variations visible in the geocenter time series derived by SLR almost completely.

Keywords

Geophysical models • Loading displacement • Space geodetic techniques • Terrestrial reference frame

O. Roggenbuck (✉)
Bundesamt für Kartographie und Geodäsie, Richard-Strauss-Allee 11, 60598 Frankfurt am Main, Germany

Present address: Jade Hochschule Wilhelmshaven/Oldenburg/Elsfleth Ofener Straße 16/19, 26121 Oldenburg, Germany
e-mail: ole.roggenbuck@jade-hs.de

D. Thaller • G. Engelhardt • S. Franke
Bundesamt für Kartographie und Geodäsie, Richard-Strauss-Allee 11, 60598 Frankfurt am Main, Germany
e-mail: daniela.thaller@bkg.bund.de

R. Dach
Universität Bern, Astronomisches Institut, Sidlerstraße 5, CH-3012 Bern, Switzerland

P. Steigenberger
Technische Universität München, Arcisstraße 21, 80333 München, Germany

Present address: Deutsches Zentrum für Luft- und Raumfahrt e.V. (DLR), Münchener Straße 20, 82234 Weßling, Germany

T. van Dam (ed.), *REFAG 2014*, International Association of Geodesy Symposia 146,
DOI 10.1007/1345_2015_214

1 Introduction

The International Terrestrial Reference System (ITRS) and its realization, the International Terrestrial Reference Frame (ITRF) is used throughout the geoscientific community. Instantaneous station positions are described as the sum of a linear model and time dependent displacements. In the recent realization, the ITRF2008, non-tidal displacements caused by mass redistribution in the atmosphere, oceans and continental water storage are unconsidered. Systematic variations of the estimated parameters are the consequence. The Global Geophysical Fluid Center (GGFC), under the auspices of the International Earth Rotation and Reference System Service (IERS), provides information and models describing mass changes and their effects. Modeled displacement vectors can be used for the reduction of geodetic observations and gravity field changes can improve the determination of precise satellite orbits.

In the past few years, many studies about the impact of loading displacements onto global space geodetic techniques have been performed. Mostly only one geophysical component, i.e. non-tidal atmospheric loading (NTAL), non-tidal ocean loading (NTOL) or continental water storage loading (CWSL), was focused in these studies (Petrov and Boy 2004; Van Dam et al. 2012; Fritsche et al. 2012; Eriksson and MacMillan 2014).

In this study we studied how the usage of all three loading components (i.e. NTAL, NTOL, CWSL) impacts global VLBI, SLR and GNSS solutions. The loading displacement is added at the observation level. Dach et al. (2011) showed that by introducing NTAL on the observation level the repeatability of weekly GNSS station height estimations can be improved by 10–20%. The importance of using NTAL in SLR analysis at the observation level was shown by Sośnica et al. (2013). After an overview of available models and differences of two different model sets in Sect. 2, the processing setup is described in Sect. 3. The results are shown and discussed in Sect. 4.

2 Geophysical Fluid Models

2.1 Background

A brief overview of gridded geophysical fluid models (available in June 2015) is given in Table 1. The displacement was calculated by most participants, whereas the impact on the Earth's gravity field was only computed by a few institutions. Comparisons between models are necessary since they have different characteristics: spatial- and temporal resolution, background models that provide the input data for the convolution, the treatment of tidal loading, use of Green's function,

Table 1 Compilation of loading models for NTAL, NTOL and CWSL which can be found at the GGFC and the International Mass Loading Service (IMLS) web-pages (status 06.2015)

	NTAL Geom./Grav.	NTOL Geom./Grav.	CWSL Geom./Grav.
Uni Luxembourg[a]	×/×	×/–	×/–
NASA GSFC[b]	×/–	×/–	×/–
TU Vienna[c]	×/×	–/–	–/–
Uni Strasbourg[d]	×/×	–/–	×/×
GFZ[e,f,g]	–/×	–/×	×/×
IMLS[h]	×/×	×/×	×/×

Only the gridded versions are included. ×: grids available. –: no grids available

[a]http://geophy.uni.lu/
[b]http://lacerta.gsfc.nasa.gov/+[./aplo_grid_ncl./oclo/l./hydlo/]
[c]http://ggosatm.hg.tuwien.ac.at/LOADING/
[d]http://loading.u-strasbg.fr
[e]GFZ = German Research Center for Geosciences
[f]http://isdc.gfz-potsdam.de
[g]http://www.gfz-potsdam.de/sektion/erdsystem-modellierung/service/hydl/
[h]http://massloading.net

and the land-sea mask. Some other features, like the handling of the oceanic response in case of NTAL or the treatment of permafrost areas, are different as well.

The RMS of the radial displacement, computed from the NASA Goddard Space Flight Center (GSFC) models, is shown in Fig. 1. The north and east surface displacements were not shown in this study due to their relatively small size. The CWSL and NTOL grids had to be detrended because the background models are not mass conserving. An offset and a long term trend per grid point were determined and removed using a minimum of 19 years of data. NTAL shows the biggest RMS of about 8 mm over Eurasia (see Fig. 1a), whereas CWSL shows strong variations in South America, central Africa and South East Asia and India. The RMS of the NTOL is much smaller and it only exceeds 5 mm in local areas such as the Sea of Japan, the North Sea and the northern Pacific Ocean. These RMS patterns suggest that the impact on global geodetic solutions will be much smaller for NTOL than it is for NTAL and CWSL.

2.2 Model Comparisons

Comparisons were made between the models from the University of Luxembourg (ULux) and the GSFC. In both cases the displacement fields for NTAL, NTOL, and CWSL are provided on a regular longitude/latitude grid. Both institutions use the NCEP Reanalysis pressure fields (Kalnay et al. 1996) for their NTAL computation of 2.5° grids with a 6 h sampling. NTOL displacement grids are calculated with pressure data from the ocean circulation model ECCO (Kim et al. 2007). ULux generates grids with a 2.5°, the GSFC

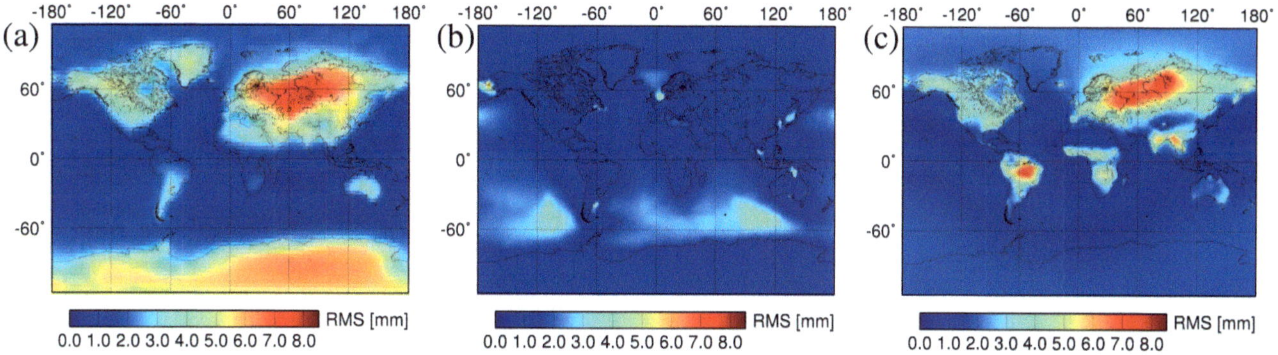

Fig. 1 RMS of the radial component of loading displacement grids from NASA GSFC. The RMS was calculated after subtracting a long term trend and an offset for the NTOL and CWS models. (**a**) Atmospheric non-tidal loading (NTAL). (**b**) Non-tidal ocean loading (NTOL). (**c**) Continental hydrology loading (CWSL)

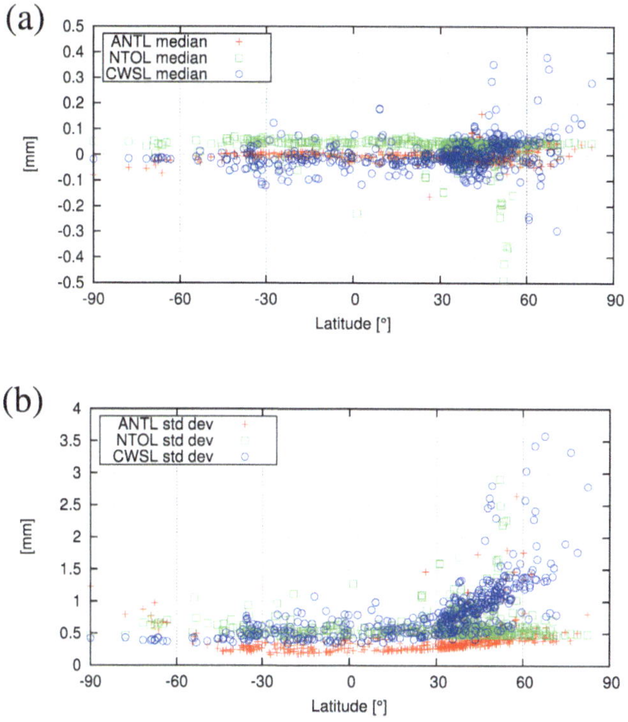

Fig. 2 Comparison between models provided by GSFC and University of Luxembourg. (**a**) Median of model differences per station. Sorted by station latitude. (**b**) Standard deviation of model differences per station. Sorted by station latitude

with a 1.0° resolution and a 12 h sampling. GLDAS/NOAH (Rodell et al. 2004) data is the basis for the generation of 2.5° (ULux) and 1.0° (GSFC) grids with a monthly temporal resolution.

The loading displacement for a set of 611 globally distributed stations was computed with a bilinear interpolation procedure. No explicit evaluation of interpolation errors was performed. The radial component was studied, using 32 years of NTAL, 19 years of NTOL, and 33 years of CWSL data. NTOL and CWSL time series were detrended before calculating the differences (GSFC-ULux). The median differences (see Fig. 2a) indicate a small offset of 0.05 mm between the NTOL models. According to Van Dam (personal communication, 2015) the offset in NTOL could not be caused by different Green's function but are the result of different mean pressure fields removed from the data. Differences in grid spacing and the resulting interpolation errors may explain the big differences occurring at stations located along the Northern Sea coast, at approximately 50° latitude. The NTAL time series agreed well, and only a few stations exceeded an absolute value of 0.05 mm. The CWLS comparison median values were more scattered, due possibly to different spatial resolution of the predicted displacements grids. Additionally, stations in areas with permafrost showed consistently significant differences. GSFC excluded such areas by their loading convolution, whereas ULux just set the snow component to zero at latitudes above 60°. Because the aforementioned differences are systematic, the standard deviations (see Fig. 2b) are considered a measurement of variability. The biggest differences in values occur at stations in permafrost areas or near coastlines. Nearly all stations with a standard deviation higher than 1 mm are coastal stations. In such areas a high grid resolution of the predicted displacements is important. Stations with a standard deviation higher than 1.5 mm are mainly located in permafrost areas.

3 SLR, GNSS and VLBI Processing

In this study, we investigated the impact of non-tidal loading displacement models on VLBI, GNSS, and SLR solutions. Modeled displacements were introduced at the observation level. If the corrections were applied a posteriori, only station coordinates, but no other parameters, were affected. The Bernese GNSS Software Version 5.3 (BSW) was used for GNSS and SLR processing. VLBI data was processed with the Calc/Solve (Solve release 2010.05.21/Calc 10) software. Both software packages are well tested and have been in

Table 2 Overview about the GNSS, VLBI and SLR data processed

	VLBI	SLR	GNSS
Data processed	R1/R4 sessions	LAGEOS 1/2 Etalon 1/2	GPS/GLONASS CODE ITRF2013
Time span	2001–2010	2001–2011	2005–2012
Software	Calc/Solve	BSW 5.3	BSW 5.3
Stations	32	51	345
Solution type	Session wise	Weekly	Daily
Displacement	Time series	Gridded	Gridded

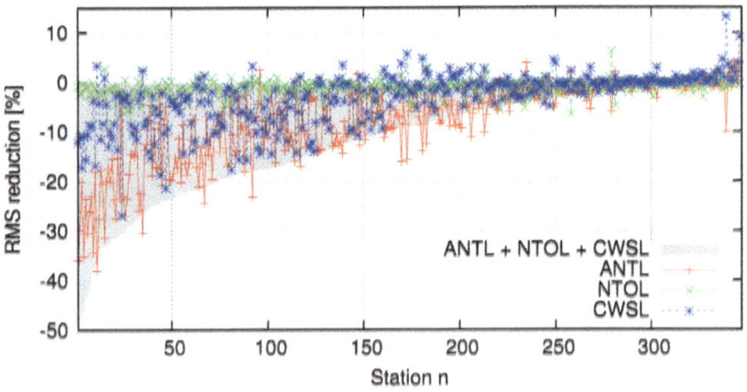

Fig. 3 Sorted reductions of rms in estimated GNSS station height time series after subtracting the linear trends and fixing for jumps

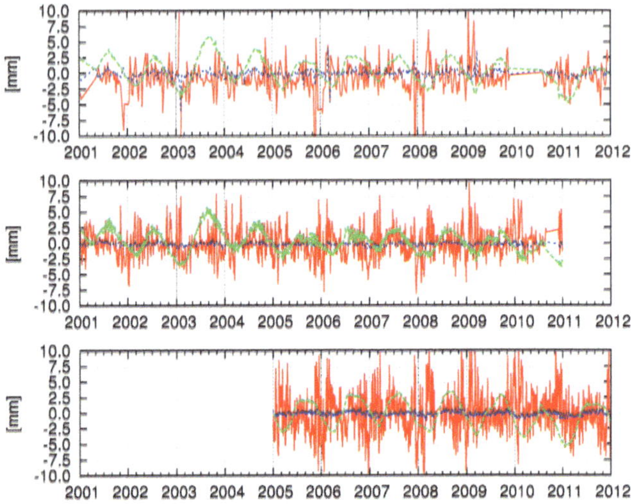

Fig. 4 Impact of NTAL (*red*), NOTL (*blue*) and CWSL (*green*) on estimated station height in Wettzell. *Top*: SLR, *Middle*: VLBI, *Bottom*: GNSS

use for the daily routine processing at IGS, ILRS, and IVS analysis centers, at BKG and CODE for a significant amount of time. An overview of the data used is given in Table 2. In addition to technique-specific parameters like the range bias in SLR, we set up common parameters such as geocenter coordinates (GCC) for SLR and GNSS and Earth Orientation Parameter (EOP) for all techniques.

The non-tidal portion of the loading displacement was taken from grids produced by the GSFC. We used displacement fields referring to the center of mass of the total Earth system (Blewitt 2003). The displacement time series for each individual grid point were detrended (NTOL/CWSL) and reformatted into a BSW readable grid format. BSW then interpolates the displacement at each station first bilinearly, and then linearly in time. In contrast, Calc/Solve uses one displacement time series per station, so in order to be consistent with the BSW analysis, these series were calculated using the same algorithms as implemented in BSW. Five solutions per space geodetic technique were generated: without non-tidal models, with all models applied i.e. NTAL+NTOL+CWSL, with only one model applied.

4 Results

4.1 Station Coordinates

The global GNSS network provides long and spatially dense time series. The relative RMS reduction of the height component was calculated for all 345 GNSS stations included in the processing. Figure 3 shows the resulting reduction for each loading component, sorted by the reduction of the solution, with all three models applied. Using all models reduces the RMS for 93% of the stations, with improvements primarily a result of NTAL followed by CWSL and NTOL. Individually, the station in Novosibirsk, Russia shows the biggest improvement of 50%. Collectively, 29% of the stations show an increasing RMS when including CWSL. Some of these stations, such as Fairbanks (FAIR), Alaska (rms +5.8%) are located in areas where local effects and/or permafrost effects play a major role. Other stations like CGGN in Toro, Nigeria (rms +9.1%, 56 sessions) were rarely observed. The biggest increase can be seen for station Parkes (PARK), Australia (rms +13.4%).

Coordinate time series from SLR and VLBI were omitted because they are noisier and not as dense as GNSS series. In order to see how different models impact individual solutions, stations where all techniques are located were studied. The model impact on the station height of Wettzell was calculated by subtracting each solution from a reference solution where no models were applied, as depicted in Fig. 4. Despite the different sampling in the GNSS, SLR and VLBI solutions, the underlying pattern of each technique looks similar. The impact magnitude of NTAL and CWSL varied between −5 mm and 5 mm, whereas NTOL did not change the height significantly because of its great distance from the ocean. The mean values and standard deviations can be seen in Table 3. Looking at the standard deviation, the mean values are not significant, but the variations from epoch to epoch are not negligible if mm accuracy of the terrestrial reference frame is desired. Nevertheless, SLR is the only

Table 3 Impact of individual loading components on station height in Wettzell

	NTAL mean/std dev [mm]	NTOL mean/std dev [mm]	CWSL mean/std dev [mm]
SLR	−0.57/2.64	0.07/0.69	0.65/2.03
VLBI	0.46/1.71	0.00/0.25	0.10/2.49
GNSS	−0.09/2.22	0.06/0.32	0.08/3.30

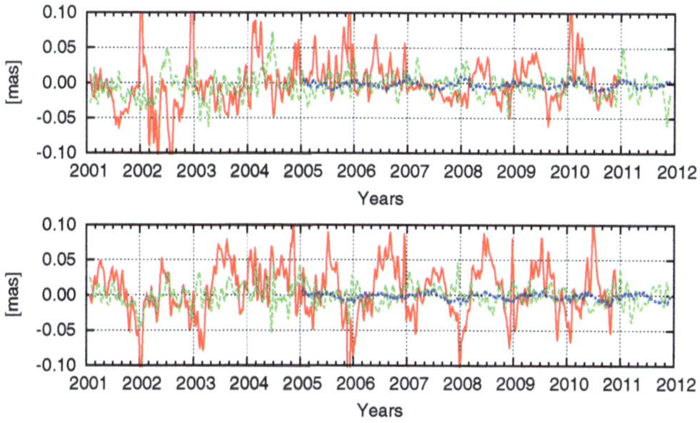

Fig. 5 Impact of the sum of NTAL, NOTL and CWSL on the estimation of the pole coordinates. *Green*: SLR, *red*: VLBI, *blue*: GNSS. *Top*: X-pole, *Bottom*: Y-pole

technique where the mean for the NTAL impact is negative (−0.57 mm). This negative mean is visible at most of the other SLR stations, as well. SLR can only operate when the sky is not cloud covered what often coincide with high pressure fields. This results are in good agreement with the studies by Otsubo et al. (2004) and Sośnica et al. (2013) regarding the Blue-Sky effect.

4.2 Earth Orientation Parameters

Figure 5 shows how non-tidal loading affects the estimated pole parameters. X-pole as well as Y-pole derived by VLBI are affected most, showing a yearly signal with an amplitude of 0.1 mas for Y-pole, and slightly smaller variations of 0.07 mas for X. SLR estimates do not show a yearly signal, and the differences vary between −0.04 mas and 0.04 mas, which may lead to the assumption that the sparse networks of these two techniques and the individual station positions are responsible for this behavior. The GNSS estimation seems nearly unaffected, with only a small yearly signal measuring an amplitude of 0.01 mas observed.

4.3 Geocenter Coordinates

In the SLR and GNSS analysis coordinates of the Earth's center of mass in the terrestrial reference frame were set up. The datum was realized by applying a no-net-translation constraint. The estimates of both SLR and GNSS derived GCC solutions were affected in a similar way by using non-tidal loading displacements. Including NTAL or CWSL led to changes of up to ±2.5 mm in the Z-component. In contrast to this, NTOL impact was much smaller (no plot shown here). The estimated time series from GNSS and SLR were transformed into the frequency domain using a Fast Fourier Transformation (FFT). The spectra of the Z-coordinate derived by SLR (see Fig. 6a) is dominated by a 4 mm yearly signal. This annual variation can be fully explained by the sum of all three loading components. Firstly, the amplitude was reduced to 2.1 mm by including CWSL and to 2.75 mm with NTAL. Secondly, NTOL did not reduce the signal significantly. Thirdly, aside from annual variations, the Z-component of GNSS-based GCC estimations show strong signals at the draconitic period (GPS = 351.2 days) and fractions therein (see Fig. 6b). These signals are caused by orbit modeling issues and are mainly visible in the Z-component reaching values of 5 mm. The additional modeling of non-tidal loading displacement showed damping in the yearly amplitude values of 0.5 mm in X and 1.3 mm in Y, but showed what appears to be an increase from 4.4 mm to 7.1 mm in Z. The processed time span is too short to separate the draconitic and the yearly period using an FFT. Overall, the amplitude of the resolvable semiannual frequency is reduced by 50% when using all three loading components. A combined SLR and GNSS solution could improve the GCC estimation (Thaller et al. 2014).

5 Summary and Conclusions

The focus of this work was to study the impact of loading displacement on space technique derived parameters. Two sets of loading displacement models consisting of NTAL, NTOL and CWSL provided by ULux and GSFC were compared. Interpolated time series were analysed. For NTAL 96% of the absolute median differences are smaller than 0.05 mm. The median differences of NTOL indicate an offset 0.05 mm. Considering these offset 87% of the absolute median differences are smaller than 0.05 mm. The comparison of CWSL models showed bigger differences reaching a 0.4 mm, which are mainly caused by different handling of permafrost areas and grid spacing of the predicting displacement values. No special evaluation of interpolation errors was done. Extending these comparisons with other models is highly desired.

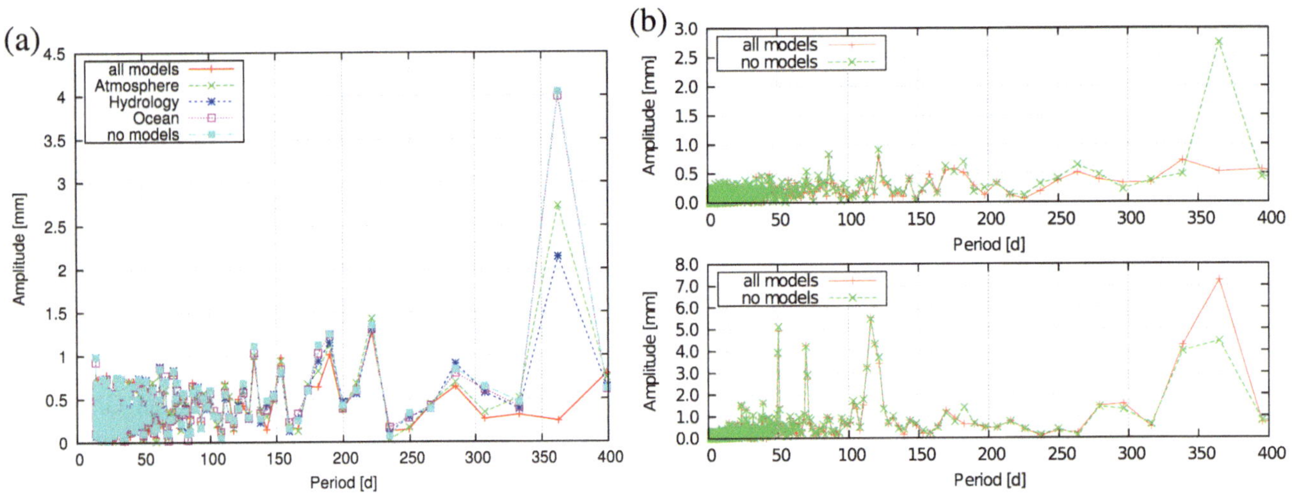

Fig. 6 Frequency domain of the GCC derived by SLR and GNSS analysis. (**a**) SLR GCC Z-component. (**b**) GNSS GCC frequency domain. *Top*: X-component, *Bottom*: Z-component

The GSFC displacement grids were introduced into global VLBI, SLR and GNSS analysis. Displacements were added a priori at the observation level to study the combined and individual impact on the estimated parameters. In contrast to SLR and VLBI, GNSS has a dense station network yielding stable, long term coordinate series with a high temporal resolution. GNSS is therefore a good technique to compare and evaluate models by investigating their effects on station coordinates. In case of VLBI and SLR, other parameters, like geocenter coordinates and Earth orientation parameters were studied. The impact of individual loading components on the station height for station Wettzell is similar for all three techniques. It was also shown that the SLR solution could especially benefit by using atmospheric loading because of the systematic Blue-Sky effect. For the GNSS stations, the RMS of the height component was significantly reduced at 93% of the stations. It was shown that the seasonal variations in the SLR based GCC can fully be explained by the sum of NTAL, NTOL, and CWSL. This is of special interest as the origin of our recent reference frame (ITRF2008) stems solely from SLR (Altamimi et al. 2011). The model impact on the geocenter estimations from GNSS and SLR are comparable. Similar studies with other models than the GSFC models would help to compare the estimated parameters.

Acknowledgements This work was funded by the German Research Foundation (DFG) as a part of the Research Project (FOR1503): Space-Time Reference Systems for Monitoring Global Change and for Precise Navigation in Space.

We would like to thank the Global Geophysical Fluid Center for providing information and assistance regarding the non-tidal loading models.

References

Altamimi Z, Collilieux X, Métivier L (2011) Itrf2008: an improved solution of the international terrestrial reference frame. J Geodesy 85(8):457–473. doi:10.1007/s00190-011-0444-4. http://dx.doi.org/10.1007/s00190-011-0444-4

Blewitt G (2003) Self-consistency in reference frames, geocenter definition, and surface loading of the solid Earth. J Geophys Res Solid Earth 108(B2). doi:10.1029/2002JB002082

Dach R, Böhm J, Lutz S, Steigenberger P, Beutler G (2011) Evaluation of the impact of atmospheric pressure loading modeling on GNSS data analysis. J Geodesy 85(2):75–91. doi:10.1007/s00190-010-0417-z

Eriksson D, MacMillan DS (2014) Continental hydrology loading observed by VLBI measurements. J Geodesy 88(7):675–690. doi:10.1007/s00190-014-0713-0

Fritsche M, Döll P, Dietrich R (2012) Global-scale validation of model-based load deformation of the Earth's crust from continental watermass and atmospheric pressure variations using GPS. J Geodyn 59–60:133–142. doi:10.1016/j.jog.2011.04.001

Kalnay E, Kanamitsu M, Kistler R, Collins W, Deaven D, Gandin L, Iredell M, Saha S, White G, Woollen J, Zhu Y, Chelliah M, Ebisuzaki W, Higgins W, Janowiak J, Mo KC, Ropelewski C, Wang J, Leetmaa A, Reynolds R, Jenne R, Joseph D (1996) The NCEP/NCAR 40-year reanalysis project. Bull Am Meteorol Soc 77:437–470

Kim SB, Lee T, Fukumori I (2007) Mechanisms controlling the interannual variation of mixed layer temperature averaged over the Niño-3 region. J Climate 20:3822–3843. doi:10.1175/JCLI4206.1

Otsubo T, Kubo-oka T, Gotoh T, Ichikawa R (2004) Atmospheric loading "Blue-Sky" effects on SLR station coordinates. In: Proceedings from the fourteenth international workshop on laser ranging instrumentation, San Fernando

Petrov L, Boy JB (2004) Study of the atmospheric pressure loading signal in very long baseline interferometry observations. J Geophys Res Solid Earth 109(B3). doi:10.1029/2003JB002500

Rodell M, Houser PR, Jambor U, Gottschalck J, Mitchell K, Meng CJ, Arsenault K, Cosgrove B, Radakovich J, Bosilovich M, Entin

JK, Walker JP, Lohmann D, Toll D (2004) The global land data assimilation system. Bull Am Meteorol Soc 85:381–394

Sośnica K, Thaller D, Dach R, Jäggi A, Beutler G (2013) Impact of loading displacements on SLR-derived parameters and on the consistency between GNSS and SLR results. J Geodesy 87:751–769. doi:10.1007/s00190-013-0644-1

Thaller D, Sośnica K, Dach R, Jäggi A, Beutler G, Mareyen M, Richter B (2014) Geocenter coordinates from GNSS and combined GNSS-SLR solutions using satellite co-locations. In: Rizos C, Willis P (eds) Earth on the edge: science for a sustainable planet. International association of geodesy symposia, vol 139. Springer, Berlin/Heidelberg, pp 129–134. doi:10.1007/978-3-642-37222-3_16

Van Dam T, Collilieux X, Wuite J, Altamimi Z, Ray J (2012) Nontidal ocean loading: amplitudes and potential effects in GPS height time series. J Geodesy 86(11):1043–1057. doi:10.1007/s00190-012-0564-5

The International Mass Loading Service

Leonid Petrov

Abstract

The International Mass Loading Service computes four loadings: (a) atmospheric pressure loading; (b) land water storage loading; (c) oceanic tidal loading; and (d) non-tidal oceanic loading. The service provides to users the mass loading time series in three forms: (1) pre-computed time series for a list of 884 space geodesy stations; (2) pre-computed time series on the global 1° × 1° grid; and (3) on-demand Internet service for a list of stations and a time range specified by the user. The loading displacements are provided for the time period from 1979.01.01 through present, updated on an hourly basis, and have latencies 8–20 h.

Keywords

Crustal deformation • Geodesy service • Mass loading

1 Introduction

Loading is a crustal deformation caused by a redistribution of air or water mass. Darwin (1882) was the first who considered this effect and made first rough estimates of its magnitude. However, only in 1980s it became feasible to compute loading quantitatively. One of the first evidence that displacements due to loading caused by ocean tides noticeably affect observations was obtained by Schuh and Moehlmann (1989). Van Dam et al. (1994) and independently MacMillan and Gipson (1994) reported detection of the atmospheric pressure loading signal in VLBI data. Later, van Dam et al. (2001) reported detection of the land water storage signal in GPS data. Ten years later Williams and Penna (2011) reported detection of the non-tidal ocean loading signal in GPS data.

The focus of loading work in 1990s was a detection of the effect. After it was recognized that unmodeled loading is a source of systematic errors, the focus in 2000s was shifted to computation of loading effects as a regular service. H.-G. Scherneck established the ocean loading service around 2000 (Baker and Bos 2003). In December 2002 the atmospheric pressure loading service (Petrov and Boy 2004) at NASA Goddard Space Flight Center was established that provides 3D displacements caused by loading for a number of sites and at a regular grid. By 2015, a number of similar services were running. The list of their URLs can be found at http://massloading.net/links.

Analysis of performance of the atmospheric loading services revealed a number of shortcomings: (1) loading at the coast and islands is affected by the dynamic response of the ocean to atmospheric forcing; (2) loading in mountainous areas has larger errors because the model data sampling is too coarse (van Dam et al. 2010); (3) latency of the loading product, 3–4 days, prevented using the service for rapid data analysis; (4) loading caused by land water storage was not modeled; (5) there is a need to have loadings for a new site not listed in the service. Interpolating the loading at a low resolution grid gave unsatisfactory results and introduced errors at a level of 15% at coastal areas.

In order to mitigate these shortcomings, a new service was launched with the name International Mass Loading Service. This paper provides the outline of the current capabilities of this service.

L. Petrov (✉)
ADNET Systems, Inc, Falls Church, VA 22043, USA
e-mail: Leonid.Petrov@lpetrov.net

T. van Dam (ed.), *REFAG 2014*, International Association of Geodesy Symposia 146,
DOI 10.1007/1345_2015_218

2 The Use of High Resolution Models for Loading Computation

The original atmospheric pressure loading service used the 2D NCEP Reanalysis surface pressure field (Kalnay et al. 1996) at a regular grid with a spatial resolution 2.5° × 2.5°. Modern models have much higher resolutions: for instance, the GEOS-FP model has resolution 0.3125° × 0.25°. The traditional approach for loading computation at a point with coordinate **r** involved a numerical evaluation of the integral of a convolution type (Farrell 1972):

$$\mathbf{u}_r(\mathbf{r},t) = \iint\limits_{\Omega} L(\phi',\lambda')\,\Delta P(\mathbf{r}',t)\,G_{\mathrm{R}}(\psi)\cos\phi'\,d\lambda'\,d\phi'$$
$$\mathbf{u}_h(\mathbf{r},t) = \iint\limits_{\Omega} \mathbf{q}(\mathbf{r},\mathbf{r}')\,L(\phi,\lambda)\,\Delta P(\mathbf{r}',t)\,G_{\mathrm{H}}(\psi)\cos\phi'\,d\lambda'\,d\phi', \quad (1)$$

where $\Delta P(\mathbf{r}',t)$ is the pressure caused by mass redistribution, $L(\phi,\lambda)$—is the land-sea mask, the share of land in an elementary cell, and $G(\psi)$ are the Green's functions.

The problem is that this algorithm has complexity $O(d^4)$, where d is the spatial grid size, i.e. it grows very rapidly with an increase of spatial resolution. It becomes impractical to use convolution for loading computation using models with a high spatial resolution. The alternative is to use the spherical harmonic transform approach. The algorithm involves the following steps:

1. forming the pressure difference with respect to the average;
2. transforming the surface pressure field to the regular grid with a higher resolution (upgridding): $2(D+1)+1 \times 4(D+1)$ over latitude and longitude, where D is degree of the expansion;
3. multiplying the surface pressure field with the land-sea mask defined as a share of land in a cell;
4. spherical harmonic transform of degree/order D in accordance with sampling theorem of Dricoll and Healy (1994).
5. scaling the output of the spherical harmonic transform with Love numbers h'_n and l'_n of the corresponding degree n:

$$V_n^m(t) = \frac{1}{\bar{\rho}_{\oplus}\,g_0}\frac{3h'_n}{2n+1} \iint\limits_{\Omega} L(\phi,\lambda)\,\Delta P(t,\phi,\lambda)\,Y_n^m(\phi,\lambda)\cos\phi\,d\phi\,d\lambda$$
$$H_n^m(t) = \frac{1}{\bar{\rho}_{\oplus}\,g_0}\frac{3l'_n}{2n+1} \iint\limits_{\Omega} L(\phi,\lambda)\,\Delta P(t,\phi,\lambda)\,Y_n^m(\phi,\lambda)\cos\phi\,d\phi\,d\lambda, \quad (2)$$

where $\bar{\rho}_{\oplus}$ is the mean Earth's density and g_0 is the equatorial gravity acceleration. The expression under the integral is the spherical harmonics $()_n^m$ of the pressure field with the land-sea mask applied.

6. inverse spherical harmonic transform:

$$D_U(\phi,\lambda) = \sum_{i=0}^{i=m}\sum_{j=-n}^{j=n} V_j^i\,Y_j^{i\,*}(\phi,\lambda)$$
$$D_E(\phi,\lambda) = \sum_{i=0}^{i=m}\sum_{j=-n}^{j=n} H_j^i\,\frac{\partial Y_j^{i\,*}(\phi,\lambda)}{\partial\lambda}. \quad (3)$$
$$D_N(\phi,\lambda) = \sum_{i=0}^{i=m}\sum_{j=-n}^{j=n} H_j^i\,\frac{\partial Y_j^{i\,*}(\phi,\lambda)}{\partial\phi}$$

This algorithm is equivalent to Eq. (1) when $D \longrightarrow \infty$, but it has complexity $O(d^3)$. It outperforms the convolution algorithm when $D > 30$. Numerical tests showed that in order to have errors in loading computation everywhere on the Earth less than 0.15 mm, degree/order 1023 is usually sufficient. Wall time of computing displacements for 10,000 stations at a single core of XEON E5-2660 v3 processor is 33 s per epoch.

3 Mass Redistribution Models

Three numerical weather models developed at the NASA Global Modeling and Assimilation Office (GMAO) are used for loading computation:

- MERRA (Modern-Era Retrospective analysis for Research and Applications) (Rienecker et al. 2011). Resolution: 0.67° × 0.5° × 72 layers × 6^h, runs from 1979.01.01 through present, latency 20^d–60^d. This model is frozen and it is considered the most stable.
- GEOS-FP (Global Earth Observing System Forward Processing) (Molod et al. 2012). Resolution: 0.3125° × 0.25° × 72 layers × 3^h, runs from 2011.09.01 through present, latency 6^h–15^h. This is the operational model, updated approximately once a year.
- GEOS-FPIT (Global Earth Observing System Forward Processing Instrumental Team) (Rienecker et al. 2008). Resolution: 0.625° × 0.5° × 72 layers × 3^h, runs from 2000.01.01 through present, latency 6^h–25^h. In terms in stability this model is intermediate between MERRA and GEOS-FP, but it has a low latency.

The surface pressure is computed from a 3D model. This process involves several steps. Firstly, each column of the output at the native, *irregular*, terrain-following grid is interpolated to the column at a new regular grid that is formally extrapolated down to −1000 m and up to 90,000 m. Then the atmospheric pressure at a given epoch is expanded into the tensor product of B-splines over the entire Earth. Using the expansion coefficients, the pressure on the surface at resolution D1023 (0.088° × 0.088°) is computed. The height of the surface is derived from 30″ × 30″ GTOPO30 model[1] by averaging over cells of the D1023 grid. Using the expansion coefficients, the atmospheric pressure on that surface is computed.

Three land water storage models are used for loading computation:

- GLDAS NOAH025 (Global Land Data Assimilation System) (Rodell et al. 2004). Resolution: $0.25° \times 0.25° \times 3^h$, runs from 2000.01.24 through present, latency 35^d–70^d.
- MERRA TWLAND (Reichle et al. 2011). Resolution: $0.67° \times 0.5° \times 6^h$, runs from 1979.01.01 through present, latency 35^d–60^d. This model is considered the most stable.
- GEOS-FPIT TWLAND. Resolution: $0.625° \times 0.5° \times 1^h$, runs from 2000.01.01 through present, latency 6^h–25^h. It was found that hourly time resolution is excessive for loading computation. The resolution was reduced to 3 h.

Upgridding involves refining the pressure field according to the fine land-sea mask. If a cell at the new $0.088° \times 0.088°$ grid falls in the area that was ocean in the old grid, the pressure of the water equivalent of soil moisture and/or snow cover is computed by interpolation from surrounding cells that are land in the original grid with applying Gaussian smoothing.

Non-tidal ocean loading is computed from the Ocean Model for Circulation and Tides (OMCT) (Dobslaw et al. 2013). The original resolution of the model is $1° \times 1° \times 6^h$, latency: 10^d–60^d.

Two models of ocean tidal loading are used: the GOT4.8 (Ray 2013) and FES2012 (Carrere et al. 2012). They are upgridded to degree/order 2047 in a similar way as it was done for land water storage, except reversal of land and sea cells.

4 Processing Pipeline

The two servers of the International Mass Loading Service that work independently check every hour whether new data appeared. If the new data appeared, they are downloaded, decoded, up-gridded, and the surface pressure anomaly at the D1023 grid is computed by subtracting a model that includes the mean surface pressure value, sine and cosine amplitudes of pressure variations in a range of frequencies in the diurnal, semi-diurnal, ter-diurnal and four-diurnal bands. Parameters of this model were computed by the LSQ over a long period of time. Then the spherical harmonic transform of degree/order 1023 of the pressure field anomaly is computed and scaled by Love numbers of the corresponding order. The coefficients V_n^m and H_n^m in Eq. (2) are stored. They are used for loading computations in three ways:

1. Computing loading at the D359 grid ($0.25° \times 0.25°$). This is done in the following way: the spherical harmonic transform of degree/order D1023 is padded with zeroes to degree/order D1079. The coefficients V_n^m, H_n^m are underwent the inverse spherical harmonic transform and produce the loading field in local Up, East, and North direction at the D1079 grid ($1/12° \times 1/12°$). Every 3rd element of the intermediate D1079 grid is written in the output file.
2. Computing loading for a set of 884 commonly used GNSS, SLR, DORIS, and VLBI stations.
3. Computing loading on-demand for the set of stations supplied by the user. A user fills the Web form where he or she specifies the model, the range of dates and the list of stations with their Cartesian coordinates. When the loading computation is finished, a user can retrieve the files with results.

[1]https://lta.cr.usgs.gov/GTOPO30.

Table 1 Estimates of admittance factors for Up, East, and North components for three different loading models from the global least squares solution using geodetic VLBI data

Atm GEOS-FPIT	UP	0.963 ± 0.023
Atm GEOS-FPIT	EA	0.609 ± 0.049
Atm GEOS-FPIT	NO	1.027 ± 0.041
Lws GEOS-FPIT	UP	0.955 ± 0.016
Lws GEOS-FPIT	EA	0.804 ± 0.029
Lws GEOS-FPIT	NO	0.886 ± 0.024
Lws NOAH025	UP	1.220 ± 0.013
Lws NOAH025	EA	0.660 ± 0.030
Lws NOAH025	NO	0.826 ± 0.033

For explanation of meaning of admittance factors, see Petrov and Boy (2004)

The loading displacements are computed using the Love numbers defined in the coordinate system with the origin at the center of mass of the total Earth: the solid Earth and the fluid under consideration. For some applications displacements with respect to the center of mass of the solid Earth are desirable. The International Mass Loading Service computes the differential loading displacements between these two systems. When this differential displacement is added to the displacement with respect to the center of the total mass, the sum is the displacement with respect to the center of mass of the solid Earth.

5 Validation

VLBI observations for the period of 2001.01.01–2014.07.01 were used for loading validation. The same technique was applied as we used for loading validation in Petrov and Boy (2004): the global admittance factors were estimated from the data together with estimation of site positions, velocities, the Earth orientation parameters, source coordinates and nuisance parameters such as clock functions and atmosphere path delays in zenith direction (see Table 1). The partial derivative for admittance factors was the contribution of the

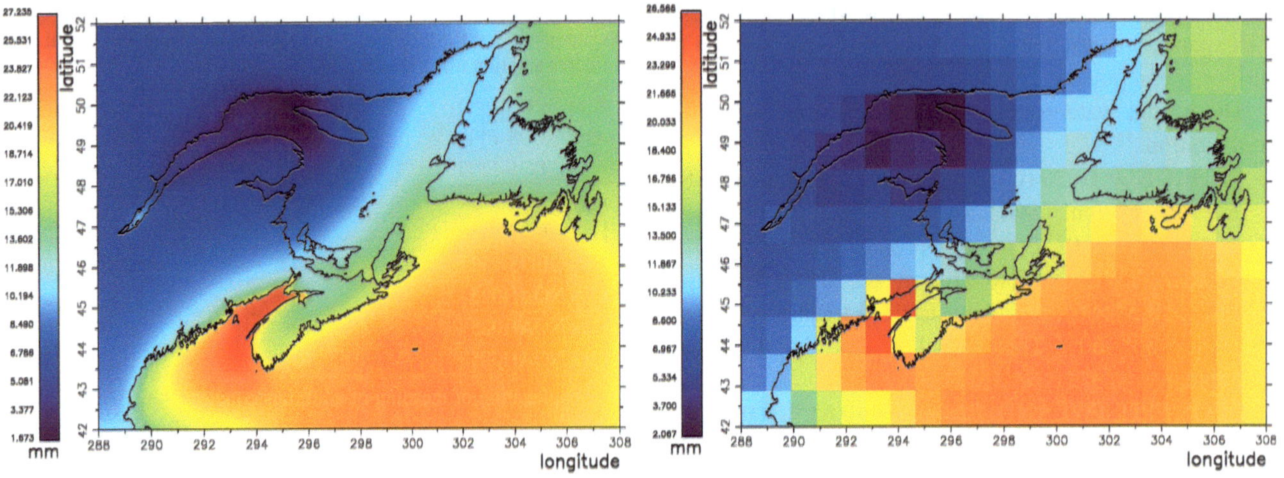

Fig. 1 Mass loading caused by the M_2 ocean tide near Newfoundland island computed with two resolutions: (*left*) 0.01° × 0.01° grid and (*right*) 1.0° × 1.0° grid. Spherical harmonic transform degree 8999 was used for computing loading at the 0.01° × 0.01° grid

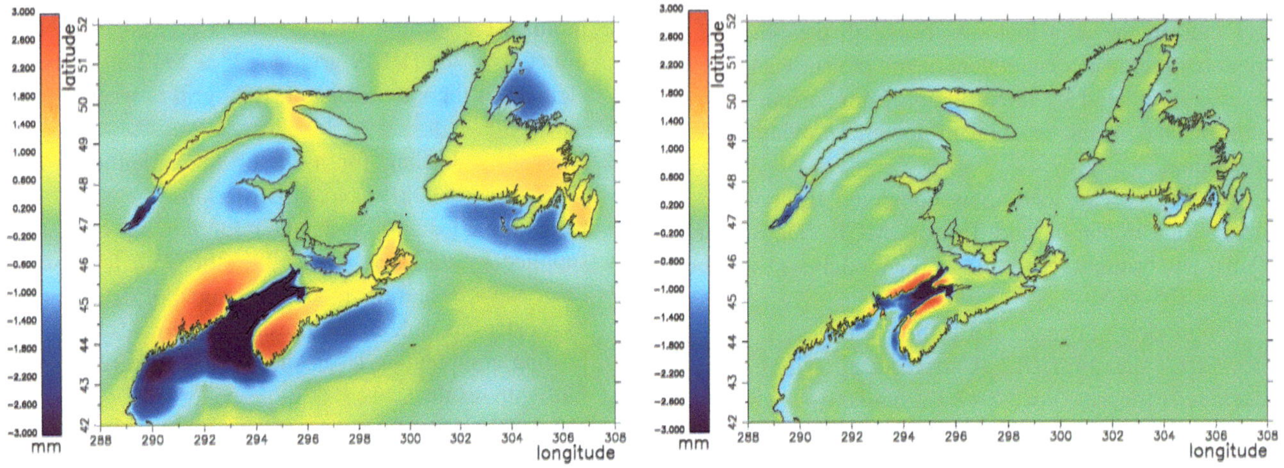

Fig. 2 The difference of mass loading caused by the M_2 ocean tide computed with two resolutions: (*left*) 1.0° × 1.0° grid versus 0.01° × 0.01° grid and (*right*) 0.25° × 0.25° grid versus 0.01° × 0.01° grid. Bilinear interpolation was used

loading displacement into path delay. If the model is perfect, the admittance factor will approach to unity.

6 Using the International Mass Loading Service

The gridded loading displacements are useful for visualization of the loading field and for computation of integrals over the area. However, a user should be aware that the field of loading displacement near the coast is not smooth. Therefore, using gridded loading for data reduction by bilinear interpolation the displacement field to the position of a given station may cause significant errors. This problem is illustrated in Figs. 1 and 2 for a case of ocean loading near Newfoundland. The M_2 ocean loading displacement has the vertical amplitude ~30 mm, but interpolation errors exceed 15% within 100 km of the coastal area when the 1.0° × 1.0° grid is used. The errors are in excess of 15% within 30 km from the coast when the 0.25° × 0.25° grid is used. They fall below 1 mm only when the grid with a resolution 0.05° × 0.05° or finer is used.

Gridded loading at 1° × 1° or 0.25° × 0.25° resolutions *should not be used* for data reduction. The International Mass Loading Service computes loadings for 884 fundamental GNSS, DORIS, SLR, and VLBI stations directly without the use of interpolation. Loading displacements for other stations are computed using Web on-demand interface.

7 Summary

At present, the International Mass Loading Service offers to the geodetic community computation of 3D displacements caused by the atmospheric pressure loading, land water storage loading, tidal and non-tidal ocean loading, free of charge, 24/7 with a latency from 8 h (atmospheric and land water storage loading) to 30 days (non-tidal loading). The URL of the primary server is http://massloading.net, the URL of the secondary server is http://alt.massloading.net. Loading bibliography and links to similar services can be found there. The loading displacement were validated against the dataset of global VLBI observations for 2001–2014.

This project was supported by NASA Earth Surface and Interior program, grant NNX12AQ29G.

References

Baker TF, Bos MS (2003) Validating Earth and ocean tide models using tidal gravity measurements. Geophys J Int 152(2):468–485

Carrere L, Lyard F, Cancet M, Guillot A, Roblou L (2012) FES2012: a new global tidal model taking advantage of nearly 20 years of altimetry. In: Proceedings of meeting "20 Years of Altimetry", Venice

Darwin GH (1882) On variations in the vertical due to elasticity of the Earth's surface. Phil Mag Ser 5 14(90):409–427

Dobslaw H et al (2013) Simulation and observation of global ocean mass anomalies. J Geophys Res 118(7):3704–3711

Dricoll JR, Healy DM (1994) Computing fourier transforms and convolutions on the 2-sphere. Adv Appl Math 15:202–250

Farrell WE (1972) Deformation of the Earth by surface loads. Rev Geophys Space Phys 10(3):751–797

Kalnay EM et al (1996) The NCEP/NCAR 40-year reanalysis project. Bull Am Meteorol Soc 77:437–471

MacMillan DS, Gipson JM (1994) Atmospheric pressure loading parameters from very long baseline interferometric observations. J Geophys Res 99(B9):18081–18087

Molod A, Takacs L, Suarez M, Bacmeister J, Song I-S, Eichmann A (2012) The GEOS-5 atmospheric general circulation model: mean climate and development from MERRA to Fortuna, NASA/TM–2012, 104606, vol 28

Petrov L, Boy J-P (2004) Study of the atmospheric pressure loading signal in VLBI observations. J Geophys Res 109:B03405. doi:10.1029/2003JB002500

Ray R (2013) Precise comparisons of bottom-pressure and altimetric ocean tides. J Geophys Res 118:4570–4584

Reichle RH, Koster RD, De Lannoy GJM, Forman BA, Liu Q, Mahanama SPP, Touré A (2011) Assessment and enhancement of MERRA land surface hydrology estimates. J Clim 24:6322–6338

Rienecker MM et al (2008) The GEOS data assimilation system — Documentation of Versions 5.0.1, 5.1.0, and 5.2.0, NASA/TM–2008–104606

Rienecker MM et al (2011) J Clim 24:3624–3648

Rodell MP et al (2004) The global land data assimilation system, Bull Am Meteorol Soc 85(3):381–394

Schuh H, Moehlmann L (1989) Ocean loading station displacements observed by VLBI. J Geophys Res 16(10):1105–1108

van Dam TM and Herring TA (1994) Detection of atmospheric pressure loading using very long baseline interferometry measurements. J Geophys Res 99:4505–4518

van Dam T, Wahr J, Milly PCD, Shmakin AB, Blewitt G, Lavallee D, Larson K (2001) Crustal displacements due to continental water loading. Geophys Res Lett 28:651–654

van Dam TM, Altimimi Z, Collilieux X, Ray J (2010) Tropospherically induced height errors in predicted atmospheric loading signal. J Geophys Res 115:B07415

Williams SDP, Penna NT (2011) Nontidal ocean loading effects on geodetic GPS heights. Geophys Res Lett 38:L09314

Pre-combined GNSS-SLR Solutions: What Could be the Benefit for the ITRF?

D. Thaller, K. Sośnica, P. Steigenberger, O. Roggenbuck, and R. Dach

Abstract

In standard combination approaches (e.g., for the International Terrestrial Reference Frame), the space-geodetic techniques are connected by the Earth rotation parameters and by the station coordinates at co-located sites, using the so-called local ties. These local ties are usually derived from terrestrial measurements together with GNSS measurements for linking to a global reference system. The local ties often differ from the coordinate differences derived from space-geodetic observations. However, the sources for the discrepancies are not always clear, and a validation within the combination is difficult as long as the local ties are needed for the combination.

We provide an alternative combination method by using the co-location of GNSS and SLR observations onboard GNSS satellites. As the local ties do not need to be applied in this approach, the resulting station coordinates are consistently estimated, but independent of the local ties. This allows us to evaluate the agreement of the terrestrial local ties and the space-geodetic coordinate differences derived from a 12-years solution. The 61 co-locations investigated in a multi-year solution show an agreement better than 1 cm in the horizontal and height components for 41 and 27 co-locations, respectively. Co-locations showing big discrepancies can be explained by the shortness of the data set included, or the rare distribution over the time span. Only the co-location site San Fernando shows unexplained differences of several centimeters.

When using satellite co-locations, the counterpart of local ties at stations are space ties at satellites. The offsets of the microwave satellite antenna form one component of the space tie, with the offset of the laser retro-reflector array (LRA) forming the second part. We show that corrections to the space ties can be estimated from combined GNSS-SLR solutions. The corrections to the LRA offsets are only 5–6 mm, whereas the corrections estimated for the microwave antennas can exceed 1 dm, with a mean correction of −86.1 mm and −110.4 mm for GPS and GLONASS, respectively.

D. Thaller (✉) • O. Roggenbuck
Bundesamt für Kartographie und Geodäsie (BKG),
Richard-Strauss-Allee 11, 60598 Frankfurt a M, Germany
e-mail: daniela.thaller@bkg.bund.de

K. Sośnica • R. Dach
Astronomical Institute, University of Bern, Bern, Switzerland

P. Steigenberger
Technische Universität München, München, Germany

Present address: DLR, Oberpfaffenhofen, Weßling, Germany

T. van Dam (ed.), *REFAG 2014*, International Association of Geodesy Symposia 146,
DOI 10.1007/1345_2015_215

The corrections to the microwave satellite antennas cause a scale difference for the GNSS ground network of 0.67 ppb. We show that this scale is consistent to the SLR scale, thus, the scale of SLR is properly transferred to the GNSS network via the co-location at the GNSS satellites.

Keywords
GNSS • Local ties • Reference frame • Satellite co-location • Scale • SLR

1 Introduction

In standard combination approaches (e.g., for computations of the International Terrestrial Reference Frame, e.g., ITRF2008, see Altamimi et al. 2011), the space-geodetic techniques are connected by the Earth rotation parameters (ERPs) and by the station coordinates at co-located sites using the so-called local ties. The local ties are the three-dimensional vectors between the reference points of the individual space-geodetic techniques. These vectors are usually derived from terrestrial measurements together with GNSS measurements for linking to a global reference system. However, the local ties often differ from the coordinate differences derived from space-geodetic observations. The discrepancies seen in our studies (Thaller et al. 2011) or by the ITRF combination centers (Altamimi et al. 2011; Seitz et al. 2012) are at the level of several millimeters up to few centimeters. The sources for the discrepancies are not always clear and a validation within the combination is difficult as long as the local ties are needed for the combination. Ray and Altamimi (2005) tried to find out the best set of local ties by introducing different subsets of local ties into combined solutions and comparing the results. However, some co-locations are found to be essential for connecting the space techniques due to their location, although there might be larger discrepancies. Thaller et al. (2007) used the ERPs of combined solutions in comparison with the single-technique solutions as an indicator for discrepancies between local ties and space-geodetic techniques.

In all cases, additional parameters common to more than one technique are highly valuable for connecting space-geodetic techniques, because they have the potential to (partially) replace local ties. Conversely, an independent validation becomes possible if the local ties themselves were not introduced into the combined solution. Krügel et al. (2007) used the troposphere parameters at co-located GNSS-VLBI sites for validation. This showed that for the 2-week CONT02 time span, the troposphere ties have the potential to replace the local ties at some co-located sites. This procedure was extended to a TRF multi-year solution by Rothacher et al. (2011).

Another possibility for connecting the satellite techniques GNSS and SLR is to make use of co-locations at satellites. Figure 1 shows the types of co-locations schematically. The left panel shows the standard approach with station co-locations, i.e., the GNSS receiving antenna and the SLR telescope has to be co-located on the ground. The local ties have to be known in this approach. Regarding the satellite co-location, two different types must be distinguished:

- Low Earth Orbiting (LEO) satellites as co-location platform [Fig. 1 (middle)]: the LEO carries a GNSS receiving antenna and a laser retro-reflector array (LRA) for SLR measurements;
- GNSS satellites as co-location platform [Fig. 1 (right)]: the satellite carries the GNSS transmitting antenna and an LRA.

In this study, the second approach is used.

When we speak about satellite co-locations, it implies that the satellite orbit parameters are estimated from both types of measurements, i.e., GNSS microwave and SLR data. In this approach, the orbits provide a strong connection between the two space-geodetic techniques, thus, local ties do not need to be applied in the combination process (Thaller et al. 2011). As a consequence, the resulting station coordinates are consistently estimated, but independent of the terrestrial measured local ties. This allows us to evaluate the agreement of the terrestrial local ties and the space-geodetic coordinate differences. The local ties can be compared to the space-geodetic coordinate differences derived from a multi-year TRF solution (see Sect. 4.1).

There is a counterpart to the local ties at the co-located stations for the satellite co-locations, too: The location of the reference points of the GNSS microwave antenna and the LRA w.r.t. the center of mass of the satellite (where the orbit refers to) must be known accurately. We call it "space ties", in analogy to the local ties. The vectors for the GNSS satellite antenna offsets (SAO) and the LRA offsets can be estimated within a combined GNSS-SLR multi-year solution and compared to the official values. The results for such an estimation are shown in Sect. 4.2.

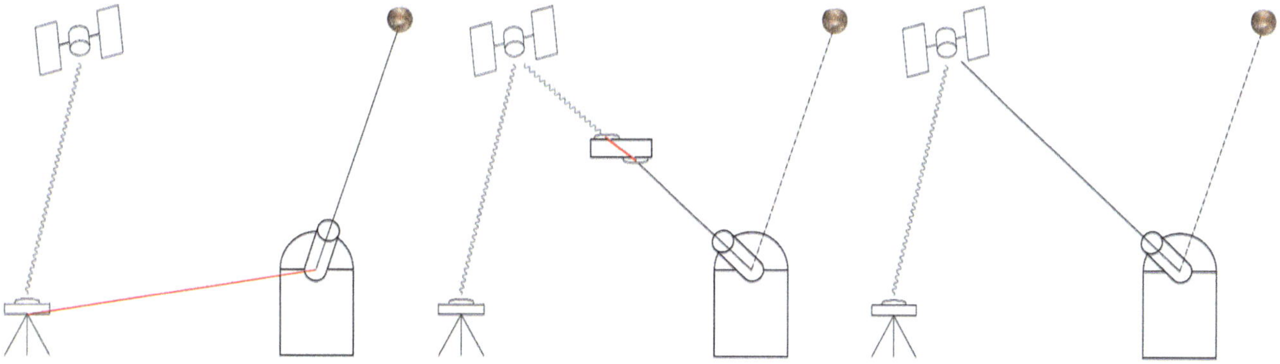

Fig. 1 Possibilities for combining GNSS and SLR. *Left*: co-location at stations; *Middle*: satellite co-location at LEO; *Right*: satellite co-location at GNSS

As the z-component of the SAO is correlated with the scale (see Zhu et al. 2003), an estimation of the z-component for the GNSS antenna offset allows the combined GNSS-SLR solution to adopt the scale of the SLR contribution (see Sect. 4.2).

2 Observations Available for GNSS-SLR Space Ties

The basic requirements for making satellite co-location usable for combined analysis are threefold: the satellites need to be equipped with an LRA, the SLR stations need to track these satellites, and the microwave GNSS data must be usable for a GNSS analysis.

The latter requirement is especially important for GLONASS in the early years, because the small network, together with only a few satellites, makes it impossible to have a good microwave GLONASS analysis. Therefore, we include GLONASS only starting 2002.

On the SLR side, unfortunately, only two GPS satellites are equipped with laser reflectors, thus, usable for co-location. In principal, all GLONASS satellites carry laser reflectors, but only three to six of them were tracked by the International Laser Ranging Service (ILRS, see Pearlman et al. 2002) in the early years up to 2009. Figure 2 shows a timeline of available satellites. Herstmonceux started to track the full GLONASS constellation in December 2009, and some other stations followed in the recent years. Figure 3 (top) shows the number of GPS and GLONASS satellites that were tracked by SLR each day. One can immediately see the time when Herstmonceux started to track all GLONASS (i.e., around 2010), as the number of satellites per day doubled. One can also see that around mid-2011 a lot of other stations tracked the full GLONASS constellation, which also caused a significant increase of the number of satellites available each day for co-location. Since that time, we have not had a day without any SLR tracking data to GLONASS.

All in all, it becomes clear from Fig. 3 (top) that GLONASS is the key element for combined GNSS-SLR solutions using co-location at GNSS satellites.

Figure 3 (bottom) shows the number of SLR stations per day that tracked any of the GPS and GLONASS satellites. In the early years, typically less than ten stations provided SLR data. Because the number of stations steadily increased, we currently receive data from 10 to 20 SLR stations.

The total amount of daily SLR observations of GNSS is shown in Fig. 4. The increased number of satellites tracked [compare Fig. 3 (top)] plays the major role here, of course. The amount of data is well below 50 during the 1990s where only GPS tracking is used. Nowadays, we have more than 100 normal points per day, and there could be up to 500 normal points in the immediate future. This number is remarkable if we compare it with the amount of LAGEOS tracking data (i.e., typically 2000–3000 normal points per week for both LAGEOS), since GNSS satellites are complicated targets due to their orbital altitude.

3 Procedure for Pre-combined GNSS-SLR Solutions

Three different data sets are used in the combination:

- Microwave data from GPS and GLONASS satellites,
- SLR data from LAGEOS and Etalon satellites,
- SLR data from GPS and GLONASS satellites.

The Bernese GNSS Software (Dach et al. 2007) (extended for SLR analysis capabilities) is used for the analysis of all space-geodetic data. Using the same software guarantees that the analysis of SLR observations from geodetic satellites (i.e., LAGEOS and Etalon), from GNSS satellites,

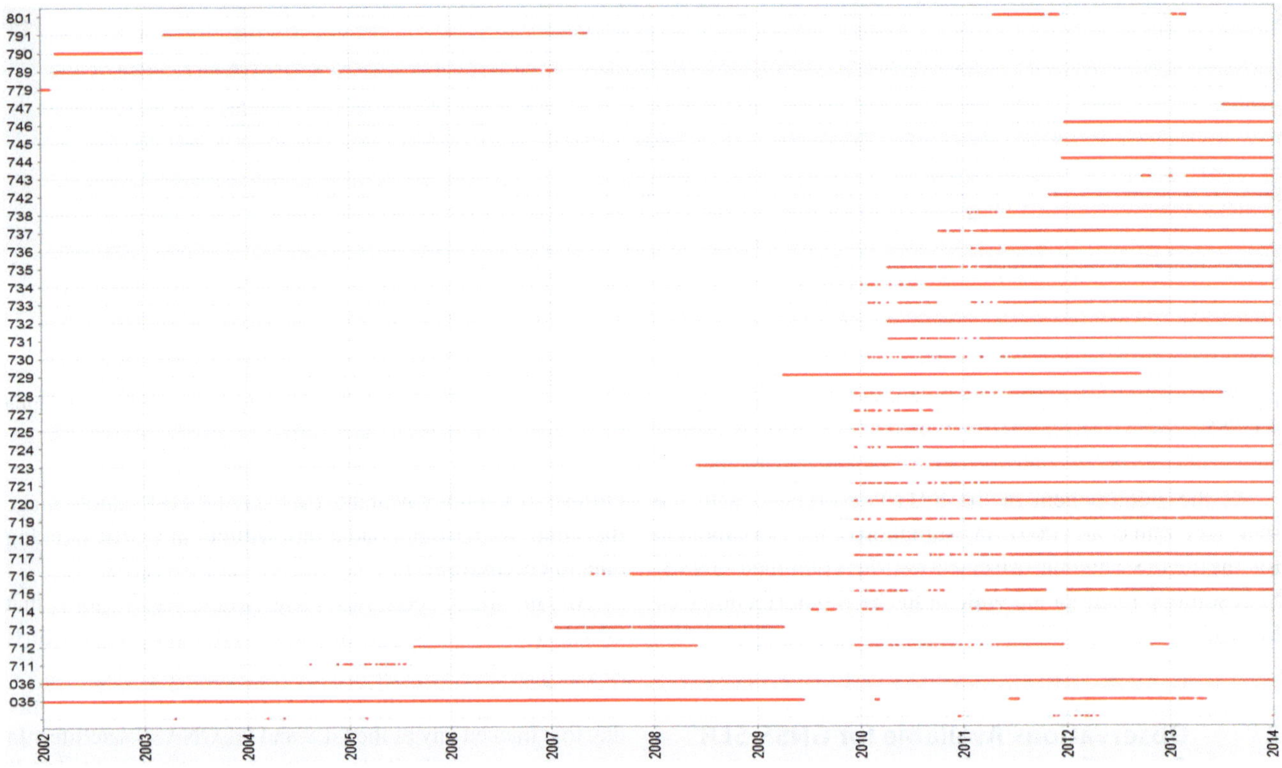

Fig. 2 Timeline of GPS and GLONASS satellites tracked by SLR. The satellites are numbered by their SVN. A list for time-dependent conversion between SVN and PRN is available, e.g., at the ftp archive for users of the Bernese GNSS software: ftp://ftp.unibe.ch/aiub/BSWUSER52/GEN/SATELLIT.I08

and microwave GNSS observations are based on the same modelling and parameterization. This procedure is identical to that described in Thaller et al. (2013) (see especially Fig. 1 therein). The analysis of the SLR data from the spherical satellites follows the procedure established at the analysis center of the ILRS, operated by BKG. In this procedure, 7-day orbital arcs for LAGEOS and Etalon satellites are estimated together with station coordinates, ERPs, and range biases (for selected non-core sites only). Additionally, geo-center coordinates are estimated for the combination studies presented here.

The analysis of the GNSS microwave data is identical to the analysis center of the International GNSS Service (IGS, see Dow et al. 2009), operated by the Center for Orbit Determination in Europe (CODE). Data from GPS and GLONASS satellites are analyzed together using identical techniques. The GLONASS satellites have been included in the analysis since 2002. For more details, we refer to Steigenberger et al. (2011).

The SLR tracking data from the GPS and GLONASS satellites are analyzed consistently with the other two data sets mentioned earlier. All SLR-specific measurement corrections (e.g. troposphere) are identical to those used for data from spherical satellites. All modeling related to GNSS orbits is identical to that used when analyzing microwave data. As there are no official values available for station-specific range biases for SLR tracking to GNSS satellites, we set up range biases for all stations to the GNSS satellites. This procedure does not necessarily lead to a weak scale definition of our pre-combined solution, because the LAGEOS and Etalon tracking data is included, and we estimate range biases only for a few non-core stations for these SLR data sets.

The SLR data from the GNSS satellites play a key role in our analysis, as they connect the two space-geodetic techniques via satellite co-locations. We do not introduce the local ties at co-located stations for generating the pre-combined solutions. For the space ties, we use the official values provided by the IGS and ILRS as a priori values, i.e., the GNSS satellite antenna offsets provided in igs08.atx[1], and the LRA offsets available from the ILRS website[2]. Corrections to both parts of the space ties are set up as parameters in our normal equations.

[1]http://igscb.jpl.nasa.gov/pipermail/igsmail/2011/006347.html.
[2]http://ilrs.gsfc.nasa.gov.

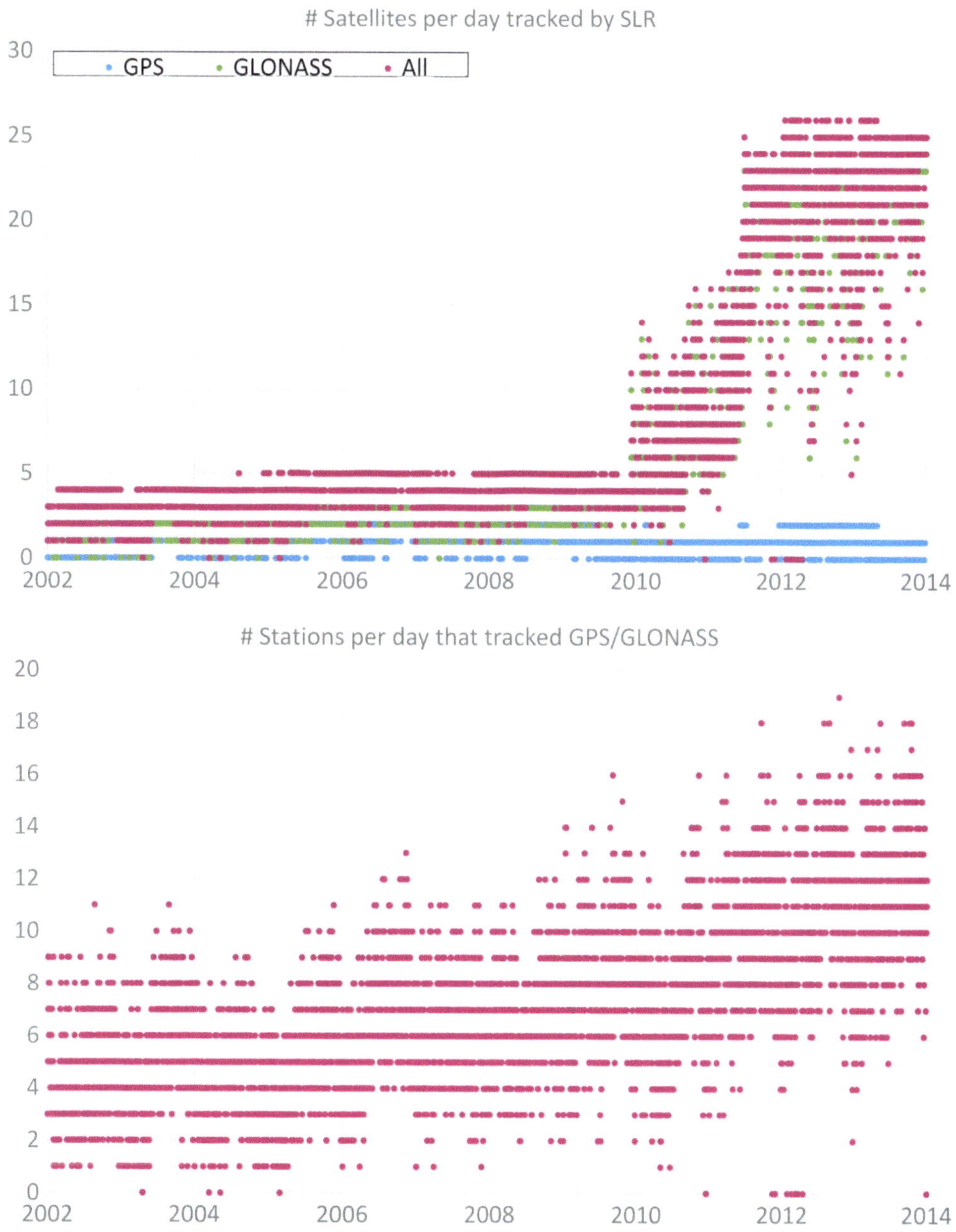

Fig. 3 Available co-locations for GPS and GLONASS satellites per day. *Top*: number of satellites tracked by SLR stations; *Bottom*: number of SLR stations tracking GPS and GLONASS

The three types of data sets mentioned at the beginning of this section are combined according to the procedure visualized in Fig. 1 of Thaller et al. (2013). The resulting weekly combined GNSS-SLR normal equations build the basis for three different types of pre-combined solutions that we then generate and study (see Fig. 5):

- Weekly solutions using the standard values for the space ties;
- Multi-year solution estimating range biases and corrections for the space ties;
- Weekly solutions using the range biases and space ties from our multi-year solution.

Corrections for the space ties can be reliably estimated from multi-year solutions only. Thus, if we compute weekly solutions, we can either use the official values, or we first have to estimate corrections in a multi-year solution, and subsequently re-introduce these corrected space ties into the weekly normal equations.

The situation for range biases from GNSS satellites is similar: the small amount of data per day (see Fig. 4) does

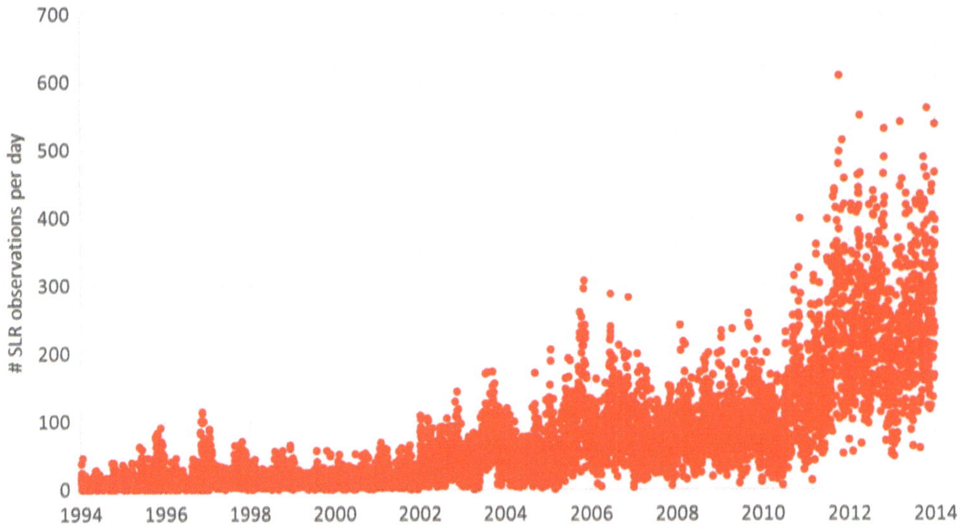

Fig. 4 Total amount of SLR tracking data (normal points) to GPS and GLONASS satellites per day

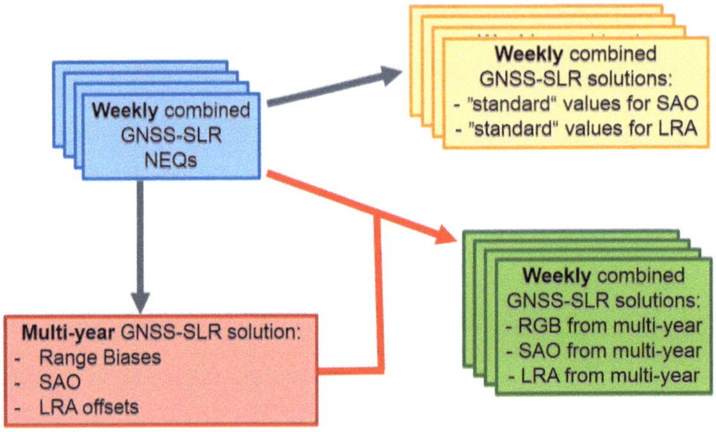

Fig. 5 Strategy for generating pre-combined GNSS-SLR solutions for range biases (RGB), microwave satellite antenna offsets (SAO), and laser retro-reflector array (LRA) offsets

not permit a reliable estimation on a weekly basis. Thus, we also estimate the GNSS range biases in a multi-year solution, and introduce these values when generating the weekly solutions.

4 Combination Results

4.1 Local Ties in a Multi-year Solution

A fully combined GNSS-SLR multi-year solution, including station coordinates and velocities and considering discontinuities where necessary (e.g., earthquakes or equipment changes), has been computed. With all possible co-locations and all sub-intervals due to discontinuities, there are 61 comparisons possible, but only those 55 possibilities with co-locations smaller than 1 km in distance are considered. Most of the co-locations are less than 200 m apart.

The resulting coordinate differences at co-located sites (e.g., between a GNSS and an SLR site) are compared to the local ties. The overall three-dimensional differences are shown in Fig. 6. We can see that the overall agreement between the coordinates estimated from the space-geodetic techniques and the terrestrial local ties are at the level of 1–2 cm, with 17 co-locations better than 1 cm, and another 13 co-locations better than 2 cm. We also see, however, a few outliers where the discrepancies amount to several centimeters. Three major reasons can explain most of these big differences:

– the time span of at least one of the co-location sites is very short,

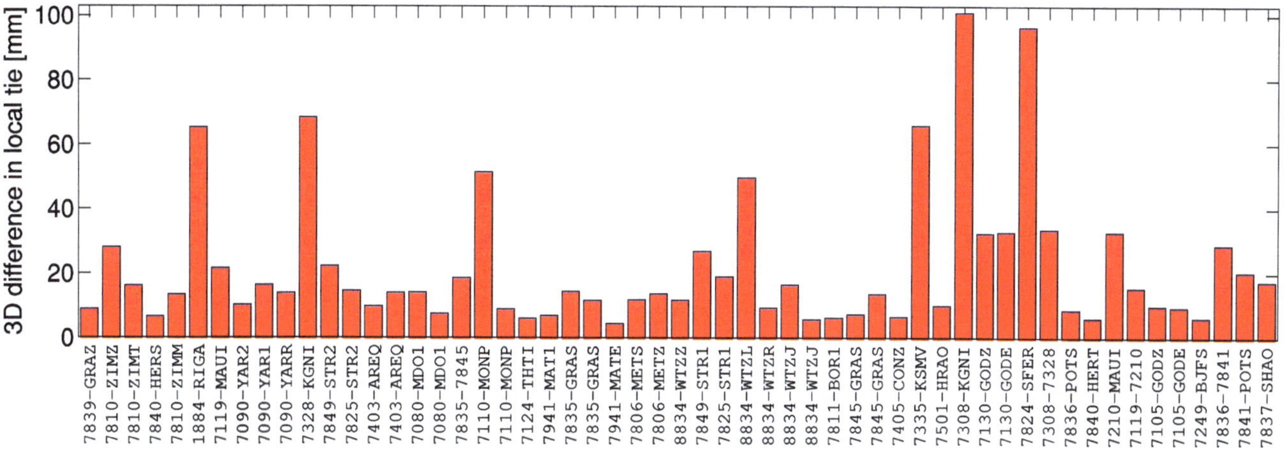

Fig. 6 Three-dimensional differences between space-geodetic coordinate estimates and local ties. Co-locations are sorted in ascending order of distance between the reference points

- the two instruments in one co-location site were never observing in parallel,
- the SLR station is included in only a few weeks, resulting in weak coordinate estimates.

To account for the first reason, we set the critical time span to 3 years of observations. This is a reasonable limit if we take into account that at least 2.5 years are needed for reliable velocity estimation (see, e.g., Blewitt and Lavallée 2002). A total of seven co-location intervals do not meet this requirement in our solution, with all of them showing large discrepancies in Fig. 6, i.e., larger than 2 cm:

- 7810-ZIMZ: ZIMZ covers only 1.5 year. The other co-locations at Zimmerwald (7810-ZIMT, 7810-ZIMM) show very good agreement, i.e., about 1.5 cm.
- 7849-STR2 covers only 3 years. The second co-location with STR2 (7825-STR2) covers about 6 years and fits at the level of 1.5 cm.
- 7849-STR1 is similar to the previous co-location at Mount Stromlo, but covers almost 4 years.
- 7110-MONP: the first interval covers only 2 years. The second interval covers 8 years, and space-geodetic techniques fit the local ties better than 1 cm.
- 8834-WTZL intersects for only 1 year. All other co-locations at Wettzell (8834-WTZZ, 8834-WTZR, 8834-WTZJ both intervals) show an agreement between space-geodetic coordinates and local ties well below 1 cm.
- 7130-GODZ and 7130-GODE intersect in our solution for half a year only. The co-location intervals with the other SLR site (i.e., 7105) is about 7.5 years and 11 years for 7105-GODZ and 7105-GODE, respectively. The agreement with local ties is well below the 1 cm level for these longer time spans.

The second requirement mentioned above, i.e., that both co-located sites observed in parallel for a certain time span, is not met by two additional co-locations shown in Fig. 6: 7328-KGNI and 7335-KSMV. Both co-locations show discrepancies between space-geodetic coordinates and local ties of about 7 cm.

When considering the Koganei station, we see big discrepancies for the second SLR-GNSS co-location as well. The coordinate difference 7308-KGNI differs by almost 10 cm from the local tie, even though the observation spans over 6 years. The reason for this discrepancy may be seen in the very weak SLR coordinate estimation, because the station is included in only 97 weeks spread over the 6 years. The SLR-SLR co-location at Koganei (7308-7328) suffers from infrequent observations as well, which explains the 3.5 cm discrepancy w.r.t. the local tie.

Riga (1884-RIGA) is another co-location showing a significant discrepancy of about 6.5 cm in Fig. 6. This is likely due to weak SLR coordinate estimation.

After considering all these facts, there are three significant discrepancies larger than 2 cm remaining:

- 7824-SFER: The reason for this discrepancy is not clear (10 years of co-location with 361 weeks).
- 7210-MAUI: The (still) short co-location time span of 3.5 years might be an explanation.
- 7836-7841: Only the SLR-SLR co-location shows about 3 cm difference. The coordinate differences to the GNSS site POTS fits well to both SLR sites.

Looking at the individual components shows that the horizontal agreement between space-geodetic coordinates and local ties is clearly better than the agreement in the height component. This is expected, as especially GNSS height

Table 1 Differences between space-geodetic coordinates and local ties at co-located sites. Numbers in brackets are without counting the co-locations separated more than 1 km

Difference in local ties	Number of co-locations					
	Height		North		East	
0–3 mm	5		21		24	(23)
3–5 mm	10	(9)	14		7	
5–10 mm	12	(11)	15	(14)	16	
10–15 mm	12		2		4	
15–20 mm	8	(7)	3	(2)	1	
20–25 mm	2		0		2	
25–30 mm	2		0		1	
30–40 mm	2		1		1	(0)
40–50 mm	1		1		0	
>50 mm	7	(4)	4	(0)	5	(1)

is not as well determined as the horizontal components. The big outliers for the overall 3D agreement discussed above appear in different components. 1884-RIGA, 7328-KGNI, and 7308-KGNI differ mainly in height, whereas the horizontal components fit at a few millimeter level. This could be an indication that the range biases estimated for the SLR observations of GNSS satellites are not reliably determined due to only a few observations. Contrary for other sites, 7110-MONP (first co-location interval) and 8834-WTZL mainly differ in the horizontal component, i.e., in East and North, respectively. In both cases the time span of data is less than 2 years, indicating that this interval is too short for a well-established velocity estimation. Finally, the big discrepancies for the co-locations 7335-KSMV and 7824-SFER cannot be attributed to only one component, but rather a generally bad agreement between space-geodetic coordinate estimates and local ties. As mentioned earlier, the reasons for these big discrepancies are not fully clear.

Table 1 lists, in detail, the number of co-locations with an agreement better than 3 mm, 5 mm, etc. for the individual components in a histogram-like style. If we take the threshold of 1 cm, we see that almost half of the co-locations investigated reach this level of agreement for the height component. Additionally, 50 and 47 out of 61 co-locations reach the 1 cm level for the North and East component, respectively.

4.2 Space Ties from Multi-year Solution and Their Impact on the Scale

As described in Sect. 3, the space ties are estimated from a multi-year combined GNSS-SLR solution. We estimate corrections only for the z-component.

The corrections estimated for the LRA offsets are very small: The mean correction for the two GPS satellites is -5.5 mm, and the mean overall correction for GLONASS satellites is -6.6 mm.

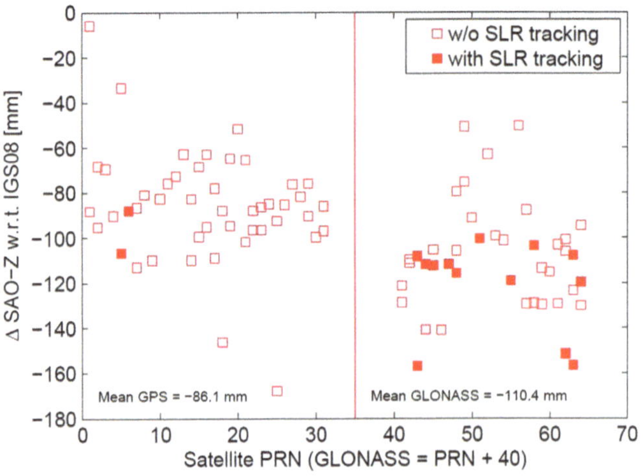

Fig. 7 Corrections for the satellite antenna offsets of the GNSS microwave antenna. The formal errors of the estimates (1-σ-level) are several mm for GPS, and 1–2 cm for GLONASS. Satellites with short observation intervals have sigmas of a few centimeters

The results look different for the microwave transmitting antenna. Figure 7 shows the corrections for the z-component for all GPS and GLONASS satellites. The corrections are at the level of several centimeters for all satellites and can reach up to 160 mm. Contrary to the LRA offsets, there is a clear difference visible between the GPS and GLONASS satellites: on average, the correction for the microwave antenna of the GPS satellites is -86.1 mm, and the mean correction for the GLONASS satellites is -110.4 mm. This difference implies that the SAO values provided in igs08.atx are not fully consistent for GPS and GLONASS. Taking into account that the GPS and GLONASS values in igs08.atx do not stem from the same solution, it is not astonishing that we see differences when estimating corrections for both together.

Since the corrections to the SAO are so large, the question arises what is the effect on the scale? According to Zhu et al. (2003), the mean z-corrections for GPS and GLONASS should lead to a scale difference of 0.67 ppb and 0.86 ppb, respectively. Figure 8 shows the scale derived from a Helmert transformation of the weekly combined GNSS-SLR solutions when using the space tie estimated from the multi-year solution. The upper part shows the scale for the sub-network of GNSS stations, and the lower part for the sub-network of SLR stations. In comparison with a GNSS-only solution (using the standard SAO values from igs08.atx) we can see a clear bias, with a mean value of 0.67 ppb, i.e., exactly the scale bias caused by the SAO corrections for the GPS satellites. On the other hand, the SLR network scale of the combined solution does not differ significantly from an SLR-only solution based on LAGEOS and Etalon data only

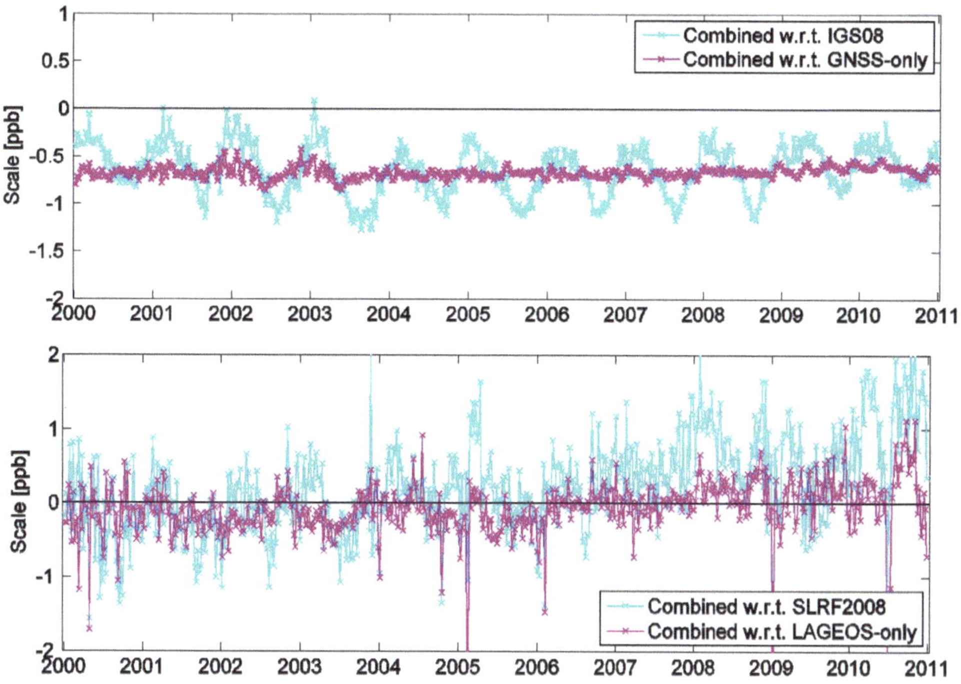

Fig. 8 Scale of the GNSS (*top*) and SLR (*bottom*) sub-networks from the combined solution w.r.t. single-technique solutions (*magenta*) and the technique-specific realizations of ITRF2008 (*cyan*)

[Fig. 8 (bottom)]. Additionally, the SLR network scale of the combined solution does not differ from the actual reference frame SLRF2008, apart from the well-known degradation since about 2008 (i.e., after the last observations contributed to SLRF2008; see e.g. the reports by the ILRS Analysis Working Group[3].) In summary, this behavior implies that:

- the SAO corrections estimated from the combined GNSS-SLR solution propagate into the GNSS network scale;
- the SLR scale is retained in the combined solution;
- the estimated SAO corrections are fully consistent with the SLR scale.

5 Conclusions and Outlook

The use of satellite co-locations instead of co-locations on the ground has been introduced with the main focus on GNSS satellites as co-location platforms. It has been shown that the number of GNSS satellites usable as a co-location platform has greatly increased since 2010, as many SLR stations are tracking the full GLONASS constellation. As a consequence, the amount of SLR normal points from GNSS satellites has markably increased, too.

We computed combined GNSS-SLR solutions using the SLR tracking data from the GPS and GLONASS satellites as the main connection between the two space-geodetic techniques. In this way, the GNSS satellites are providing the co-location platform instead of the co-located stations. The local ties derived from terrestrial measurements are not introduced into the combination, thus, the station coordinates of the co-located sites are independently estimated. This fact allows us to compare and validate the space-geodetic coordinate differences w.r.t. the terrestrial local ties.

In general, we can summarize that most of the space-geodetic co-locations agree at the level of few millimeters up to 1 cm with the terrestrial local ties. For the north, east, and height component, 50, 47 and 27 co-locations show discrepancies below 1 cm, respectively.

Most of the co-location intervals that show discrepancies of 2–10 cm between the space-geodetic coordinate estimates and the local ties can be explained by either short time intervals in our solution, sparse SLR observations, or no parallel observations of the two co-located sites. Only the SLR-GNSS co-location at San Fernando shows a discrepancy of almost 10 cm that we cannot explain at the moment. All remaining co-locations investigated in our studies show an overall 3-dimensional agreement between space-geodetic coordinates and terrestrial local ties better than 2 cm.

It must be pointed out, however, that the reason for discrepancies cannot be solely attributed to the terrestrial measurements for deriving the local ties. It is more likely that there are still several modeling issues in the space-geodetic data analysis that propagate into the station coordinate estimates.

[3]http://ilrs.gsfc.nasa.gov/science/awg/awgActivities/index.html.

The offsets of the microwave GNSS antennas have been found to be one source of space tie discrepancies. The combined GNSS-SLR solutions allow us to estimate corrections for the antenna offsets provided by the IGS. The estimated offsets are fully consistent with the SLR scale. However, the estimation reveals discrepancies of igs08.atx values and the SLR scale at the level of 0.67 ppb. In view of increasing the consistency of the space-geodetic techniques within future ITRF computations, the satellite co-locations thus provide a valuable method for deriving consistent scale for the GNSS and SLR networks.

Another advantage of using these types of combined GNSS-SLR solutions within future ITRF computations is the fact that a significant amount of additional SLR tracking data will be included into the ITRF. Firstly, this would increase the contribution of SLR to the ITRF due to the increased number of observations. Secondly, this would give more credit to the SLR station operators who contribute a significant amount of effort into tracking the high GNSS satellites.

Acknowledgements The work is partly funded by the German research foundation (DFG) within the sub-project PN5 of the research group FOR1530. The space geodetic stations providing data through the ILRS and IGS are acknowledged, as well as the ILRS, IGS and IERS themselves for their support. The SLR stations providing tracking data to the GNSS satellites are especially acknowledged for their effort.

References

Altamimi Z, Collilieux X, Métivier L (2011) ITRF2008: an improved solution of the international terrestrial reference frame. J Geodesy 85(8):457–473. doi:10.1007/s00190-011-0444-4

Blewitt G, Lavallée D (2002) Effect of annual signals on geodetic velocity. J Geophys Res 107(B7). doi:10.1029/2001JB000570

Dach R, Hugentobler U, Fridez P, Meindl M (eds) (2007) Bernese GPS Software Version 5.0. Astronomical Institute, University of Bern, Bern, Switzerland

Dow JM, Neilan RE, Rizos C (2009) The International GNSS Service in a changing landscape of global navigation satellite systems. J Geodesy 83(3–4):191–198. doi:10.1007/s00190-008-0300-3

Krügel M, Thaller D, Tesmer V, Rothacher M, Angermann D, Schmid R (2007) Tropospheric parameters: combination studies based on homogeneous VLBI and GPS data. J Geodesy 81(6–8):515–527. doi:10.1007/s00190-006-0127-8

Pearlman MR, Degnan JJ, Bosworth JM (2002) The International Laser Ranging Service. Adv Space Res 30(2):125–143

Ray J, Altamimi Z (2005) Evaluation of co-location ties relating the VLBI and GPS reference frames. J Geodesy 79(4–5):189–195. doi:10.1007/s00190-005-0456-z

Rothacher M, Angermann D, Artz T, Bosch W, Drewes H, Gerstl M, Kelm R, König D, König R, Meisel B, Müller H, Nothnagel A, Panafidina N, Richter B, Rudenko S, Schwegmann W, Seitz M, Steigenberger P, Tesmer S, Tesmer V, Thaller D (2011) GGOS-D: homogeneous reprocessing and rigorous combination of space geodetic observations. J Geodesy 85(10):679–705. doi:10.1007/s00190-011-0475-x

Seitz M, Angermann D, Bloßfeld M, Drewes H, Gerstl M (2012) The 2008 DGFI realization of the ITRS: DTRF2008. J Geodesy 86(12):1097–1123. doi:10.1007/s00190-012-0567-2

Steigenberger P, Hugentobler U, Lutz S, Dach R (2011) CODE contribution to the first IGS reprocessing campaign. Technical Report 01/2011, Astronomical Institute, University of Bern, Bern

Thaller D, Krügel M, Rothacher M, Tesmer V, Schmid R, Angermann D (2007) Combined Earth orientation parameters based on homogeneous and continuous VLBI and GPS data. J Geodesy 81(6–8):529–541. doi:10.1007/s00190-006-0115-z

Thaller D, Dach R, Seitz M, Beutler G, Mareyen M, Richter B (2011) Combination of GNSS and SLR observations using satellite co-locations. J Geodesy 85(5):257–262. doi:10.1007/s00190-010-0433-z

Thaller D, Sośnica K, Dach R, Jäggi A, Beutler G, Mareyen M, Richter B (2013) Geocenter coordinates from GNSS and combined GNSS-SLR solutions using satellite co-locations. In: Rizos C, Willis P (eds) Proceedings of the IUGG General Assembly 2011, IAG Symposia Series, Vol. 139. doi:10.1007/978-3-642-37222-3_16

Zhu S, Massmann F, Yu Y, Reigber C (2003) Satellite antenna phase center offsets and scale errors in GPS solutions. J Geodesy 76(11–12):668–672. doi:10.1007/s00190-002-0294-1

GGOS-SIM: Simulation of the Reference Frame for the Global Geodetic Observing System

Harald Schuh, Rolf König, Dimitrios Ampatzidis, Susanne Glaser, Frank Flechtner, Robert Heinkelmann, and Tobias J. Nilsson

Abstract

The accuracy and stability requirements for the International Terrestrial Reference Frame (ITRF) postulated by the Global Geodetic Observing System (GGOS) are not met so far. The GGOS–SIM project builds a software tool that by simulating the space geodetic infrastructure allows to assess the impact of technique upgrades, new sites, new satellites, local ties, and space ties on the ITRF accuracy and stability. As also the procedure for the combination of the techniques plays a fundamental role in the generation of an ITRF, we discuss peculiarities of current day approaches and draw conclusions relevant for this project. As the assessment of the accuracy of an ITRF is needed for checking against the GGOS requirements, we compile actual methods and present here a new measure of stability which is exemplarily applied to recent ITRFs.

Keywords

GGOS • Global geodetic observing system • ITRF • International terrestrial reference frame

1 Introduction

It is one of the main tasks in geodesy to define and realize Terrestrial Reference Frames (TRFs) precisely enough to correctly interpret and determine products of geometry, rotation, and gravity field of the Earth. In the framework of the Global Geodetic Observing System (GGOS) (Plag et al. 2009), the flagship component of the International Association of Geodesy (IAG), the requirement has been postulated for a global TRF with an accuracy of 1 mm and a stability of 0.1 mm/year. The International Earth Rotation and Reference Systems Service provides the International Terrestrial Reference Frame (ITRF), the most recent release being the ITRF2008 (Altamimi et al. 2011). It was derived from a combination of four main space geodetic techniques: Very Long Baseline Interferometry (VLBI), Satellite Laser Ranging (SLR), Global Navigation Satellite Systems (GNSS), and Doppler Orbitography and Radiopositioning Integrated by Satellite (DORIS). Local tie vectors at co-located sites were used to connect the station coordinates of the different techniques. Starting with ITRF2005 also the Earth Orientation Parameters (EOPs) have been included in the combination serving as a "global tie" (Altamimi et al. 2007). The global distribution of the station networks of the techniques involved within the ITRF2008 determination is given in Fig. 1.

For many geodetic applications and for the requirements of GGOS the accuracy and stability achieved so far is not

H. Schuh • F. Flechtner
Institute of Geodesy and Geoinformation Science, Technische Universität Berlin, Straße des 17. Juni 135, 10623 Berlin, Germany

GFZ German Research Centre for Geosciences, Telegrafenberg, 14473 Potsdam, Germany
e-mail: harald.schuh@tu-berlin.de

S. Glaser (✉)
Institute of Geodesy and Geoinformation Science, Technische Universität Berlin, Straße des 17. Juni 135, 10623 Berlin, Germany
e-mail: susanne.glaser@tu-berlin.de

R. König • D. Ampatzidis • R. Heinkelmann • T.J. Nilsson
GFZ German Research Centre for Geosciences, Telegrafenberg, 14473 Potsdam, Germany

T. van Dam (ed.), *REFAG 2014*, International Association of Geodesy Symposia 146,
DOI 10.1007/1345_2015_217

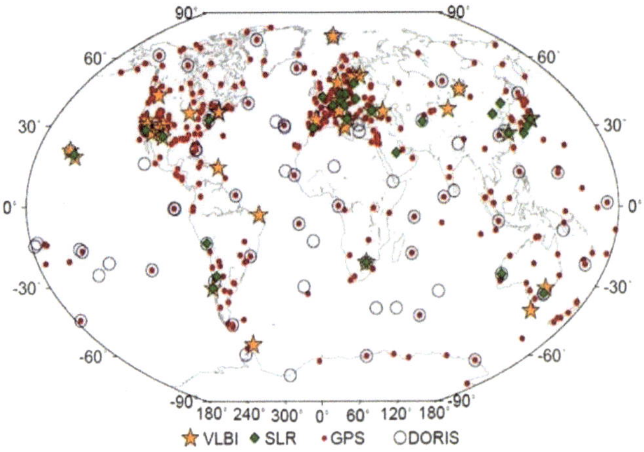

Fig. 1 The ITRF2008 station network (Altamimi et al. 2011)

sufficient. There is a need to investigate error sources of the ITRF and to give recommendations for future improvements of the ITRF. This is the intention of our recently started project GGOS-SIM funded by the German Research Foundation (DFG).

In this paper we will describe the GGOS-SIM project tasks and the way the project objectives will be reached. Further we will discuss the usually applied methods for the combination of the space geodetic techniques and we will shed some light on their respective advantages and disadvantages. We will also present an alternative methodology for the assessment of the temporal evolution of TRFs and give an example based on existing TRFs.

2 Tasks and Objectives of the GGOS-SIM Project

Project GGOS-SIM has been implemented to improve the ITRF and to test the accuracy requirements of GGOS. In GGOS-SIM we will simulate observations based on the infrastructure of the IAG Technique Services: the International VLBI Service for Geodesy and Astrometry (IVS, Schuh and Behrend 2012), the International Laser Ranging Service (ILRS; Pearlman et al. 2002), the International GNSS Service (IGS; Dow et al. 2009), and the International DORIS Service (IDS; Willis et al. 2010). The overall objectives of GGOS-SIM will be to create a tool to simulate

- the complete current ground network of all space geodetic techniques,
- the effects of technical improvements of the space geodetic techniques on the ITRF,
- the impact of number and accuracy of local ties and space ties on the ITRF, and
- the impact of new co-location sites on the ITRF.

As a main result GGOS-SIM will offer the ultimate tool for testing and improving state-of-the-art and future ITRFs. Consequently, the simulations to be carried out within GGOS-SIM will help planning and coordinating an efficient development of the international geodetic infrastructure as backbone of GGOS.

As a basis the simulated observations of all the space geodetic techniques will be tested thoroughly to match the real world closely. For simulation of the VLBI infrastructure the VieVS@GFZ software (Böhm et al. 2012; Nilsson et al. 2015) is used. For simulation of the SLR, GNSS, and DORIS infrastructures the EPOS-OC software (see Zhu et al. 2004) is suited. The combination of the solutions will be done on the normal equation level also performed with the EPOS package in order to get a solution with highest consistency.

The simulation of the VLBI infrastructure is founded on the current infrastructure of the IVS. We simulate the standard rapid turnaround sessions within the time span of 2008–2014 using a global network of 28 stations. Prominent random error sources are the refraction of the radio waves in the troposphere and the station clock corrections. Therefore we perform Monte Carlo simulations where we model the zenith wet delays and clock values at each station and add observation noise simulated as white noise per baseline.

The simulation of the SLR infrastructure will be done according to the current infrastructure of the ILRS. Prominent systematic error sources of SLR are range and time biases. In order to reach millimeter accuracy the range bias parameters and the coordinate parameters need to be separated. The separation gets better the more low elevation observations become available. Hence particular emphasis will be given to multi-satellite constellations where the probability for a better separation increases. Therefore, the current constellation of ILRS, i.e. the LAGEOS and ETALON satellites will be enhanced by the LARES, AJISAI, STARLETTE, and STELLA satellites.

The simulation of the GNSS infrastructure will initially concentrate on GPS with the space and ground segment as currently adopted by the IGS. As the GPS environment already consumes rather exhaustive computational resources the other GNSS constellations like GLONASS or GALILEO will gradually be included later on. Systematic error sources arise from mismodelling of radiation pressure, tropospheric corrections, and ambiguities. There is also a stochastic error source induced by colored observation noise. The future enhancement of the constellation inherently leads to a more reliable and accurate TRF solution but comes along with unknown inter-system biases. Anyway, it can be expected that deficiencies in modeling and parametrization can be reduced.

The simulation of the DORIS infrastructure will follow the general lines of the IDS. Systematic errors arise from the frequency instability of the receivers which result in scale deviations of the TRF depending on the solution strategy or in systematic deviations of coordinate time series of individual stations. Similar to GNSS there is also an issue with colored observation noise. By augmenting the space segment one can predict a more accurate and more reliable TRF solution.

All the technique-specific simulations will be tested against real observations. From the simulated observations, technique-specific TRFs will be generated and tested against available solutions. Once realistic simulations are achieved the technique solutions are combined to a simulated ITRF which is then tested against current ITRFs.

Also, the impacts of technique upgrades, new ground stations, and new satellites in space can be assessed. Of particular interest for the combination of station positions of single-technique solutions are the local tie vectors between the reference points of the instruments at co-located sites. Here, the accuracy and number of local ties play a fundamental role. For the improvement of the ITRF new co-location sites could be essential. For the current ITRF space ties are not used; they however can fulfill the same role as local ties, therefore their impact will also be investigated at the end of the project.

3 On the Combination of Space Geodetic Techniques

The combination of the technique-specific solutions is a crucial point in generating an ITRF. The Institut National de l'Information Géographique et Forestière (IGN, formerly Institut Géographique National) and the Deutsches Geodätisches ForschungsInstitut of the Technische Universität München (DGFI-TUM) published the recent ITRF2008 (Altamimi et al. 2011) and the DTRF2008 (Seitz et al. 2012) where the adjacent procedures and methods were described to some extent. A fruitful discussion about the different combination techniques applied by the two groups is given in Seitz (2015). Here we compile just some interesting points of concern which will be investigated in project GGOS-SIM.

- The concept of "minimum constraints" as used in the generation of the ITRFs do rather not just cure the rank deficiency of the NEQ. A more appropriate term is probably "non-distorting constraints" as used by Angermann et al. (2004a).
- IGN eliminates scale and origin information from GPS and DORIS as far as possible from the beginning. Therefore, the ITRF solution gets dependent on the loosely constrained SLR solution.
- DGFI and IGN apply different handling of velocities within the combination of the single-technique solutions to determine a TRF. IGN uses the 14 parameter similarity transformation of the single-technique solutions to determine positions and velocities of the TRF (seven parameters for positions, seven for velocities). DGFI introduces velocities by expanding the NEQs.
- The global polyhedron of ground stations used when generating the ITRFs is too sparse to allow for an estimation of a plate tectonic model. The recent IGN approach is to minimize the deviations with respect to the previous realization to reach a high consistency between the ITRFs. Projecting this back in time, the temporal change of the orientation of ITRF2008 is given by the NNR-NUVEL-1A (DeMets et al. 1994) plate tectonic model. One should examine also the usefulness of more recent plate tectonic models.

We aim for a combination method as follows after studying the ones usually applied:

- Data: We are going to generate simulated, weekly (in case of VLBI session-wise), datum-free normal equations from the four techniques. Thus, there is no effect coming from the datum information included. Indeed, in reality, IGS and IDS provide normal equations with minimal constraints, ILRS with loose constraints, and IVS being the only technique to provide unconstrained normal equations. DGFI f.i. removes the datum effect on the combination by expanding the normal equations with seven Helmert parameters.
- Epoch-wise normal equations: the velocity parameters are set up from the beginning, so there is no need to artificially expand the normal equations later on.
- Preliminary TRFs for each technique: a solution is possible by applying No-Net-Rotation (NNR), No-Net-Translation (NNT) and No-Net-Scale (NNS) conditions for GNSS and DORIS, NNR for SLR, and NNR and NNT for VLBI to the normal equations accumulated per technique. Solving for epoch-wise normal equations under the same datum conditions and comparing them to the preliminary TRFs is useful for quality assessment. Also the time series can be checked for discontinuities, outliers, and non-linear behaviour.

- Integration of the local ties under various aspects: Usage of the full covariance matrix of the co-locations from the terrestrial measurements; empirical weight factors for the local ties; detection of gross errors of local ties; or constraints on the velocities of co-located ground stations.
- Datum: scale from VLBI and/or SLR, origin from SLR, orientation with respect to the most recent ITRF or with respect to a plate tectonic model. Following ITRF2008 and DTRF2008, we will also apply "non-distorting constraints" and shall investigate different ways in constraint handling as well.

GGOS-SIM will not realize a new global TRF from real data. It will investigate methodologies to improve accuracy and stability under various scenarios:

- Station distribution: e.g. uniform coverage.
- Datum definition: constraint options as e.g. described by Davies and Blewitt (2000); GNSS and DORIS observations in addition to SLR for origin definition; time dependent NNR with respect to a recent ITRF or with respect to recent plate tectonic models.
- Technique specific biases and systematics: f.i. impact of historic data.

4 The Assessment of the TRF by an Alternative Methodology

The ITRF is based on various conventions for its origin, scale, and orientation and their rates respectively (Altamimi et al. 2002, 2007; Petit and Luzum 2010; Seitz et al. 2012). In general, a reference frame should be stable over time so that the projection of the station coordinates to past or future epochs can be done reliably. Therefore, the stability of a TRF is directly connected to its velocity field. Consequently, we focus here on the evaluation of the temporal evolution of a TRF by using the station velocities.

Origin and scale of the ITRF are well defined from the SLR and SLR/VLBI techniques respectively (e.g. Lambeck 1987), but none of the space geodetic techniques provide orientation information for the TRF (Angermann et al. 2004b; Altamimi and Dermanis 2009). VLBI senses the orientation with respect to the Celestial Reference Frame (Ma et al. 1998). Thus, the orientation and especially its rate are defined by external sources (Altamimi et al. 2002). For the initial ITRF realizations till ITRF2000, the plate tectonic models AM0-2 (Minster et al. 1974) or NNR–NUVEL–1A were used; see e.g. Altamimi et al. (2002). From ITRF2005 onwards, the orientation and its rate were realized in a way that they coincide with the previous ITRF version (Altamimi et al. 2007, 2011). Hence, the orientation and its rate are dependent on the choice of the core sites for the NNR condition and of course on the quality of the estimated velocities. Mathematically, the deficiencies of orientation and its rate are treated with the NNR conditions applied to carefully selected points, both for coordinates and velocities (e.g. Altamimi et al. 2002; Angermann et al. 2004b). The NNR condition for velocities has some interesting physical interpretations: they remove the relative angular momentum of the frame and they minimize its associated relative kinetic energy of the stations included in the condition (Altamimi et al. 2002).

Many investigations were performed to analyze the impact of the NNR implementation in ITRF realizations (Dermanis 2001; Altamimi et al. 2003, 2011; Kreemer et al. 2006). Furthermore, various studies were published on the stability of the ITRF origin with respect to the reference frame which is realized by SLR measurements (Dong and Feng 2007). Seitz et al. (2012) present a methodology where they assess the impact of using VLBI only or both, SLR and VLBI, for the scale definition in global reference frames. The core of the reference frame origin and scale assessment is based on the comparison of the inter-technique combined solution with the individual technique solution. On the other hand, the evaluation of the NNR condition is based mainly on the comparison of the ITRF associated velocities with some modern tectonic plate motion model (e.g. Kreemer and Holt 2001; Altamimi et al. 2003).

In this paper we present an alternative strategy to validate the global reference frames' relative temporal evolution. Let us consider that the 3-D velocity vector $\mathbf{v}_i$ of each point i contains two parts: The "unreal" quantity $\mathbf{v}_i^{tr}$ which refers to the effect of the choice of the reference system (realized by the minimal constraints) and the deformation part $\delta\mathbf{v}_i$ (including also the random errors which are relatively small, compared to the deformation itself) (Dermanis 2015, personal communication). Thus, we may define the following equation:

$$\mathbf{v}_i = \mathbf{v}_i^{tr} + \delta\mathbf{v}_i. \quad (1)$$

If we pass to the coordinate connection between two epochs t_0 and t, we get:

$$\begin{aligned} \mathbf{x}_i(t) &= \mathbf{x}_i(t_0) + (t - t_0)\mathbf{v}_i \quad (2) \\ &= \mathbf{x}_i(t_0) + (t - t_0)\mathbf{v}_i^{tr} + (t - t_0)\delta\mathbf{v}_i \\ &= \mathbf{x}_i(t_0) + \delta\mathbf{x}_i(t) + (t - t_0)\delta\mathbf{v}_i \end{aligned}$$

where the term $\delta\mathbf{x}_i(t)$ corresponds to a change of the reference frame between two epochs and thus it can be written -pointwise- as a classical Helmert type similarity transformation (e.g. Altamimi and Dermanis 2013):

$$\delta\mathbf{x}_i(t) = (t - t_0)\mathbf{v}_i^{tr} = \mathbf{E}_i\mathbf{\Theta}(t) \quad (3)$$

where $\mathbf{E}_i$ is the Jacobian matrix (cf. e.g. Eq. (A6) in Altamimi et al. 2002) and $\mathbf{\Theta}(t) = [T_x(t), T_y(t), T_z(t), \delta s(t), R_x(t),$

$R_y(t), R_z(t)]^T$ are the time-dependent Helmert parameters (T_x, T_y, T_z stands for the translation, ds the scale and R the orientation terms, respectively). Consequently:

$$\mathbf{x}_i(t) - \mathbf{x}_i(t_0) = \delta\mathbf{x}_i(t) + (t - t_0)\delta\mathbf{v}_i = \mathbf{E}_i\boldsymbol{\Theta}(t) + (t - t_0)\delta\mathbf{v}_i \quad (4)$$

and since $\mathbf{x}_i(t) - \mathbf{x}_i(t_0) = (t - t_0)\mathbf{v}_i$, we obtain

$$\mathbf{v}_i = \frac{1}{t - t_0}\mathbf{E}_i\boldsymbol{\Theta}(t) + \delta\mathbf{v}_i. \quad (5)$$

Adopting the linear model for the temporal evolution of the parameters, we assume that $\boldsymbol{\Theta}(t) = \boldsymbol{\Theta}_0 + \dot{\boldsymbol{\Theta}}(t - t_0)$, where the term $\dot{\boldsymbol{\Theta}}$ corresponds to the linear rates of the Helmert parameters. Assuming that $\boldsymbol{\Theta}_0 = \mathbf{0}$ so that the initial coordinates are recovered at the reference epoch t_0 and combining Eqs. (4) and (5) for all the network points, we finally get:

$$\mathbf{v} = \mathbf{E}\dot{\boldsymbol{\Theta}} + \delta\mathbf{v}. \quad (6)$$

If we implement the optimization criterion $\delta\mathbf{v}^T\mathbf{W}\delta\mathbf{v} \rightarrow min$ the classical Least Squares adjustment can be applied. A reasonable choice for the weight matrix $\mathbf{W} = \mathbf{C}_\mathbf{e}^{-1}$ where $\mathbf{C}_\mathbf{e}^{-1}$ is the error covariance matrix of the estimated velocities of the reference frame. The estimated parameters and their associated covariance matrix become:

$$\hat{\dot{\boldsymbol{\Theta}}} = (\mathbf{E}^T\mathbf{C}_\mathbf{e}^{-1}\mathbf{E})^{-1}\mathbf{E}^T\mathbf{C}_\mathbf{e}^{-1}\mathbf{v} \quad (7)$$

$$\mathbf{C}_{\hat{\dot{\boldsymbol{\Theta}}}} = (\mathbf{E}^T\mathbf{C}_\mathbf{e}^{-1}\mathbf{E})^{-1}. \quad (8)$$

Equation (7) quantifies the temporal evolution of the reference frame in terms of Helmert transformation rate parameters. This means, it can act as a diagnostic tool for the relative assessment of the reference frame temporal evolution when comparing the results of two different frames. For example, one can understand the different choices of the velocity constraints to the global networks, comparing older and newer versions of the same reference frame.

For a first glimpse into this new quality assessment method we evaluate the temporal features of the recent global reference frames ITRF2008 and DTRF2008, cf. Table 1. In order to prevent from gross errors all stations in the ITRF2008 with standard deviations of the velocities larger than 2 mm/year are excluded. As a result 595 stations are used to determine the rate parameters. For DTRF2008 the same stations are taken.

The differences of the translational temporal evolution can reach 1.4 mm/year for the y-axis. This also can be an indication of different origin constraint handling between the two frames. Regarding the orientation rates, they do not exceed 0.9 mm/year. Finally, the scale rate difference is found at 0.3 mm/year. In mean the consistency between the two frames is at the level of 1 mm/year. Seitz et al. (2012) claim a consistency of 0.5 mm/year; however they compare each space technique results separately (w. r. t. DTRF2008). For example, they found a discrepancy of 0.85 mm/year for the DORIS reference frames.

Table 1 Helmert transformation rate parameters for ITRF2008 and DTRF2008

Parameter	ITRF2008	DTRF2008	Differences
$\dot{T}_x$ (mm/year)	−14.0 ± 0.02	−13.6 ± 0.02	0.4 ± 0.03
$\dot{T}_y$ (mm/year)	11.0 ± 0.02	9.6 ± 0.02	−1.4 ± 0.03
$\dot{T}_z$ (mm/year)	11.6 ± 0.02	10.6 ± 0.02	−1.0 ± 0.03
$\dot{\delta s}$ (mm/year)	−1.4 ± 0.02	−1.7 ± 0.02	−0.3 ± 0.03
$\dot{R}_x$ (mm/year)	−3.7 ± 0.03	−2.8 ± 0.03	0.9 ± 0.04
$\dot{R}_y$ (mm/year)	2.2 ± 0.03	2.5 ± 0.03	0.3 ± 0.04
$\dot{R}_z$ (mm/year)	−4.6 ± 0.03	−4.0 ± 0.03	0.6 ± 0.04

5 Summary

The GGOS-SIM project aims at the improvement of the accuracy and stability of the ITRF to meet the requirements postulated by GGOS. The way is to set up a software tool for simulating the space geodetic observations contributing to the generation of an ITRF. As a basis the infrastructures as provided by the IAG Technique Services IVS, ILRS, IGS, and IDS are simulated first closely to reality. Then all options supposed to improve the ITRF accuracy as technique upgrades, new sites, and new co-locations can be quantified. The combination strategy applied to the individual techniques affects the ITRF and its accuracy. Therefore, within GGOS-SIM the impact of the combination will be tested. We also outline a new approach for judging the stability of a TRF, and give an example of the results from ITRF2008 and DTRF2008. It turns out, that both realizations are still a factor of ten above the GGOS requirement. This shows the need for further improvements.

Acknowledgements GGOS-SIM is funded by the German Research Foundation (DFG, SCHU 1103/8-1). We are grateful to Athanasios Dermanis and two anonymous reviewers for their valuable comments which helped to improve the quality of the manuscript.

References

Altamimi Z, Dermanis A (2009) The choice of reference system in ITRF formulation. In: Sneeuw et al (eds) VII Hotine-Marussi symposium on mathematical geodesy. International Association of Geodesy Symposia Series, vol 137. Springer, Berlin, pp 329–334. http://link.springer.com/bookseries/1345

Altamimi Z, Dermanis A (2013) Theoretical foundations of ITRF determination. The algebraic and the kinematic approach. In: Katsampalos KV, Rossikopoulos D, Spatalas S, Tokmakidis D (eds) On measurements of lands and constructions. Volume in honor of Prof. D. Vlachos, Publication of the School of Rural & Surveying Engineering, Aristotle University of Thessaloniki, pp 331–359

Altamimi Z, Sillard P, Boucher C (2002) ITRF2000: a new release of the international terrestrial reference frame for Earth science applications. J Geophys Res 107(B10):2214

Altamimi Z, Sillard P, Boucher C (2003) The impact of a no-net-rotation condition on ITRF2000. Geophys Res Lett 30(2):1064

Altamimi Z, Collilieux X, Legrand J, Garayt B, Boucher C (2007) ITRF2005: a new release of the international terrestrial reference frame based on time series of station positions and Earth orientation parameters. J Geophys Res 112:B09401

Altamimi Z, Collilieux X, Metivier L (2011) ITRF 2008: an improved solution of the international terrestrial reference frame. J Geod 85(8):457–473. http://link.springer.com/article/10.1007%2Fs00190-011-0444-4

Angermann D, Drewes H, Kelm R, Krügel M, Meisel B, Tesmer V (2004a) IERS combination research centre at DGFI. IERS Annual Report 2003, 112–113, Verlag des Bundesamtes für Kartographie und Geodäsie, Frankfurt a. M., 2004

Angermann D, Drewes H, Krügel M, Meisel B, Gerstl M, Kelm R, Müller H, Seemüller W, Tesmer V (2004b) ITRS combination centre at DGFI - A terrestrial reference frame realization 2003. Deutsche Geodätische Kommission, Reihe B, 313, Muenchen

Böhm J, Böhm S, Nilsson T, Pany A, Plank L, Spicakova H, Teke K, Schuh H (2012) The new Vienna VLBI software VieVS. In: Geodesy for planet Earth, International Association of Geodesy Symposia Series, vol 136. Springer, Berlin, pp 1007–1011

Davies P, Blewitt G (2000) Methodology for global geodetic time series estimation: a new tool for geodynamics. J Geophys Res 105(B5):11083–11100

DeMets C, Gordon RG, Argus DF, Stein S (1994) Effect of recent revisions to the geomagnetic reversal timescale on estimates of current plate motions. Geophys Res Lett 21:2191–2194

Dermanis A (2001) Global reference frames: connecting observation to theory and geodesy to geophysics. IAG 2001 scientific assembly vistas for geodesy in the New Milennium 2–8 Sept. 2001, Budapest. http://der.topo.auth.gr/DERMANIS/PDFs/Dermanis_Budapest.pdf. Accessed Feb 2015

Dong D, Feng P (2007) ITRF origin: diagnosis of current realization. EGU General Assemply, Vienna, 15–20 Apr. 2007. http://itrf.ensg.ign.fr/doc_ITRF/egu2007/Dong-egu_070417.pdf. Accessed Feb 2015

Dow JM, Neilan RE, Rizos C (2009) The International GNSS Service in a changing landscape of Global Navigation Satellite Systems J Geod 83(3-4):191–198. http://link.springer.com/article/10.1007%2Fs00190-008-0300-3

Kreemer C, Holt WE (2001) A no-net-rotation model of present-day surface motions. Geophys Res Lett 28:4407–4410

Kreemer C, Lavallée DA, Blewitt G, Holt WE (2006) On the stability of a geodetic no-net-rotation frame and its implication for the International Terrestrial Reference Frame. Geophys Res Lett 33:L17306, doi:10.1029/2006GL027058

Lambeck K (1987) Geophysical Geodesy: The Slow Deformations of the Earth. Clarendon Press, Oxford http://onlinelibrary.wiley.com/doi/10.1029/GM060p0007/summary

Ma C, Arias EF, Eubanks TM, Fey AL, Gontier A-M, Jacobs CS, Sovers OJ, Archinal BA, Charlot P (1998) The International Celestial Reference Frame as Realized by Very Long Baseline Interferometry. Astron J 116:516–546. http://iopscience.iop.org/article/10.1086/300408/meta;jsessionid=2B5FF1BD4EF0F88AA55F9F1A9C08220A.c4.iopscience.cld.iop.org

Minster JB, Jordan TH, Molnar P, Haines E (1974) Numerical modelling of instantaneous plate tectonics. Geophys J R Astron Soc 36:541–576

Pearlman MR, Degnan JJ, Bosworth JM (2002) The International Laser Ranging Service. Adv Space Res 30(2):135–143. http://www.sciencedirect.com/science/article/pii/S0273117702002776

Petit G, Luzum B (eds) (2010) IERS Conventions. IERS Technical note 36, Verlag des Bundesamts für Kartographie und Geodäsie, Frankfurt am Main. ISBN 3-89888-989-6. http://www.iers.org/IERS/EN/Publications/TechnicalNotes/tn36.html

Plag H-P, Beutler G, Gross R, Herring T, Rizos C, Rummel R, Sahagian D, Zumberge J (2009) Global Geodetic Observing System. Springer, Berlin/Heidelberg. Chapter Introduction:1–13

Seitz M, Angermann D, Bloßfeld M, Drewes H, Gerstl M (2012) The 2008 DGFI realization of the ITRS: DTRF2008. J Geod 86(12):1097–1123

Seitz M (2015) Comparison of different combination strategies applied for the computation of terrestrial reference and geodetic parameter series. In: The 1st International Workshop on the Quality of Geodetic Observation and Monitoring Systems (QuGOMS'11). International Association of Geodesy Symposia Series, vol 140. Springer, Berlin, pp 57–64. http://www.springer.com/us/book/9783319108278

Schuh H, Behrend D (2012) VLBI: a fascinating technique for geodesy and astrometry. J Geodyn 61:68–80

Nilsson T, Soja B, Karbon M, Heinkelmann R, Schuh H (2016) Application of Kalman filtering in VLBI data analysis. Earth Planets Space 67:9 pp (Article ID 136). doi:10.1186/s40623-015-0307-y

Willis P, Fagard H, Ferrage P, Lemoine FG, Noll CE, Noomen R, Otten M, Ries JC, Rothacher M, Soudarin L, Tavernier G, Valette J-J (2010) The International DORIS Service (IDS): Toward Maturity. Adv Space Res 45(12):1408–1420. http://www.sciencedirect.com/science/article/pii/S0273117709007212

Zhu S, Reigber C, König R (2004) Integrated adjustment of CHAMP, GRACE and GPS data. J Geodyn 78(1-2):103–108

Overview of the ILRS Contribution to the Development of ITRF2013

V. Luceri, E.C. Pavlis, B. Pace, D. König, M. Kuzmicz-Cieslak, and G. Bianco

Abstract

Satellite Laser Ranging (SLR) data have been fundamental over the past three decades for the realization of the International Terrestrial Reference Frame (ITRF), which is based on an inter-technique combination of the geodetic solutions obtained from an intra-technique combination strategy performed at each IERS Technique Centre. This approach provides an opportunity to verify the internal consistency for each technique and a comparison of Analysis Center (AC) adherence to internal procedures and adopted models.

The International Laser Ranging Service (ILRS) contribution is based on the current IERS Conventions 2010 as well as on internal ILRS ones, with a few documented deviations.

The main concern in the case of SLR is monitoring systematic errors at individual stations, accounting for undocumented discontinuities, and improving the satellite target signature models. The SLR data re-analysis for ITRF2013 extends from 1983 to the end of 2013 and was carried out by 8 ACs according to the guidelines defined by the ILRS Analysis Working Group (AWG). These individual solutions have been then combined in the official solution by the ILRS Combination Center.

This work allows point-wise monitoring of the quality of the SLR contribution and a thorough investigation on the time behaviour of its characteristic products, i.e. origin and scale of ITRF. The stability and consistency of these products are discussed for the individual and combined SLR time series. The critical issues from this analysis will be presented to highlight the key points that SLR should take into account to contribute in the best possible way to the present and future ITRF realizations.

Keywords

Geocenter • International laser ranging service • International terrestrial reference frame • Satellite laser ranging • Scale

1 Introduction

The next realization of the ITRF will follow the same approach already adopted for ITRF2008 (Altamimi et al. 2011). It will be constructed using time series of station positions and Earth Orientation Parameters (EOPs) from the four space geodetic techniques (SLR, VLBI, GNSS, DORIS). The IERS Technique Centers were requested to provide time series that are as long as possible and preferably

V. Luceri (✉) • B. Pace
e-GEOS S.p.A., Centro di Geodesia Spaziale, Matera, Italy
e-mail: cinzia.luceri@e-geos.it

E.C. Pavlis • D. König • M. Kuzmicz-Cieslak
GEST/UMBC, Baltimore, MD, USA
e-mail: epavlis@umbc.edu

G. Bianco
Agenzia Spaziale Italiana, Centro di Geodesia Spaziale, Matera, Italy
e-mail: giuseppe.bianco@asi.it

T. van Dam (ed.), *REFAG 2014*, International Association of Geodesy Symposia 146,
DOI 10.1007/1345_2015_153

cover the full history of observations of their technique. Thus, the Technique Combination Centers' role implies the responsibility of generating an official single-technique solution, merging in an optimal way, all of the available AC solutions.

The role of SLR in the realization of the ITRF is fundamental, both for its temporal data coverage, starting at the beginning of the eighties, and its specific sensitivity to the terrestrial origin and scale. As in the case of ITRF2008, it is expected that SLR will realize the ITRF origin and along with VLBI the ITRF scale.

The official ILRS (Pearlman et al. 2002) contribution to the new ITRF2013[1] is generated by the Primary ILRS Combination Center (CC) at the Space Geodesy Center of the Italian Space Agency (ASI/CGS) and is designated as ILRSA. A backup solution time series (named ILRSB) is computed at the Goddard Earth Sciences and Technology Center (GEST/UMBC), the backup CC.

The ILRS contribution to ITRF2013 is a time series of weekly station coordinates and daily EOPs: X-pole, Y-pole and excess Length-Of-Day (LOD), estimated over 7-day arcs aligned with calendar weeks (Sunday to Saturday) from January 1993 to December 2013. For the earlier period, January 1983 to December 1992 15-day arcs had to be used generating station coordinates as biweekly averages and EOP estimated as 3-day averages since the network size and the data from a single satellite do not allow stable solutions with the higher resolution. Each weekly combination is obtained through the weighted averaging of the weekly solutions submitted by the official ILRS Analysis Centers. Both the individual and combined solutions follow strict standards agreed upon within the ILRS AWG to provide products of the highest possible quality.

2 Individual Solutions

The individual solutions are computed by the official ILRS ACs (see Table 1) using the SLR data acquired from the worldwide network that observed the satellites LAGEOS, LAGEOS-2, Etalon-1 and Etalon-2. From 1983 to 1992 the dataset made up of LAGEOS data only is called from now on the "historical period". This dataset is complemented with the LAGEOS-2 and ETALON satellites starting from 1993 (see Fig.1). The main difference in the data amount is due to the LAGEOS-2 data; the amount of the ETALON data is roughly one tenth of the data from the two LAGEOS, and have a practically negligible impact on the results despite an increase of data during specific campaign periods organized by the ILRS requesting intensive tracking.

Table 1 Current ILRS Analysis Centers

ASI	Agenzia Spaziale Italiana	Italy
BKG	Bundesamt für Kartographie und Geodäsie	Germany
DGFI	Deutsches Geodätisches Forschungs Institut	Germany
ESA	European Space Operation Center	Europe
GFZ	GeoForschungsZentrum Potsdam	Germany
GRGS	Groupe de Recherche de Géodésie Spatiale – Observatoire de la Cote d'Azur	France
JCET	Joint Center for Earth Systems Technology – NASA&UMBC	USA
NSGF	NERC Space Geodesy Facility	UK

The SLR observations are retrieved from the CDDIS and/or EDC data archive facilities and analyzed to generate the individual EOP and station position solutions. The measurements are processed in intervals of 7 days (15 days in 1983–1992) to generate a loosely-constrained solution for station coordinates and EOP. The EOPs (X_p,Y_p and LOD) are all computed as daily averages since 1993 and as 3-day averages when only LAGEOS data are available during the "historical period". Daily UT parameters are also solved for, but they are of course considered as weakly-determined parameters by any satellite technique and are not included in the analysis product that is submitted to the combination centers.

The product quality is affected by different factors including the adopted conventions, application/estimation of systematic errors, applied satellite center-of-mass (CoM) corrections, data coverage and hidden constraints. These factors were addressed in the past years within the ILRS AWG in order to give the ACs some guidelines. Analysis contributors are generally free to follow their own computation model and/or analysis strategy, but a number of constraints must be followed for consistency:

- The computational models follow the prevalent IERS Conventions as closely as possible (with documentation of any exceptions).
- Daily series of the Mean Pole (MP) coordinates and their rates based on the interpolated/extrapolated IERS MP series are adopted, instead of the fixed polynomial version in the IERS 2010 Conventions.
- As requested by the ITRF2013 Call for Participation, the non-tidal atmospheric loading effects are not modeled.
- The stations are included in the weekly analysis if the number of observed LAGEOS plus LAGEOS 2 ranges is greater than 10. Data weighting is applied according to the analyst's preference. However, the AWG has agreed to down-weight "non-core" sites significantly.

[1]At the time of the REFAG 2014 meeting, the ITRS Call for Participation for ITRF2013 was still valid. The decision to extend the series through 2014 and go for the realization of ITRF2014, came later. The description of the ILRS contribution is applicable to ITRF2014 with the extension of the series to 2014.

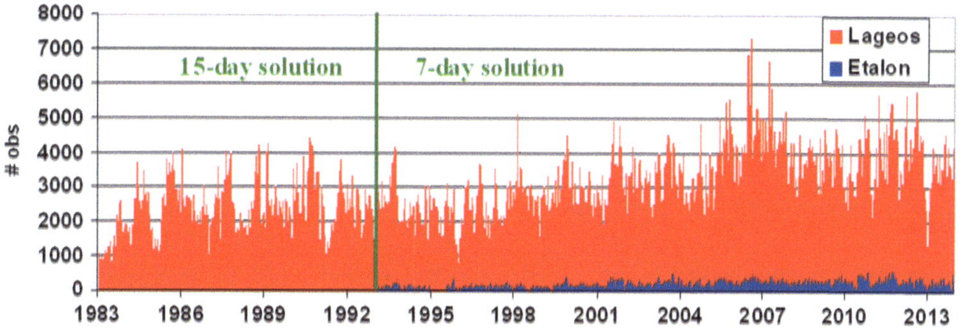

Fig. 1 Input observation dataset for the ILRS AC solutions

- The center-of-mass correction for each satellite is applied following the site- and time-specific tables provided by Graham Appleby (Appleby and Otsubo 2013) that take account of the various laser station technologies and tracks the changes over the years.
- Range corrections were modeled or estimated for a number of sites, based either on engineering reports from these sites or long-term analysis of their systematic behavior. All of the applied corrections are documented in the ILRS Data Handling file available at http://ilrs.dgfi.tum.de/fileadmin/data_handling/ILRS_Data_Handling_File.snx.
- The weekly solutions are loosely constrained with an a priori standard deviation on station coordinates of ~1 m and the equivalent of at least 1 m for EOPs.

The individual AC solutions are stored at CCDIS and EDC, available for the ILRS CC only.

3 ILRSA Intra-Technique Combination

The eight ACs submitted several versions of their SLR SSC/EOP solutions in order to give their best time series, in strict cooperation with the CCs. One of the key strengths of the contributing solutions is the use of different software (s/w) for the individual AC data reduction. This "s/w diversity" provides a safeguard against erroneous implementations of the required models and algorithms, and allows us to check internally the accuracy of our products. Beyond the quality of the individual AC solutions, the final combined product quality is also affected by the weight applied to the contributing AC solutions and the application of outlier editing.

The first step in the combination process is the rigorous check of each individual AC time series in terms of looseness in the applied constraints, the application and/or estimation of systematic corrections for the set of sites over specific periods, the deletion of data when they are not recoverable, all according to the adopted AWG guidelines. This phase is in general time consuming, requiring a close interaction between the ACs and the CCs, and can take several months especially when reprocessing of the time series is required.

Once the input AC solutions are fully checked, the CCs are ready to start the final combination process. The official ILRS combined solution is produced by the Primary Combination Center, ASI/CGS, and named ILRSA; a backup combined solution (ILRSB) is computed at GEST/UMBC, the backup CC.

The ILRSA solution has been obtained by a direct combination of the loosely constrained solutions, taking advantage of the fact that loosely constrained solutions, although they possess an ill-defined datum, they still preserve the relative geometry of the station polyhedron figure.

The combination is based on the method described in (Davies and Blewitt 2000) and allows handling input solutions easily, with no inversion problems for the solution variance-covariance matrix, no need to know a priori values for the estimates and no need to estimate or remove relative rotations between the reference frames before combining the solutions.

Each contributing solution (and related variance-covariance matrix) is treated as an 'observation' whose misclosure with respect to the combined solution must be minimized in an iterative Weighted Least Square approach. Each solution is stacked using its full covariance matrix rescaled by an estimated scale factor. A scaling of the covariance matrix of the ith solution is required because the relative weights of the contributing solutions are arbitrary. Imposing $\chi^2 = 1$ for the combination residuals and requiring that each contribution to the total χ^2 is appropriately balanced, the relative scaling factors (σ_i) are estimated iteratively together with the combined solution. If R_i represents the solution residuals (with respect to the combined product), Σ_i the solution covariance matrix and N the number of solutions, the imposed conditions are:

$$R_1^T(\sigma_1\Sigma_1)^{-1}R_1 = \cdots = R_i^T(\sigma_i\Sigma_i)^{-1}R_i \cdots =$$
$$= R_N^T(\sigma_N\Sigma_N)^{-1}R_N \quad \text{and}$$
$$\chi^2 = R_1^T\Sigma_1^{-1}R_1 + \cdots + R_N^T\Sigma_N^{-1}R_N = 1$$

Table 2 Scaling factors applied to the individual AC solutions' covariance

	ASI	BKG	DGFI	ESA	GFZ	GRGS	JCET	NSGF
Mean	4.3	4.9	11.6	3.9	7.6	4.7	5.4	10.6
Standard deviation	2.7	4.1	5.5	1.7	5.4	2.9	3.5	6.0

The first guess for the combination is obtained with $\sigma_i = 1$ for each solution.

The scale factors for each contributing AC are reported in Table 2 as mean value and standard deviation over the period 1993–2013 when the solutions are more stable with the complete 4 satellites configuration. Five ACs have similar scale factors (between 4 and 5) while 3 ACs need higher scale factors to achieve a balanced contribution in the combination, which means that they have higher residuals with respect to the combined solution.

A rigorous editing (Brockmann 1996) has been introduced to eliminate outliers with respect to the combined solution following a 5σ criterion for sites with less than 10 observations, erroneously present in the contributing solutions, sites with too large uncertainties (>1 m) and sites with coordinate residuals with respect to the a priori SLRF2008[2] (>0.5 m).

4 ILRSA Assessment

The combined time series is subject to several checks before its release.

The internal precision of the ILRSA solution is checked through the computation of the weighted root mean square (wrms) over the time series of the coordinate residuals of each input solution with respect to the combination. Thus, a cumulative 3-dimensional value of the wrms (3D wrms) is computed for each solution using the coordinate residuals in all three components, of all the sites contributing to each solution. The time series of the 3D wrms for each AC are illustrated in Fig. 2 as a yearly running average, from 1993 to 2013, in order to make more evident their mean value and their trend. The internal "agreement" is roughly 4 mm over the last years, with a higher value for three input solutions, as mentioned above.

The external precision is checked comparing the ILRSA solutions with SLRF2008 in terms of:

- the mean of the 3D wrms of the site coordinate residuals w.r.t. SLRF2008 (see Fig. 3); and,
- the translation and scale offsets of ILRSA with respect to SLRF2008.

and the EOPs with respect to the USNO final daily values (derived primarily from GNSS data).

[2]SLRF2008 is an extension of ITRF2008 including the new SLR stations active in the network after the ITRF2008 release.

The initial decade of the solution time series, the "historical period", (1983–1992) consists of less precise estimates. However, the old portion of the series is a valuable and unique contribution of the SLR technique to the long-term Terrestrial Reference Frame definition, contributing a number of sites from the early stages of space geodetic networks. The 3D wrms is computed for each single solution using the full network and using a subset of "core sites", namely those sites selected by the ILRS AWG for their stability over time, the lengthy data history and being well-modeled in SLRF2008 (see Table 3 with the mean values).

Figure 3 shows the 3D wrms values for the input solutions as well as the ILRSA combination, where each dot is the value of the single solution and the black line is a polynomial fit that highlights the trend of the ILRSA combination.

The datum stability of the ILRSA combination is assessed through the estimation and analysis of the translation and scale offsets with respect to SLRF2008. In the last ITRF realization (ITRF2008), the ITRF origin was defined by the ILRS SLR time series and the ITRF scale was obtained as an average of the VLBI and SLR scales and their rates. The next ITRF datum will be presumably made in the same way and the quality of the frame defining parameters will largely depend on SLR. The weak and noisy solutions of the "historical period" will have a small or null impact on the frame definition. The translations shown in Fig. 4 are relative to the 1993–2013 time-span. Very small offsets and drifts are visible (Table 4) while, as expected, significant seasonal variations are present. A small deviation from the trend is visible in Ty after 2010; this signature is present in all the input time series, but the reason for the deviation is not clear. Tz is noisier, as expected, driven largely by mass redistribution between the two hemispheres and a change of slope occurs around 1997, already present in previous solutions (Altamimi et al. 2011). The linear fit on the scale (Fig. 5) time series indicates clearly a negative slope (−0.37 mm/year). The yearly running average over the time series highlights a change of slope around 1997 and an anomalous signature around 2010 (similar and contemporary with what we see in Ty), unexplained at the moment. It seems that the anomaly in 2010 is due to an isolated event at that time, with the subsequent part of the time series continuing with the same linear trend observed over the main part of the series. As in the case of the translations, all the input AC solutions show the same behavior and a much more extensive investigation will be undertaken to resolve it. A summary

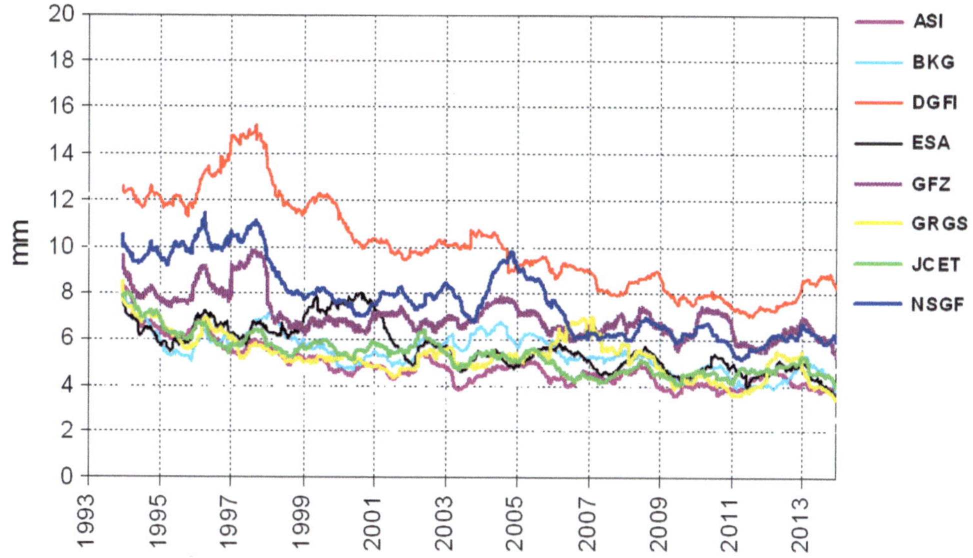

Fig. 2 3D WRMS of the coordinate residuals with respect to ILRSA (yearly running average)

Table 3 3D WRMS of the ILRSA coordinate residuals with respect to SLRF2008

Units in millimeters (mm)	1983–1992	1993–2013
All sites (mean)	15.4	7.7
Core sites (mean)	11.2	5.0

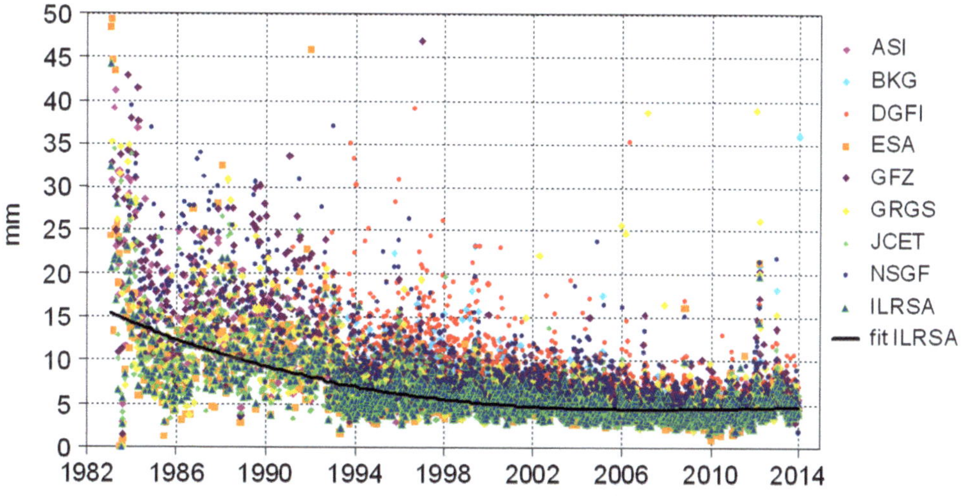

Fig. 3 3D WRMS of the core site coordinate residuals with respect to SLRF2008

Table 4 Translation and scale with respect to SLRF2008

	T_X	T_Y	T_Z	SCALE
Offset @ 2005.0 (mm)	-0.3 ± 0.1	-0.7 ± 0.1	0.7 ± 0.2	-3.76 ± 0.08
Slope (mm/year)	-0.01 ± 0.01	-0.12 ± 0.01	0.28 ± 0.03	-0.37 ± 0.01

of the Helmert transformation parameters linear trend is in Table 4.

As stated above, another external comparison is made for the EOPs with the USNO final daily values. The comparison is performed in terms of wrms of the residuals and the results agree in general with what is expected from the SLR technique: 167 μas for the X-component, 190 μas for the Y-component and 32 μs for LOD.

A further step in the assessment of the ILRSA solution will be the detection of discontinuities in the site coordinates

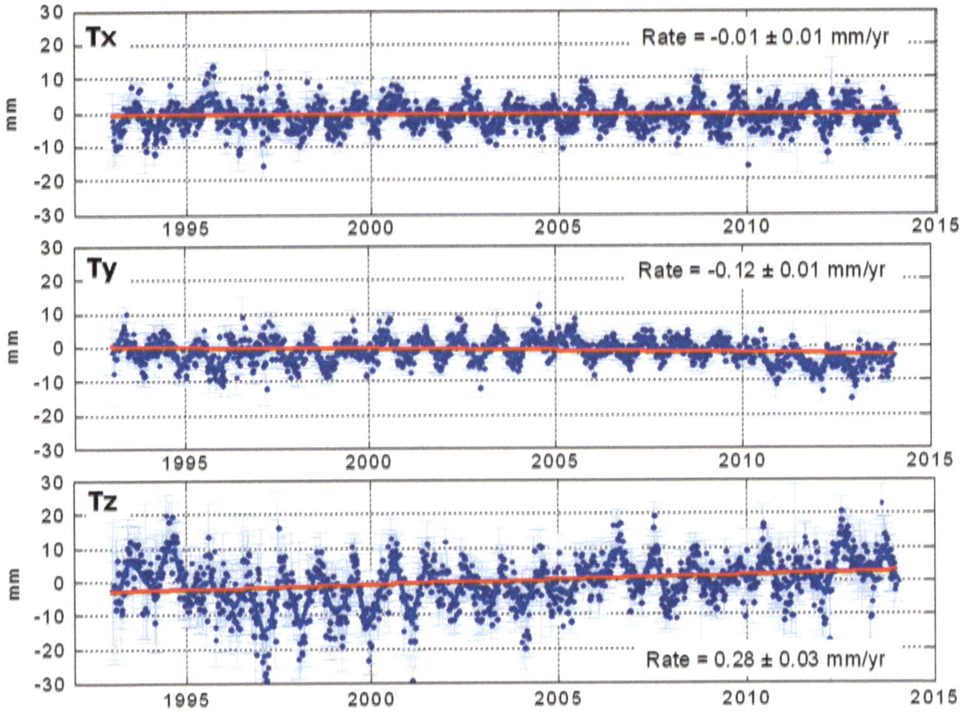

Fig. 4 ILRSA translation offsets with respect to SLRF2008

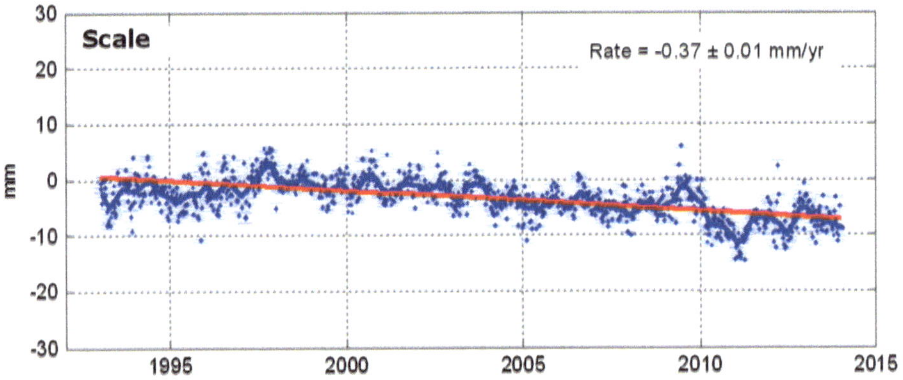

Fig. 5 ILRSA scale factor with respect to SLRF2008

time series. These discontinuities will be discussed with the IERS ITRS centre in charge of the official release of the next ITRF in order to reach an agreement on the set to be implemented and delivered with the new ITRF.

5 Conclusion

The ILRS contribution to the next ITRF has been delivered following the guidelines of the ITRS Call for Participation. The eight ILRS Analysis Centers produced time series of station coordinates and EOPs (X_p, Y_p and LOD) over the period 1983–2013 under the constraints agreed within the ILRS Analysis Working Group. The ILRS Combination Centers delivered the official ILRSA combined time series and the backup series ILRSB. Similar to the case of the ITRF2008 contribution, the ILRSA combination is computed by a direct weighted average of the loosely constrained solutions, whose contribution to the final product is balanced using a simultaneously estimated scale factor. The internal and external precision of the combined time series has been evaluated through the comparison between the input individual series and the combined one, between the combined and SLRF2008 a priori, and for the EOPs, with the USNO time series.

These comparisons show good performance of the quality parameters (site coordinates WRMS, Helmert transformation parameters time series) for the final combined solution and a remarkable coherence for the single AC solutions: the 3D WRMS of the Core Site residuals with respect to SLRF2008 reaches 5 mm in the last years, the Helmert transformation parameters time series (origin and scale) show coherence over time and very limited noise, allowing the detection of small secular and periodic components, with only Tz slightly noisier, as expected. A signature starting around February 2010 is visible in the Ty and scale series and will be further investigated as to its origin and possible remedy.

As requested by the IERS ITRS centre, the ILRS time series will be extended to include 2014; the network will benefit by a considerable number of new stations, mainly Russian, but the quality of the ILRS contribution is not expected to be substantially affected.

References

Altamimi Z, Collilieux X, Métivier L (2011) ITRF2008: an improved solution of the international terrestrial reference frame. J Geod 85(8):457–473

Appleby G, Otsubo T (2013) Centre of Mass corrections for precise analysis of LAGEOS, Etalon and Ajisai data. In: Proceedings of the 18th international workshop on laser ranging, 11–15 November 2013, Fujiyoshida, Japan

Brockmann E (1996) Combination of solutions for geodetic and geodynamic applications of the Global Positioning System (GPS), Ph.D. thesis, Astronomical Institute, University of Berne

Davies P, Blewitt G (2000) Methodology for global geodetic time series estimation: a new tool for geodynamics. J Geophys Res 105(B5):11083–11100

Pearlman MR, Degnan JJ, Bosworth JM (2002) The international laser ranging service. Adv Space Res 30(2):135–143

Part III

Regional Reference Frames

A Spatial Analysis of Global Navigation Satellite System Stations Within the Context of the African Geodetic Reference Frame

Ivan F. Muzondo, Ludwig Combrinck, Joel O. Botai, and Cilence Munghemezulu

Abstract

Permanent Global Navigation Satellite Systems (GNSS) stations that are operating within the African Geodetic Reference Frame (AFREF) are contributing data to the datum realizations at global, regional and local levels. The infrastructure supports development and administrative functions of African governments, the public and investors throughout the African continent. However, African stations with high quality and continuous data that have been acquired over several decades are limited, which result in a non-uniform network. This means that additional station investment are required in Africa for new stations to contribute to the AFREF project. An assessment of the spatial distribution and densification of GNSS stations that contribute to AFREF ensure future geometrical improvements in the network have the most impact. Established GNSS stations within AFREF network contribute data for the realization of the International Terrestrial Reference Frame (ITRF), International GNSS Service (IGS) products and local AFREF solutions.

This study contributes to AFREF through an improved understanding of the progress, identifies countries that have either adequate or inadequate permanent GNSS stations and informs planning and decisions on sites for new GNSS station installations. In turn, the result forms the basis for evaluating the impact of locations, proximity, neighbourhood and distribution of GNSS stations in AFREF static coordinates and station velocities computations. The study concludes that due to current sparse and uneven geometric distribution of GNSS stations, a strategic and planned densification of AFREF stations is essential.

Keywords

AFREF • GNSS • Voronoi diagrams

1 Introduction

The global standard in datum definition is the International Terrestrial Reference Frame (ITRF). The ITRF is an Earth-fixed and therefore Earth-centred and Earth-rotating coordinate frame. Modern ITRF computations employ four space geodetic techniques – Very Long Baseline Interferometry (VLBI), Satellite and Lunar Laser Ranging (SLR/LLR), Doppler Orbitography and Radiopositioning Integrated by Satellite (DORIS) and GNSS (Angermann et al. 2000; Altamimi et al. 2011). The IGS combines

I.F. Muzondo (✉) • L. Combrinck • J.O. Botai • C. Munghemezulu
Department of Geography, Geoinformatics and Meteorology, University of Pretoria, Private Bag X20, Pretoria 0028, South Africa
e-mail: ivan@hartrao.ac.za

L. Combrinck • C. Munghemezulu
Hartebeesthoek Radio Astronomy Observatory, P.O. Box 443, Krugersdorp 1740, South Africa

T. van Dam (ed.), *REFAG 2014*, International Association of Geodesy Symposia 146,
DOI 10.1007/1345_2015_199

global solutions from GNSS data (Dow et al. 2009). A regional datum, such as AFREF, is computed to be coincident and compliant with the ITRF. Stations within the AFREF project are installed and operated, based on IGS standards, to promote the establishment of a permanent network of GNSS stations for Africa (Combrinck 2008; Wonnacott 2012). GNSS technology is preferred, compared to VLBI, SLR/LLR and DORIS, as the installations are comparably cheaper, technically simpler, and operationally well-understood.

In this paper, we present a spatial analysis of permanent GNSS stations established on the African plates as at 30th August 2014. The spatial analysis methods include the use of Voronoi diagrams, proximity analysis, buffering and overlay techniques. The GNSS station locations are analysed in terms of the spatial distribution and the effects of the recognised ITRF2008 and IGS stations and local AFREF stations. Local AFREF stations are sites that are neither used to compute the ITRF nor recognized as IGS stations. Data from all stations used are publicly available from global, regional and local data centres. The GNSS stations that are located on or near the African plates, Nubian and Somalian, are modelled to illustrate the characteristics of the AFREF network geometry as well as African country participation. The Voronoi polygons and the statistical analysis of the final diagrams' (or tessellations') areas and diagrams' sides provide evidence of the strengths and the weaknesses of the spatial distribution and site influences of the permanent GNSS stations.

2 Theory and Background

2.1 ITRF and IGS

Permanent stations contribute GNSS data in the Receiver Independent Exchange (RINEX) format to the data centres, which in turn makes the data available to analysis centres for computations and hence contribute to the improvement of global, regional and local reference frames. Two frames are defined by the International Earth Rotation and Reference Systems Service (IERS), the International Celestial Reference Frame (ICRF) and ITRF (IUGG 2013). The ICRF is an absolute coordinate system based on a selection of quasars, while the ITRF is a global terrestrial coordinate system realized through instrumentation located on the Earth's surface (e.g. phase centre of GNSS antenna, referred to some fixed marker located on bedrock), and linked to the ICRF through the Earth Orientation Parameters (EOP). Computations of AFREF are fully compliant and consistent with the ITRF (Altamimi 2003; Farah 2009; Wonnacott 2012).

2.2 AFREF

The AFREF project is an initiative of the United Nations Economic Commission for Africa (UNECA) Committee on Development Information (CODI) (Farah 2009). It encompasses works from IGS, the International Association of Geodesy (IAG), International Federation of Surveyors (FIG) and other organisations that endeavour to ensure the availability of permanent GNSS stations data. Ideally, station location requires contiguous stations to be no greater than 1,000 km apart (Wonnacott 2012). In other words, the distance between GNSS station throughout the African continent would be overlapping on circles of radii 500 km. AFREF scientists are involved in coordinating activities to promote the establishment, maintenance and use of GNSS stations, data sharing, GNSS product distribution and application. Further, the AFREF community oversees the data and computations of the AFREF regional reference frames (Altamimi 2003; Farah 2009; Wonnacott 2012). Organisations that participate and support AFREF have established centres to archive or duplicate existing RINEX data, to compute precise GNSS products and realise station coordinates and velocity solutions. This favourable setting allows AFREF to gain tremendous opportunities to grow its network of GNSS stations.

The primary technology for AFREF and IGS is GNSS (Combrinck 2008). The densification of GNSS stations for Africa requires a coordinated approach within AFREF that ensures stations are located where they are used and required (Combrinck 2008; Wonnacott 2012). Currently, the IGS site and equipment specifications for permanent GNSS stations have been adopted as the AFREF standard. Most stations in the AFREF network are established and maintained through voluntary organisations. Institutions such as the Jet Propulsion Laboratory (JPL) – USA, Institut Geographique National (IGN) – France, the German Research Centre for Geosciences (GFZ) – Germany and the Hartebeesthoek Radio Astronomy Observatory (HartRAO) – South Africa, have contributed to the African initiative. Based on the IGS standards, GNSS stations are established with geodetic quality receivers/antennas and specifications for resolving calibration, standardisation and management challenge (Dow et al. 2009). These include dual-frequency receivers, site locations with low multipath environment and rigidly attached to bedrock. The quality of the monuments, relative spatial distribution and location (or spatial impact) is very important so as to maximise stability, improve network geometry and lessen adverse influences on the data such as multipath and GNSS signal obstruction (Combrinck and Schmidt 1998).

Operational aspects in AFREF network are often less than ideal. This originates largely from the federated organisational participation, which at times focuses on different applications. Geodesy, geodynamics, development ventures, environmental and mapping programmes, and international boundary disputes are some examples (Combrinck 2008). The inclusive approach has resulted in permanent GNSS stations being established through a diverse range of organisations that utilise a variety of equipment and often target different applications. Stations incorporate vendor biases and exhibit variable standards in station rigidity, data latency and reliability of power supplies (Wonnacott 2012). Spatial locations of sites are influenced by subjective primary purposes for permanent stations, and not objective criteria inherent in techniques including Voronoi diagrams (Fortune 1992; Klein 2014).

2.3 Voronoi Diagrams Theory

Voronoi diagrams are polygons or tessellations such that the edges indicate the limits of extent of influence or region exerted by contained site (Okabe and Suzuki 1997; Okabe et al. 2000). They are derived from Delaunay triangulations which have origins in computational geometry. Voronoi diagrams are based on work by G.F. Voronoi on quadratic forms (Fortune 1992; Okabe et al. 2000; Edelsbrunner 2014; Klein 2014). They have wide application in spatial data analysis where there is a sudden change in characteristics when changing from one region to another region (Zhang et al. 2014; Edelsbrunner 2014; Okabe et al. 2000; Edelsbrunner and Seidel 1986). According to Fortune (1992), Okabe et al. (2000), Edelsbrunner (2014) and Klein (2014), site s may be associated with surrounding space of influence, termed Voronoi regions (enclosed by polygons), which are defined from a finite set of points that exert influence for a region in a domain R^2. The elements of set S sites ($s \in S$) are distinguishable from other points of set X in their character. For each site s, an area of influence, the Voronoi region V_s, is defined by the apex points, ($x \in X$), as limits to each region that are at least as close to one site as they are to other adjacent sites. This is expressed by Edelsbrunner (2014) as:

$$V_s = \{x \in R^2 \,|\, \|x - s\| \le \|x - t\|,\ \forall t \in S\,|\}, \quad (1)$$

where the sites $s = (s_1, s_2)$ are distinguishable from the other points on the surface due to their *influence* characteristic and the apex points $x = (x_1, x_2)$ are any discernible location. The maximum influence distance, $\|x - s\|$, is expressed as:

$$\|x - s\| = \left[(x_1 - s_1)^2 + (x_2 - s_2)^2\right]^{\frac{1}{2}} \quad (2)$$

The definitions of regions for each site, s, are referred to as Voronoi diagrams. The use in mathematics, computer science and Geographical Information Science (GISc) disciplines is growing (Fortune 1992; Okabe et al. 2000; Edelsbrunner 2014; Klein 2014).

The spatial analysis using Voronoi diagram defines GNSS stations as sites, s_i. Regions, V_s, are associated with sites, s_i, that are bound by apex points or vertices x_i, being limits to polygon regions. The Voronoi diagrams provide a visual display of site proximity, neighbourhood and distribution. The visual enhancement, coupled with modern computer efficiency in computing the Voronoi algorithms has made Voronoi diagrams the spatial analysis tools for systematic and objective approach to AFREF station location analysis (see Edelsbrunner and Seidel 1986; Fortune 1992; Okabe et al. 2000; Edelsbrunner 2014; Zhang et al. 2014). The permanent GNSS stations provide a finite set of sites located on African continent, the domain. The region defined for each site is represented by separate Voronoi diagrams.

The elements of a set S (ITRF, IGS or local AFREF) represented by s_1, s_2 and s_3 are assumed to have the same weight i.e. quality standard, station construction, rigidity, sampling rate and latency. In set notation, S is:

$$S = s_1 \cup s_2 \cup s_3 \quad (3)$$

Given the set of GNSS stations, S such that S_i represents the ith GNSS station location, then following on Okabe et al. (2000), the Voronoi diagram is a polygon with a variable number of sides such that:

a) Each point s_i is of known location and indicates the centre of gravity of the polygon defined around that site.
b) Each polygon contains only one s_i site, where $i \in \{1, 2, \ldots, n\}$ and n is the finite number of sites (GNSS stations) represented by S in space.
c) Each polygon side is generated from point x_i and lies equidistant to all neighbouring and surrounding sites.

Theoretically, the ideal geometry for a network of regular distribution of stations, Voronoi diagrams would display as regular pentagonal or hexagonal polygons circumscribed in a circle of constant diameter. In this study, the geometry of the Voronoi polygons is irregular shapes of different sizes and shapes across Africa.

3 Research Methodology

The methodology was designed to ensure that the geometrical relationship between permanent GNSS stations can be both qualitatively and quantitatively explored.

The study was conducted using the geodetic latitude–longitude coordinate system. The polar coordinate system, with the radius assumed to be the unit sphere enables the use of spherical trigonometry formulae for distance and angle computations, informed the geo-database designed. This required a scale change from the unit sphere to a sphere with radius based on the Weighted Mean Radius (WMR) in World Geodetic System 1984 (WGS84) ellipsoid. The origin was the centre of mass of the Earth and the primary axis coincided with the average rotation of the Earth as determined by the IERS. The secondary axis was the intersection of the equatorial and Greenwich Meridian planes while the tertiary axis completed a right-handed system. A weighted mean radius was computed based on the WGS84 ellipsoidal parameters to enable transformation to ellipsoidal distances.

Environmental Systems Research Institute (ESRI) ArcGIS 10.1 was the primary geo-database software. This was composed of shape files of the African cities, two GNSS station layers and the African countries layer as contained on or near the African plates. The GNSS stations layer contained ITRF2008 and IGS recognised points, while the second layer was composed of local AFREF points only. The reference date for all data collection was 30 August 2014 and relied exclusively on publicly available datasets from IGS and AFREF data centres. The main base layer was the polygon layer of United Nations recognised African countries. Although ITRF2008 and IGS stations are widely scattered worldwide, for this study, only stations on or neighbouring the African tectonic plates were included in generating Voronoi diagrams.

Voronoi diagrams were generated from the GNSS stations in the following order:

- ITRF2008 and IGS layer,
- AFREF layer

A union function was performed [Eq. (3)] to output a combined GNSS stations layer. The Voronoi diagrams were constructed on this combined ITRF2008, IGS and local AFREF (or combined GNSS) layer. The countries' layer was intersected with the combined GNSS layer to generate the base layer to improve the analysis and interpretation on the Voronoi diagrams and distinguish land from sea. From the combined GNSS layer, polygon areas indicating station distribution that exceeded 1,000 km were selected from lengths of the polygon sides generated using Voronoi diagrams. Similarly, regions where the radius was less than 500 km were also highlighted. Scatter plots were generated from polygon areas for further analysis.

4 Results and Discussion

The geo-database contained an African countries' layer. Africa is made up of 56 United Nations (UN) and 55 African Union (AU) recognised countries. Both official counts differ from the on-the-ground administrative units and political jurisdictions, which some unofficial reports estimate to be approximately 65. These counts differ from the generated country polygons which are 762 with some African countries having multiple polygons. The in-polygon analyses based on countries was therefore complicated.

The spatial distribution of GNSS stations may be interpreted as indicating that the locations of local AFREF points are in the vicinity of joint ITRF2008 and IGS points. African regions can therefore be classified as either adequate to more than adequate or lacking to critically lacking. Figure 1a, b both show that the northern and central African regions are lacking local AFREF stations as well as ITRF2008 and IGS stations.

Based on the stations shown in Fig. 1, the resultant Voronoi diagrams are shown in Fig. 2. In Fig. 2a, Voronoi diagrams are generated for local AFREF stations, whilst in Fig. 2b they are generated for ITRF2008 and IGS stations. The Voronoi diagrams are clearly not ideal as they appear as irregular polygons, and not the theoretical regular pentagons and hexagons as discussed in Sect. 2.4. The following analysis focuses on assessing further Voronoi polygon parameters.

The combined GNSS stations layer is represented in Fig. 3a and the corresponding Voronoi diagram shown in Fig. 3b. Based on our assessment of Fig. 3, the regions in the southern, eastern and some western countries, specifically the regions covered by the Nigeria Geodetic Network (NigNET) – Nigeria, Millennium Development Challenge (MDC) – Benin, and Africa Monsoon Multidisciplinary Analysis (AMMA) GPS – West Africa have a dense network of GNSS stations. An assessment of the polygon side lengths and polygon areas in the Voronoi diagrams is included in the attribute table. An analysis of the attribute values has resulted in a better understanding of the AFREF initiatives on a regional basis. A greater density of GNSS stations results in the polygons having shorter sides and smaller areas. We note that the Voronoi polygons with longer sides occur when the vertices originate from oceans and regions in central and northern Africa. Regions with better station densities in the southern, eastern and parts of the horn of Africa are also evident from the Voronoi diagrams.

However, the regions with better coverage need to be assessed to verify that the minimum requirements for the AFREF objectives are attained. We focus on an analysis of the polygon side lengths and polygon areas of the Voronoi diagrams as generated from the geo-database layer containing ITRF2008, IGS and local AFREF points. The ITRF2008 + IGS + local-AFREF generated Voronoi diagrams only are used for these additional analyses.

A scatter plot of polygon areas (Fig. 4) computed from the Voronoi diagrams generated from GNSS stations using the geo-database layer of combined ITRF2008, IGS and local AFREF points. The ideal theoretical polygon area that meets the minimum AFREF objectives, simplified to

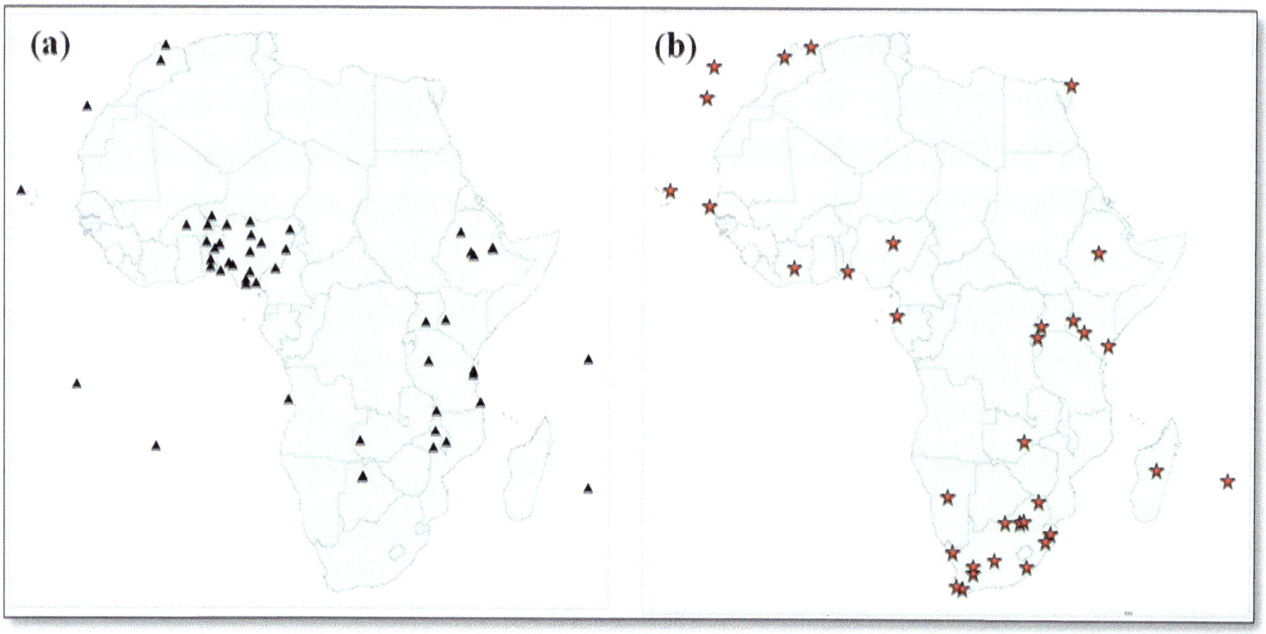

Fig. 1 (a) AFREF and (b) ITRF + IGS GNSS stations

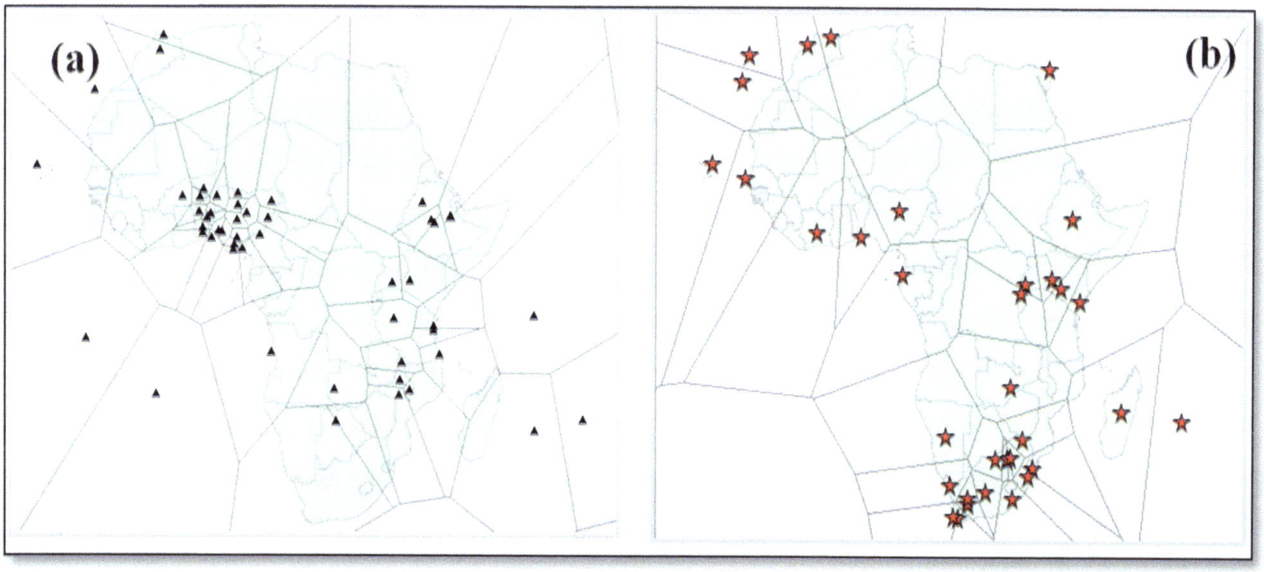

Fig. 2 (a) AFREF (b) ITRF and IGS Voronoi analysis of Africa

a regular spherical hexagonal shape, is approximately 52.5 square degrees. A polygon area of 52.5 square degrees is synonymous with GNSS station density corresponding to a circle of radius 500 km (see Sect. 2.2 for details). This assessment resulted in many polygon areas failing to meet the criterion, with the worst polygon area being 2,983 square degrees. An analysis of the attribute table for polygon areas shows that the majority of these extremely large polygons are on the perimeter and contained the ocean. While installations of GNSS station at sea or ocean is not feasible, new stations maybe installed on land regions to reduce the size of these perimeter polygons. Polygons that included the sea and ocean were considered outlier polygons and excluded in the polygon areas scatter plot (Fig. 5). The largest Voronoi area becomes 350.8 square degrees, which is greater than the AFREF recommended maximum of 52.5 square degrees. The line marked 'Max Area for AFREF Specification' is the turning point of polygon areas that comply with AFREF requirements and those that exceed the minimum specifications.

Figure 6a highlights the excluded polygons with some give-and-take approximations that were necessary for a

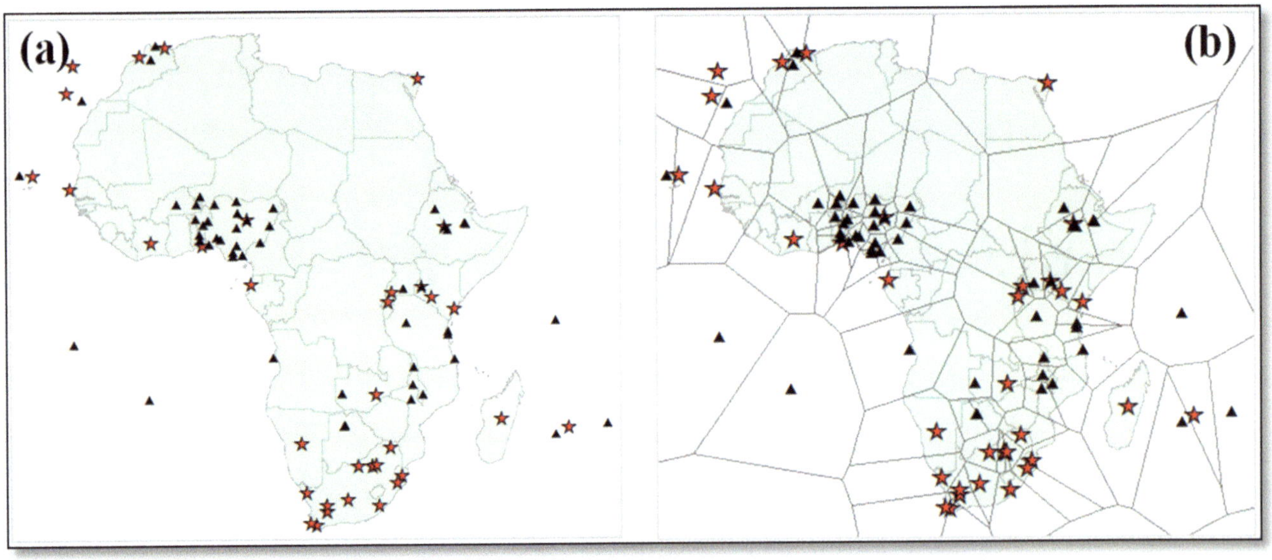

Fig. 3 (**a**) ITRF, IGS and local AFREF stations (**b**) Voronoi diagrams from ITRF, IGS and local AFREF stations

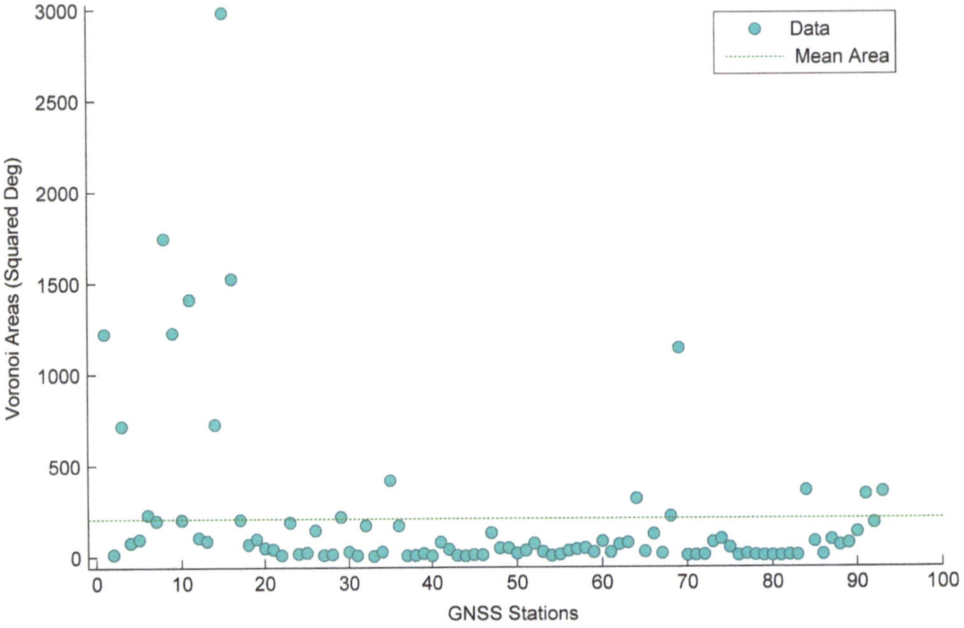

Fig. 4 Scatter plot of Voronoi Areas generated from GNSS Stations

continued assessment. The polygon areas that met AFREF minimum requirements are highlighted in Fig. 6b. The countries that are intersecting the Voronoi diagrams highlighted in Fig. 6a are considered to have met the minimum requirements for AFREF.

An illustration of the resultant scatter plot after excluding these outliers indicates relatively smaller areas of the Voronoi diagrams in Fig. 5.

5 Conclusions and Recommendations

We have identified regions within the African continent for which the AFREF project must focus on when planning and locating new GNSS stations. The polygon areas scatter plots and the statistical assessments form the base data when comparing network improvements in the future, as well as

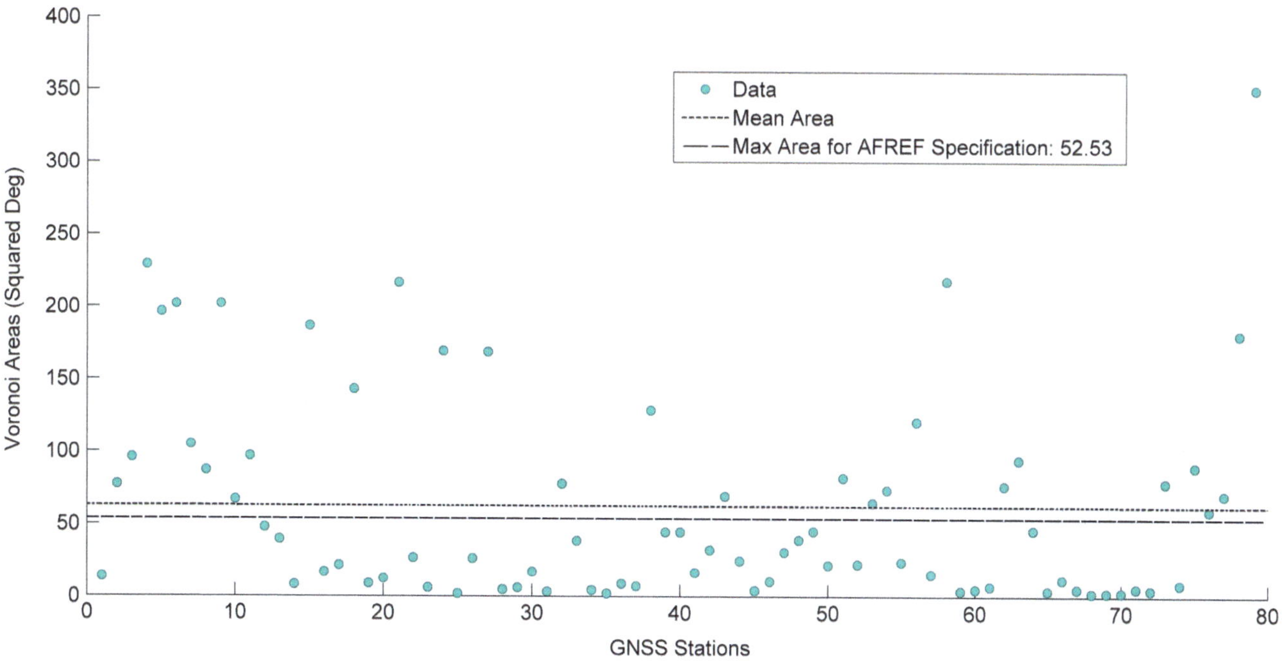

Fig. 5 Scatter plot of Voronoi Areas generated from GNSS station excluding polygons that contains ocean and sea outliers

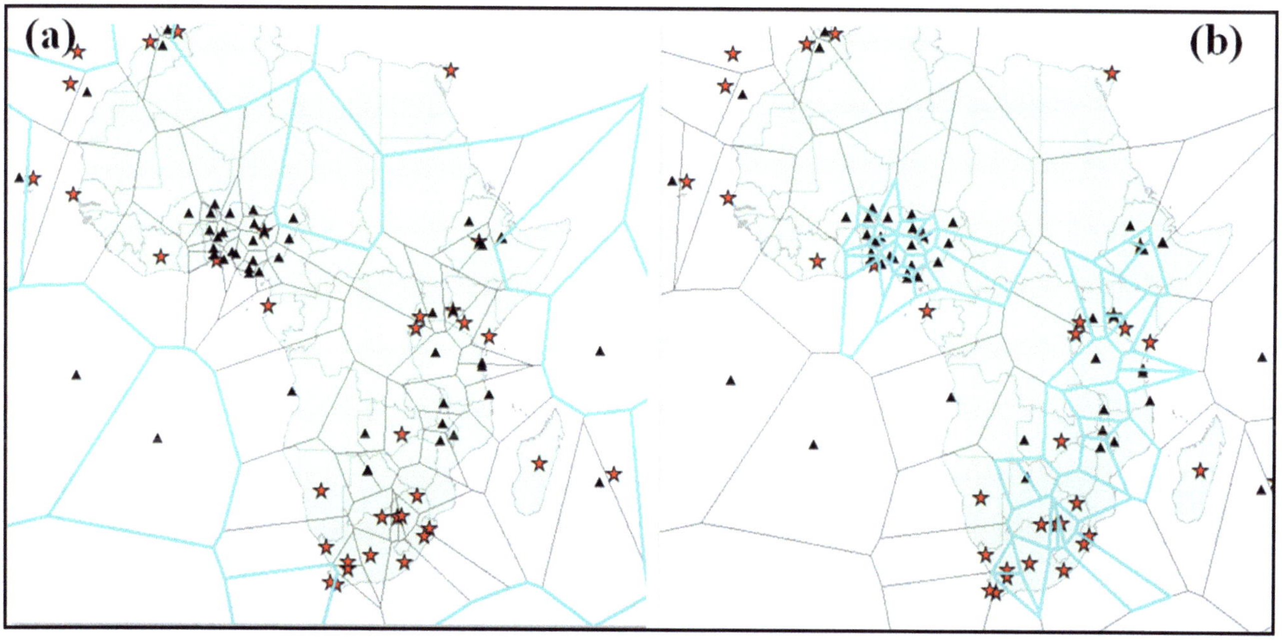

Fig. 6 (**a**) Outlier (**b**) Compliant Voronoi diagrams

the impact that new GNSS Stations would have in improving the network geometry.

In further studies, we will focus on the techniques to reduce the resultant polygon areas of Voronoi diagrams and lengths of the longer sides of the Voronoi diagrams. Additional variables, such as country selection, proximity to towns or cities and site suitability, need to be integrated in the analysis. The results of such a study would be identifying the least number of GNSS stations required to establish an optimal GNSS network in Africa.

Acknowledgements This research is being conducted with support from the Hartebeesthoek Radio Astronomy Observatory and the University of Pretoria. The ArcGIS 10.1 software package is used under the Tshwane University of Technology license.

References

Altamimi Z (2003) ITRF, GPS permanent stations and the AFREF project. In: 2nd FIG regional conference, Dec 2–5, 2003. FIG, Marrakech, Morocco

Altamimi Z, Collilieux X, Metivier L (2011) ITRF2008: an improved solution of the international terrestrial reference frame. J Geod 85:457–473

Angermann D, Klotz J, Reigber C (2000) Realization of a terrestrial reference frame for large-scale GPS networks. Int Assoc Geod Symp 121:304–309

Combrinck L (2008) Status of the African Geodetic Reference Frame (AFREF) project as at September 2008. Augmentation systems and Applications, 11–14 November 2008. Berlin

Combrinck L, Schmidt M (1998) Physical site specifications: geodetic site monumentation. In: Proceedings of the IGS network system workshop. Annapolis, USA

Dow J, Neilan R, Rizos C (2009) The International GNSS Service in a changing landscape of Global Navigation Satellite Systems. J Geod 83:191–198

Edelsbrunner H (2014) Voronoi and Delaununay diagrams. In: Edelsbrunner H (ed) A short course in computational geometry and topology. Springer, Klosterneuburg, pp 9–15

Edelsbrunner H, Seidel R (1986) Voronoi diagrams and arrangements. Discret Comput Geom 1:25–44

Farah H (2009) The African Reference Frame (AFREF) project: a fundamental geodetic tool for Africa. European Geosciences Union General Assembly, 19–24 April 2009. The Smithsonian/NASA Astrophysics Data Systems, Vienna

Fortune S (1992) Voronoi and delaunay triangulations. Comput Euclidean Geometry 193–233

IUGG (2013) International Union of Geodesy and Geophysics (IUGG) Yearbook 2013. IUGG

Klein R (2014) Voronoi diagrams an delaunay triangulations. Encycl Algorithm 1–5

Okabe A, Suzuki A (1997) Location optimization problems solved through Voronoi diagrams. Eur J Oper Res 98(3):445–456

Okabe A, Boots B, Kokichi C, Sung N (2000) Spatial tessellations: concepts and applications of Voronoi diagrams. Wiley, Chichester

Wonnacott R (2012) AFREF: concepts and progress. In: United Nations Regional Cartographic Conference for Asia and the Pacific. UNRC-CAP, Bangkok

Zhang F, Li J, Zhao H, Kang S, Li Z (2014) The generation algorithm for spherical full-feature Voronoi diagram. In: International conference on information science, electronic and electrical engineering (ISEEE). IEEE, Sapporo, pp 351–355

The Development of a Station Coordinate Estimation Program to Model Time Series from Continuous GPS Stations in New Zealand

C. Pearson, C. Crook, and P. Denys

Abstract

This paper describes models that Land Information New Zealand (LINZ) uses to model time series for the 38 stations in New Zealand's PositioNZ CORS network. A selection of the PositioNZ stations are chosen to act as reference stations for each job submitted to the PositioNZ-PP on-line GPS processing package. These models are used to estimate ITRF2008 coordinates at the epoch of observation for the reference stations. Because the estimated positions of the PositioNZ reference stations act as the control for the final coordinates that are provided to the users of the PositioNZ-PP service, they must be of sufficient accuracy.

New Zealand's tectonic situation is such that time dependent processes such as slow-slip events and post-seismic relaxation are included in our models. Slow-slip events with amplitudes of 3 cm or more are common for PositioNZ stations along the east coast of North Island and the northern end of South Island and have a significant effect on GPS time series in this region (Wallace and Beavan, J Geophys Res 115(B12), B12402, 2010). The model also includes the effect of post-seismic relaxation from the 2009 Dusky Sound earthquake, which has a significant effect on most stations in the South Island.

As a result, models of PositioNZ coordinate time series incorporate site velocity and include transient effects associated with post-seismic relaxation from large earthquakes, slow-slip events, the affect of annual seasonal coordinate variation, offsets caused by equipment changes or earthquakes and velocity changes. The post-seismic relaxation term is modelled using exponential decay function and slow-slip events are modelled as error functions.

The first realization of the station coordinate estimation model was developed by GNS Science (Beavan, GNS Science, Lower Hutt, 2008) and was recently updated to include the effects of the 2009 Dusky Sound earthquake (Beavan et al., Geophys J Int 183(3):1265–1286, 2010), 2010–2011 Christchurch earthquake sequence (Beavan et al., Bull N Z Soc Earthq Eng 43(4):228–235, 2012; Beavan et al., N Z J Geol Geophys 55(3):207–221, 2012)

C. Pearson (✉) • P. Denys
School of surveying, University of Otago, Dunedin, New Zealand
e-mail: chris.pearson@otago.ac.nz

C. Crook
Land Information New Zealand, Wellington, New Zealand

T. van Dam (ed.), *REFAG 2014*, International Association of Geodesy Symposia 146,
DOI 10.1007/1345_2015_177

and the 2013 Cook Strait earthquake sequence along with all slow-slip events that have occurred since 27 June 2008. The model needs to be continuously monitored and maintained to respond to new slow-slip events, as well as other influences such as GNSS equipment changes.

Keywords
Dynamic Datums • Geodetic time series modelling • Time dependent processes

1 Introduction

New Zealand is located on the Pacific-Australia plate boundary where up to 4–5 cm of relative plate motion is accommodated per year across the country. For a country of its size, New Zealand has a complex tectonic environment with two oppositely directed subduction zones separated by a zone of oblique convergence (see Fig. 1). The northern (Hikurangi) subduction zone is subject to frequent and slow-slip events (SSEs) causing a few centimetres displacement over periods ranging from a few weeks to a few years. As a consequence of this active tectonic deformation, New Zealand is subject to a high rate of seismic activity. Over the last 11 years New Zealand has been affected by ten earthquakes with magnitudes greater than Mw 6, the largest being the M_W (moment magnitude) 7.8 Dusky Sound earthquake of 15 July 2009. This last event has caused measurable post-seismic deformation over the southern half of South Island.

Land Information New Zealand (LINZ) uses a station coordinate estimation program to model time series for the 38 stations in the PositioNZ CORS network, which act as reference stations for the PositioNZ-PP, an on-line GPS processing package supported by LINZ (Pearson et al. 2014). Because the estimated positions act as control for the final coordinates calculated by PositioNZ-PP, they must be highly accurate. Due to the prominence of time dependent processes on the New Zealand plate boundary, the program includes models of post-seismic relaxation and SSEs. Once the model has been developed for a time series it only needs to be updated when some event affects the station environment, for example the occurrence of a large earthquake or the start of a new SSE.

The first realization of the model was developed by GNS Science (Beavan 2008) and was recently updated to include the 2009 Dusky Sound earthquake (Beavan et al. 2010), the 2010–2011 Christchurch earthquake sequence (Beavan et al. 2012a, b) and the 2013 Cook Strait earthquake sequence along with all slow-slip events since 27 June 2008.

2 Post-seismic Deformation

The 2009 M_W 7.8 Dusky Sound Earthquake (Fig. 2) has been followed by a large post-seismic signal. Measurable post-seismic relaxation extends nearly to Mt Cook, approximately half of the length of the South Island (Ross et al. 2013). Figure 2 shows the time series for selected continuous GNSS base stations in the South Island, south of latitude 44°S. Previous studies have shown that post-seismic deformation can be modelled using either exponential, power-law or logarithmic functions, each of which is associated with a different geophysical process. Ross et al. (2013) modelled the fit of the time series using all functions and concluded that, for most stations, the three functions cannot be distinguished. We found that an exponential provides an adequate fit to the time series for the purposes of estimating coordinates. Pearson et al. (2013) also used an exponential function to estimate coordinates for the Denali earthquake. Figure 2 shows the time series for all PositioNZ stations in the southern half of the South Island. While both the coseismic and post-seismic signals diminish with distance, as shown in Fig. 2, the relative importance of post-seismic relaxation to co-seismic offset increases with distance. This can be seen most clearly by comparing the north components of PYGR, which is the closest station and LEXA, which is the most distant.

3 Slow-Slip Events

Frequent SSEs occur in the eastern part of North Island and northern part of South Island, but are not observed north of HIKB or south of KAIK (Wallace and Beavan 2010; Wallace et al. 2012). Figure 3 shows the time series of selected PositioNZ stations from the northern half of New Zealand that are known to be affected by SSEs. The affect of SSEs is indicated by departures from the normal linear trend particularly in the east component. The SSEs change from relatively high frequency in GISB to a lower

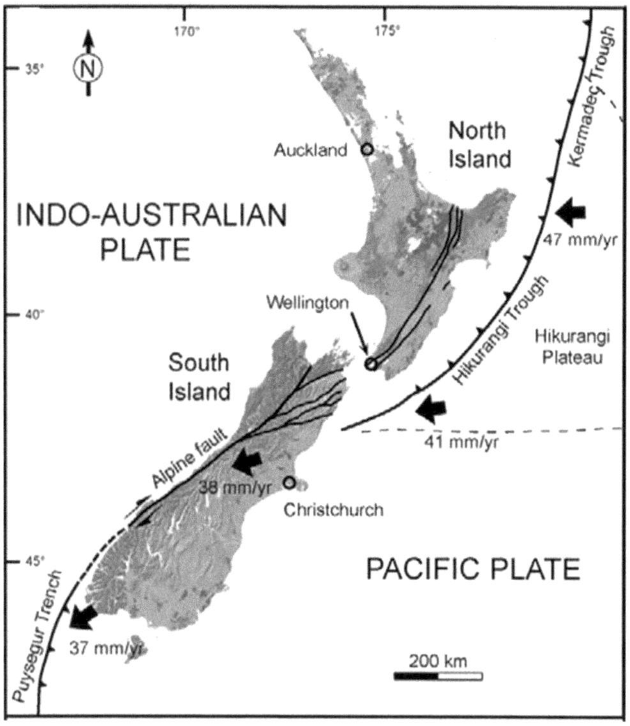

Fig. 1 Tectonic setting for New Zealand

frequency farther south. The maximum amplitude is roughly 30 mm in the east component of GISB and 20 mm for NSLN and DNVK. We parameterized the SSE as an error function, since this functional form is similar to the shapes of SSEs that are observed in New Zealand (Beavan 2008).

4 Station Predictive Model

Following Beavan (2008), we modelled the time series by fitting Eq. (1) to the east, north and up components of the time series. The equation contains the following eight terms:

1. constants, c_k, relating to the east, north, and up offsets from the nominal epoch 2000.0 latitude and longitude and height;
2. velocity, v_k, where the k index refers to the east, north, and up components;
3. seasonal (annual ($f_i = 1$) and semi-annual ($f_i = 2$)) cycles. A_{ki} and B_{ki} are the amplitudes of the in-phase and quadrature seasonal terms;
4. velocity changes at specified times. V_{kj} and t_j are the magnitudes and times of N_v velocity changes;
5. offsets at specified times of equipment changes. E_{ki} and t_{ki} are the magnitudes and times of N_C equipment offsets;
6. co-seismic offsets at specified times for relevant earthquakes. C_l and t_l are the magnitudes and times of N_E co-seismic offsets;
7. amplitudes of decaying exponential post-seismic signals starting at a specified times. P_m, t_m and K_m are the magnitudes, start times and inverse time constants of Np exponentially-decaying post-seismic signals;
8. slow-slip events (amplitude and duration) at specified times. The parameters are the duration and east, north, and up amplitudes of the error functions. The mid-point time is user defined and S_n, t_n and D_n are the magnitudes, centre times and inverse durations of N_s slow-slip events.

$$\begin{aligned} m_k(t) = c_k + v_k t + \sum_{i=1,2} \left[A_{ki}\cos\left(2\pi f_i t\right) + B_{ki}\sin\left(2\pi f_i t\right)\right] \\ + \sum_{j=1,Nv} V_{kj}\left(t - t_j\right) H\left(t - t_j\right) + \sum_{i=1,N_E} E_{ki} H\left(t - t_i\right) \\ + \sum_{l=1,N_C} C_{kl} H\left(t - t_l\right) \\ + \sum_{m=1,N_P} P_{km} H\left(t - t_m\right)\left(1 - e^{K_m(t - t_m)}\right) \\ + \sum_{n=1,N_S} 0.5 S_{kn} erf\left(D_n\left(t - t_n\right)\right) \end{aligned} \quad (1)$$

where m_k is the model of the east, north, or up time series, The k index refers to the east, north, and up components, t is the time in days from 2000.0, and $H(t - t_0)$ is the Heaviside step function. These parameters are used to interpolate or extrapolate coordinates for the PositioNZ stations.

5 Fit of GNSS Time Series

As a first step, the geocentric Cartesian coordinate estimates from the daily Bernese solutions of PositioNZ stations from 2000.0 onwards are converted to a (east, north, up) displacement time series by subtracting the nominal ITRF2008. In order to develop models for the Station Coordinate Predictive Model, non-linear least square techniques were used to model each time series using Eq. (1). The process requires manually defining the times of offset and other events, which are then refined by the optimization process. Typically, only a subset of the terms in Eq. (1) is required.

The parameters used depend on the tectonic environment of the station. All stations require the constants, the velocity, and the amplitudes of the in-phase and quadrature seasonal terms plus any offsets associated with equipment charges. For stations effected by earthquakes, we also determine the co-seismic offsets. Stations located south of Christchurch also require the amplitudes and inverse

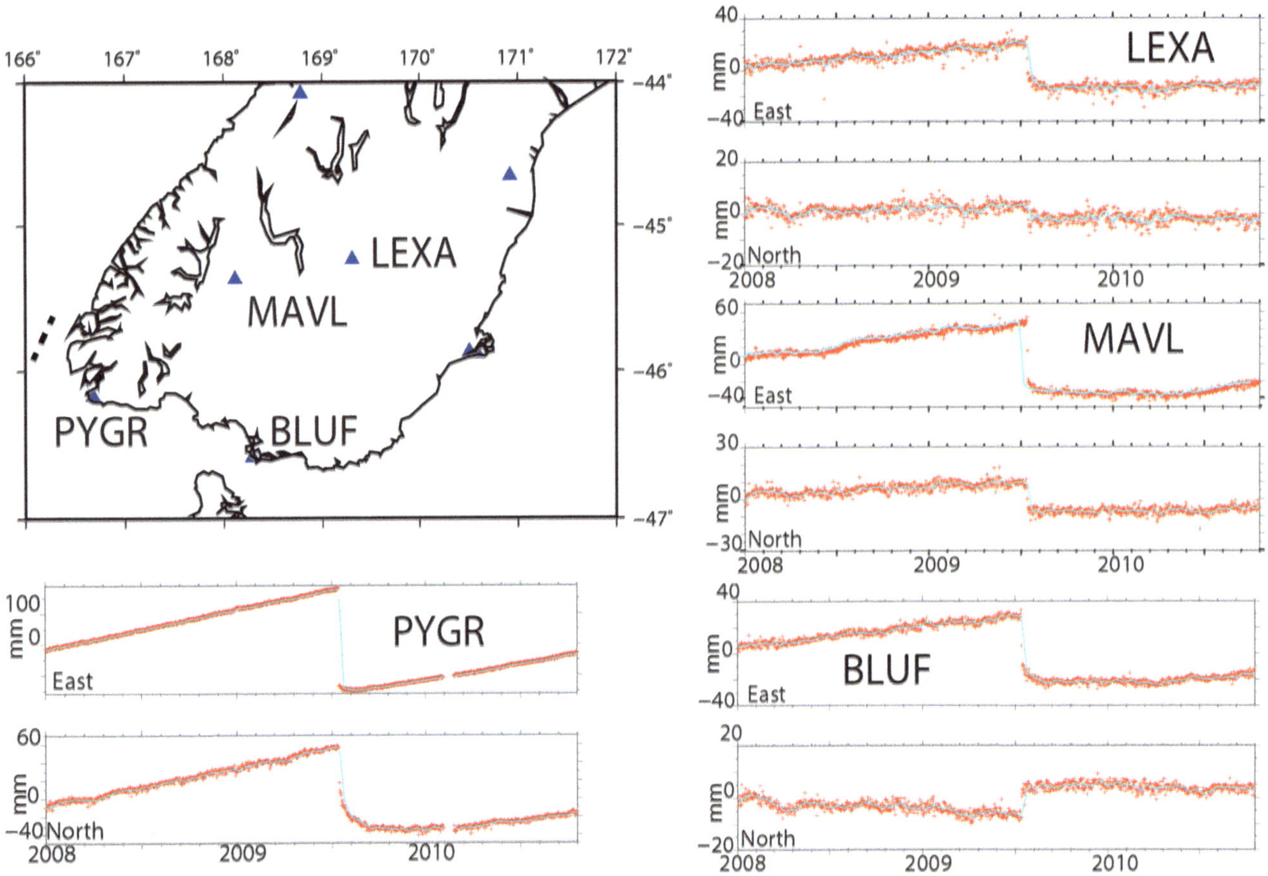

Fig. 2 E and N components of de-trended time series for four stations near the epicenter of the MW 7.8 2009 Dusky Sound earthquake. The time series show co-seismic offset and post-seismic relaxation for the earthquake. Location map shows locations of PositioNZ stations (*blue triangles*). *Dashed line* shows location of the Puysegur subduction zone. Note that Y axis for PYGR and MAVL is compressed compared to the other stations

time constants of exponentially-decaying post-seismic signals. The eastern North Island and northern South Island also require the duration and east, north, and up amplitudes of the error functions that model slow-slip events.

The median time series length was about 3,400 days with the longest being nearly 5,000 days and the shortest being a little over 1,000 days.

6 GNSS Data

The GNSS data for this study comes from the New Zealand PositioNZ CORS network of 38 stations. For all but one of these stations Zephyr Geodetic antennae were used providing a reasonably homogenous network. The GPS phase data from each session were processed in a network solution using the Bernese version 5.0 processing package (Dach et al. 2007), to determine daily estimates of coordinates and their covariance matrices. During each day's processing, IGS final orbits and associated polar motion files were held fixed, and a 7-parameter transformation was applied to fit the ITRF2008 coordinates of a set of IGS stations at the epoch of observation.

The models generally do an excellent job fitting the time series (Table 1). In all cases, the RMS residuals for the time series compared to the estimation s of the model is better than 2.5 mm for east and north and 6 mm for up (see Fig. 4). Generally the difference between the best and worst fitting model is relatively small with the range between the minimum and maximum RMS being about 1 mm for the two horizontal components and 1.8 for the vertical.

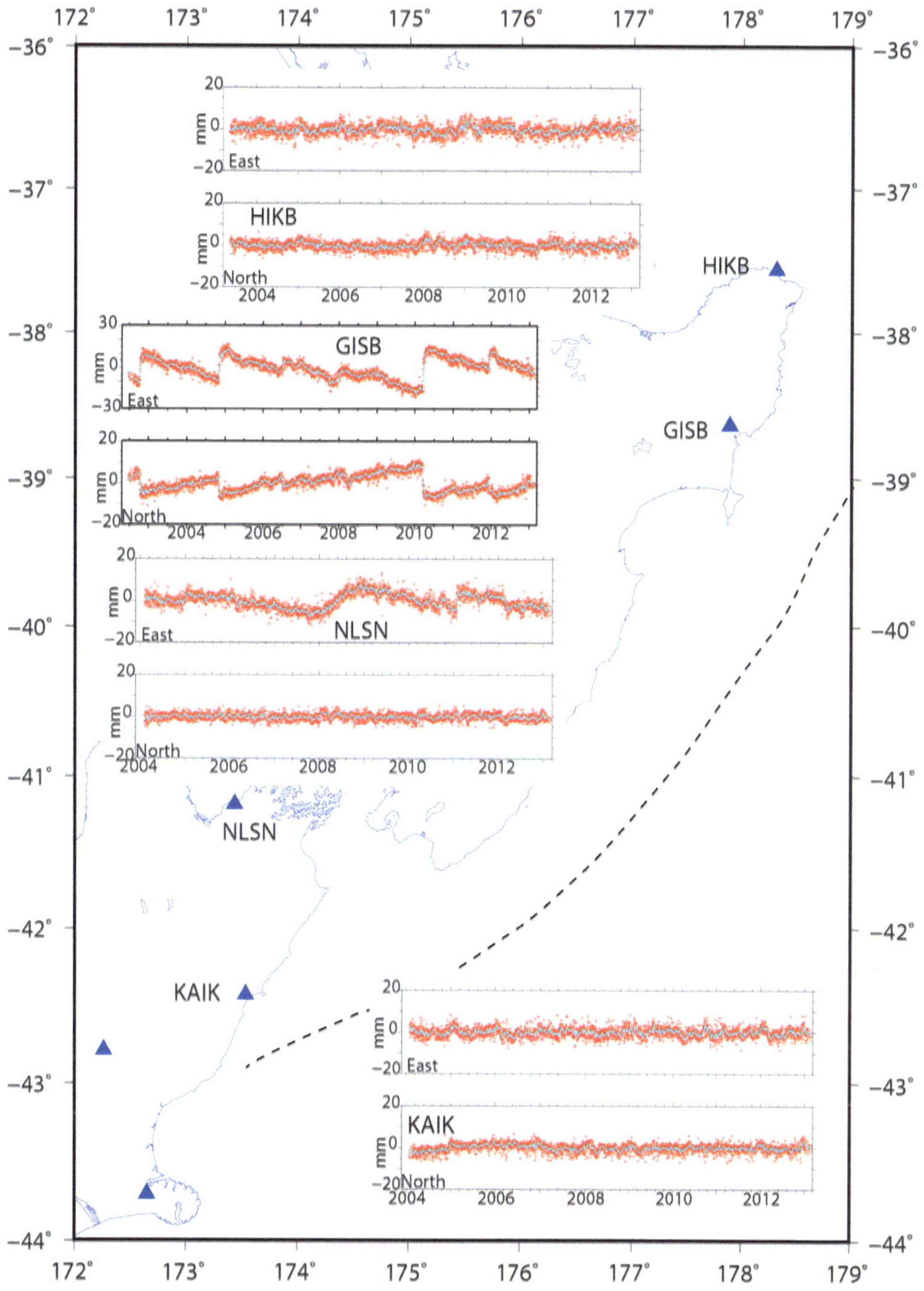

Fig. 3 E and N components of time series for four stations in Northern New Zealand. Two stations (GISB and NLSN) are from the area effected by SSEs while the other two are not. Time series are corrected for secular velocity and equipment – co-seismic offsets but not SSEs. SSEs are shown by deviations from horizontal trend. *Dashed line* shows location of the Hikurangi trench. Note that Y axis for GISB is compressed compared to the other stations

Table 1 Statistical summary for time series fits

	Length (days)	RMS E (mm)	RMS N (mm)	RMS U (mm)
Mean	1,735	1.72	1.36	4.55
Min	1,039	1.37	1.04	3.87
Max	4,923	2.35	2.1	5.69

We did not find any obvious correlation between stations containing complex time dependent processes like SSEs and post-seismic relaxation and high RMS. Indeed the stations GISB (which is the site most effected by SSEs) and PYGR (which has the largest post-seismic signal) are close to the centre of the rms ranges for each coordinate component.

7 Modelling Time Dependent Processes

In the New Zealand tectonic environment, models of SSEs and post-seismic relaxation must be incorporated into station coordinate estimation models in order to produce coordinates of sufficient accuracy for the PositioNZ-PP GPS processing application (Pearson et al. 2014).

Figure 5a, b shows examples of modelled time-series for the PositioNZ station most effected by SSEs. Figure 5a shows the modelled time series with SSEs terms included in the model. Figure 5b shows the modelled time series with no model for SSEs. Figure 5 also shows the residuals for the time-series. Excluding the SSEs causes an increase in the RMS residual to increase by a factor of two (or more) in the east component, which is the most affected. Figure 5b shows that not modelling the SSE for station GISB, which has the strongest SSE signal, can cause a 2 cm error in the estimated east coordinate. Not including the SSE model when modelling the time series also causes a significant change in the estimated station velocity (from −7.96 and 22.58 mm/year in the east and north component respectively for models with SSE incorporated to 1.13 and 18.81 mm/year without an SSE model), because the velocity term simply averages through the periodic SSE deformation rather than determining the velocity of the linear segments between the SSEs.

Figure 6a, b shows modelled time series for a PositioNZ station BLUF located about 150 km from the epicenter of the 2009 Dusky Sound Earthquake and within the region effected by post-seismic relaxation. The time series in Fig. 6a includes the exponential model for post-seismic relaxation while Fig. 6b shows the same stations with no model for post-seismic relaxation. In both cases ignoring post-seismic relaxation introduces significant errors into the positions for the time around the earthquake. It also introduces significant errors into the velocities because some of the unmodelled post-seismic deformation is incorporated into the constant velocity. Note that post-seismic relaxation is not just a problem for stations located in the epicenter region. For this earthquake, measurable post-seismic relaxation extends 500 km from the epicenter.

8 Discussion

Developing models of CORS time series in areas subject to time dependent processes like SSEs or post-seismic relaxation with sufficient accuracy (to act as control of adjustment of GNSS baselines) requires that these processes are explicitly modelled. As a result, the information contained in a SINEX file, which is used by OPUS (Soler et al. 2006) and ITRF2008 (Altamimi et al. 2011) and is restricted to the velocity and offsets, is not sufficient. In New Zealand we have developed a station predictive program to interpolate or extrapolate coordinates for use by the PositioNZ-PP online GPS processing algorithm.

SSEs or post-seismic relaxation provide quite different challenges to developing and maintaining models of GNSS time series. SSEs can occur at any time and often their onset is not clear till until the event is well established. Consequently, time series for areas subject to SSEs must be closely monitored to make sure that the model is sufficiently accurate for coordinate estimation purposes. The functional form for SSEs is not clear based on geophysical grounds, but we have found that simple error functions are usually sufficient to model the position time series.

In contrast, post-seismic relaxation only occurs after a major earthquake, so the potential onset of such a process is obvious. There is considerable discussion in the geophysical literature on the functional form that post-seismic relaxation follows with the two most common (logarithmic and exponential decay) models being associated with different tectonic processes. Experience both with the Dusky Sound earthquake in New Zealand and the Denali earthquake in Alaska (Pearson et al. 2013) show that either functional form will give a sufficiently accurate fit for the purposes of coordinate estimation.

One of the challenges in the New Zealand situation is maintaining these models, as including new offsets and seismic events requires manual intervention. The models serve two main purposes. One is to calculate reference

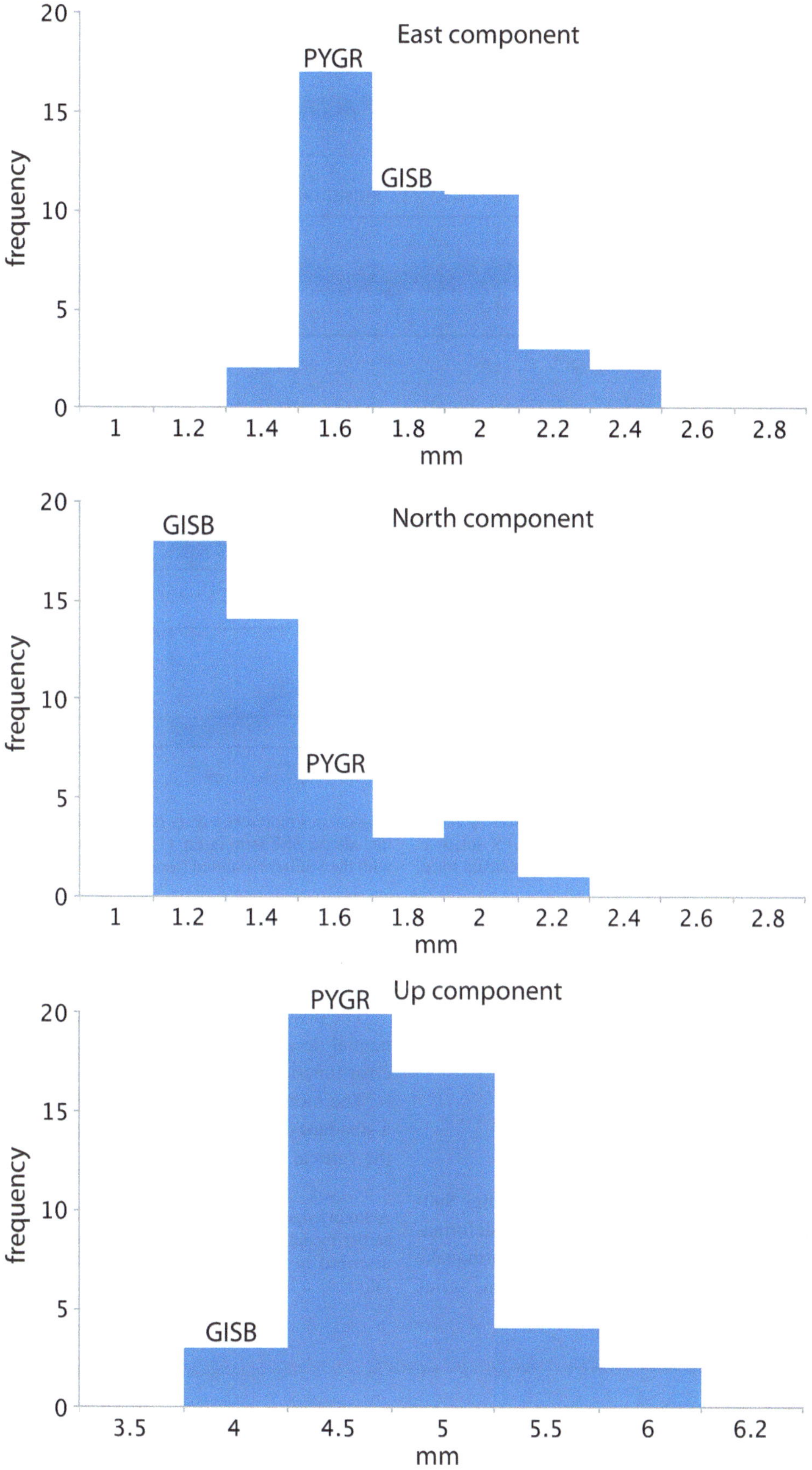

Fig. 4 Histograms of the RMS east, north and up residuals for the modelled time series at each CORS site. *Labels* indicate the bins containing the RMSE values for GISB and PYGR

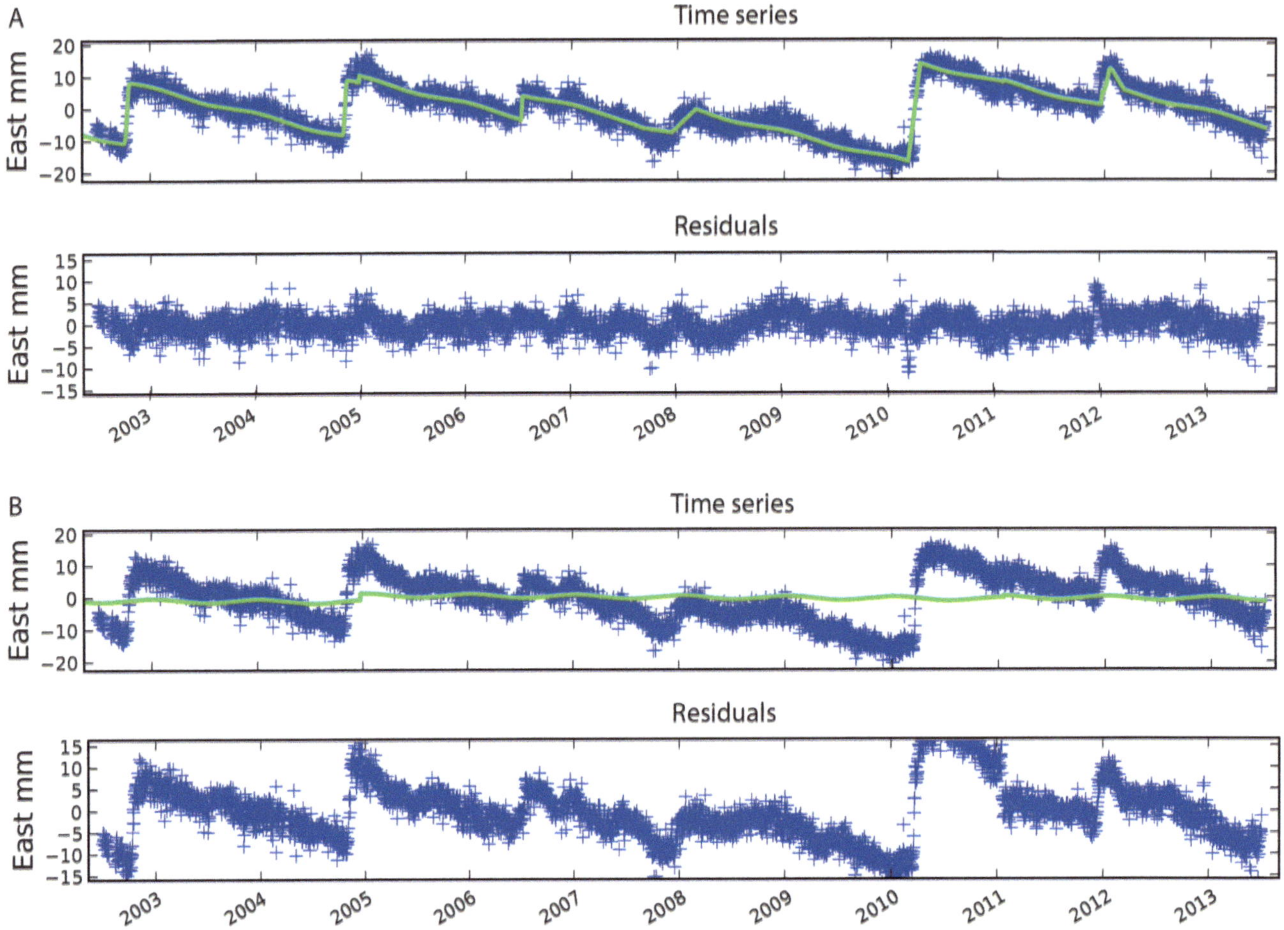

Fig. 5 Modelled time series for a station (GISB) affected by SSEs. In the time series trace *blue crosses* represent daily GPS solutions and the *green line* the Station Predictive Model. In the residual trace, *blue crosses* represent a daily residual estimate. (**a**) Shows models that include the SSE term in Eq. (1), while (**b**) shows the same time series with the SSE term omitted from the model

station coordinates for the PositioNZ-PP online GNSS post-processing service. The other is to catalogue and quantify the tectonic events affecting these stations. For the first of these requirements we are considering an alternative approaches such as using filtered versions of the time series, as these are much easier to maintain automatically.

9 Conclusions

The prevalence of slow-slip events on the Hikurangi Subduction Zone along with extensive post-seismic deformation associated with the M_W 7.8 Dusky Sound earthquake requires that models for the PositioNZ position time series incorporate terms for transient effects associated with these processes in addition to velocity, seasonal variation in the coordinates, equipment and co-seismic offsets.

The post-seismic relaxation term is modelled using exponential decay function and slow-slip events are modelled as error functions.

The models fit coordinates from the daily solutions with a standard deviation of ±2.5 mm in E and N and ±6 mm in the vertical.

Acknowledgements Much of the work described here is built on an earlier model developed by John Beavan of GNS Science. The work described in this paper was funded by a research grant from Land Information New Zealand to the University of Otago.

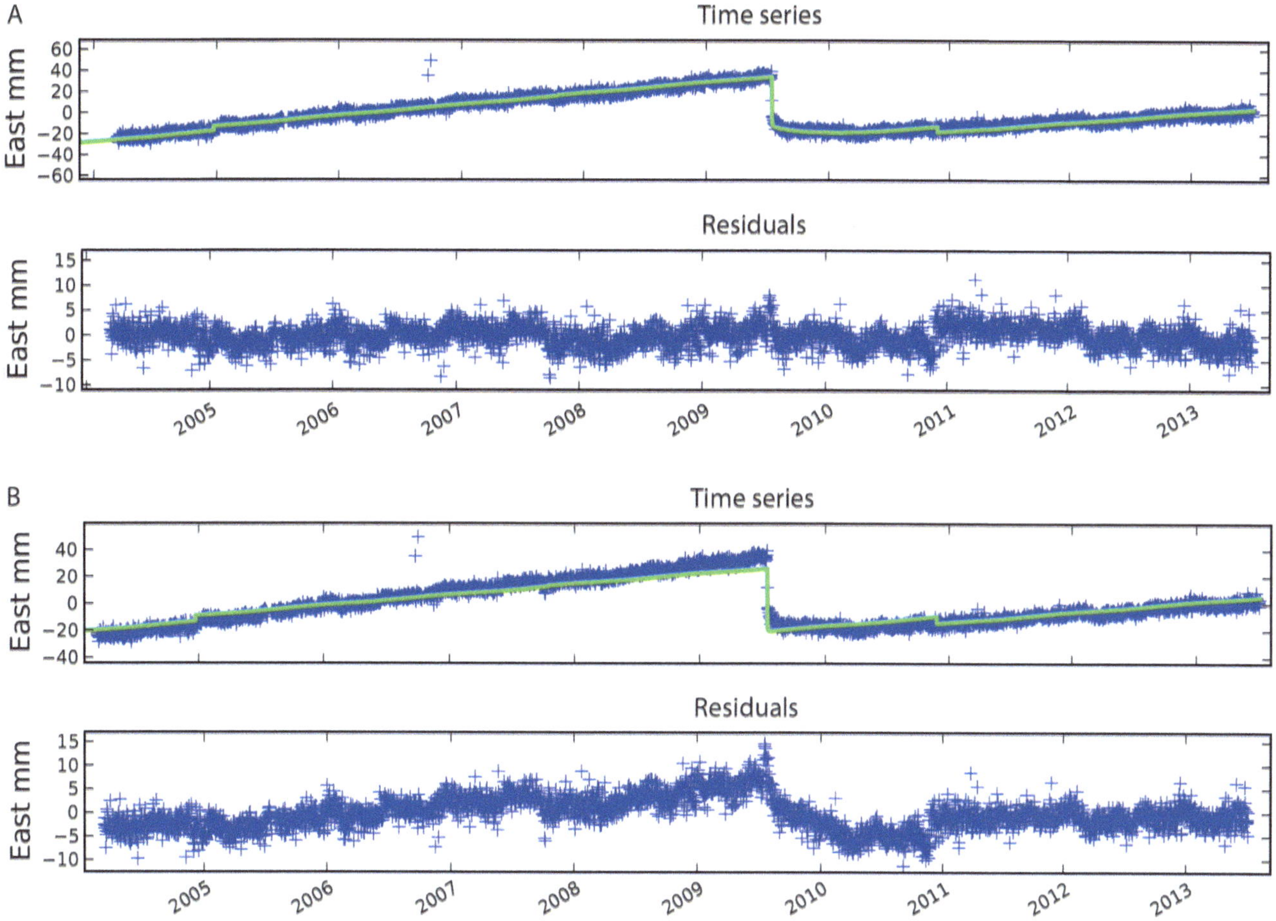

Fig. 6 Modelled time series for a station (BLUF) affected by post-seismic relaxation. In the time series trace *blue crosses* represent daily GPS solutions and the *green line* the Station Predictive Model. In the residual trace, *blue crosses* represent a daily residual estimate. (**a**) Shows models that include the post-seismic relaxation term in Eq. (1), while (**b**) shows models where the post-seismic relaxation term is neglected

References

Altamimi Z, Collilieux X, Metivier L (2011) ITRF2008: an improved solution of the international terrestrial reference frame. J Geod. doi:10.1007/s00190-011-0444-4

Beavan J (2008) Consultancy services for coordinates for PositioNZ onLine, Phase 2 (PONL-02). Consultancy Report for Land Information New Zealand. GNS Science, Lower Hutt

Beavan J, Samsonov S, Denys P, Sutherland R, Palmer N, Denham M (2010) Oblique slip on the Puysegur subduction interface in the 2009 July MW 7.8 Dusky Sound earthquake from GPS and InSAR observations: implications for the tectonics of southwestern New Zealand. Geophys J Int 183(3):1265–1286. doi:10.1111/j.1365-246X.2010.04798.x

Beavan J, Samsonov S, Motagh M, Wallace L, Ellis S, Palmer N (2012a) The Darfield (Canterbury) earthquake: geodetic observations and preliminary source model. Bull N Z Soc Earthq Eng 43(4):228–235

Beavan J, Motagh M, Fielding E, Donnelly N, Collett D (2012b) Fault slip models of the 2010–2011 Canterbury, New Zealand, earthquakes from geodetic data, and observations of post-seismic ground deformation. N Z J Geol Geophys 55(3):207–221

Dach R, Hugentobler U, Fridez P, Meindl M (eds) (2007) Bernese GPS software version 5.0. Astronomical Institute, University of Berne, Berne, Switzerland, January 2007

Pearson C, Freymueller J, Snay R (2013) Software to help surveying engineers deal with the coordinate changes due to crustal motion in Alaska. In: Zufelt JE (ed) Proceedings of the 10th international symposium on cold regions development (ISCORD): planning for sustainable cold regions [CD-ROM & Book]. American Society of Civil Engineers, Reston, VA, pp 297–307. doi:10.1061/9780784412978.031

Pearson C, Crook C, Jordan A, Denys P (2014) PositioNZ-PP an online GPS processing application for New Zealand. Int Assoc Geod Symp v 143. doi:10.1007/1345_2015_159

Ross L, Denys P, Williams CA, Pearson C, Faegh-Lashgary P, Wallace LM, Motagh M, Beavan J, Hamling IJ, Townend J (2013) Post-seismic deformation associated with recent New Zealand earthquakes. AGU Fall conference, San Francisco

Soler T, Michalak P, Weston ND, Snay RA, Foote R (2006) Accuracy of OPUS solutions for 1- to 4-h observing sessions. GPS Solut 10:45–55. doi:10.1007/s10291-005-0007-3

Wallace LM, Beavan J (2010) Diverse slow slip behavior at the Hikurangi subduction margin, New Zealand. J Geophys Res 115(B12), B12402. doi:10.1029/2010jb007717

Wallace LM, Beavan J, Bannister S, Williams C (2012) Simultaneous long-term and short-term slow slip events at the Hikurangi subduction margin, New Zealand: implications for processes that control slow slip event occurrence, duration, and migration. J Geophys Res 117(B11), B11402. doi:10.1029/2012jb

Results from the Regional AUSTRAL VLBI Sessions for Southern Hemisphere Reference Frames

Lucia Plank, James E.J. Lovell, Jamie McCallum, Elizaveta Rastorgueva-Foi, Stanislav S. Shabala, Johannes Böhm, David Mayer, Jing Sun, Oleg Titov, Stuart Weston, Sergei Gulyaev, Tim Natusch, and Jonathan Quick

Abstract

The AUSTRAL observing program is an initiative led by the Australian AuScope VLBI antennas in collaboration with radio telescopes in Warkworth, New Zealand, and Hartebeesthoek, South Africa. In 2014 the number of AUSTRAL sessions increased tremendously. Comparing recent results to the standard products achieved in global VLBI sessions regularly undertaken by the International VLBI Service for Geodesy and Astrometry (IVS), better accuracies in terms of baseline length repeatabilities are found for these regional AUSTRAL sessions. The network of (almost) identical small and fast telescopes as well as the technical equipment at all stations allows for new observing modes and improved operations, as such serving as a testbed for the future VLBI Global Observing System (VGOS). Special AUST-Astro sessions are used for dedicated astrometry of sparsely observed radio sources in the southern sky, as well as for detecting new radio sources for geodesy. In 2015, the AUSTRAL program will be further increased and final steps are now being undertaken for full VGOS compatibility of the three AuScope VLBI antennas. We present the latest results of the AUSTRAL sessions and give an overview of the multiple areas of research they support.

Keywords

AuScope • AUSTRAL • VGOS • VLBI

1 Introduction

Since 2011, more than 70 dedicated AUSTRAL VLBI sessions have been observed, providing valuable data for future realisations of the terrestrial and celestial reference frames. Observations are made with the AuScope VLBI array (Lovell et al. 2013) and antennas in Hartebeesthoek (South Africa) and Warkworth (New Zealand). The AUSTRALs are observed with a data rate of 1 Gbps, which is four times the data rate of the usual observations within the IVS (International VLBI Service for Geodesy and Astrometry; Schuh and Behrend 2012). This allows for significantly more observations per station. High cadence observing and the increased amount of data (up to 6 TB per station per day) demand for steady developments in data transfer logistics and remote operations. As such, the

L. Plank (✉) • J.E.J. Lovell • J. McCallum • E. Rastorgueva-Foi • S.S. Shabala
School of Physical Sciences, University of Tasmania, Private Bag 37, 7001 Hobart, TAS, Australia
e-mail: Lucia.Plank@utas.edu.au

J. Böhm • D. Mayer
Vienna University of Technology, Vienna, Austria

J. Sun
Shanghai Astronomical Observatory, Shanghai, China

O. Titov
Geoscience Australia, Canberra, ACT, Australia

S. Weston • S. Gulyaev • T. Natusch
Auckland University of Technology, Auckland, New Zealand

J. Quick
Hartebeesthoek Radio Astronomy Observatory, Hartebeesthoek, South Africa

T. van Dam (ed.), *REFAG 2014*, International Association of Geodesy Symposia 146,
DOI 10.1007/1345_2015_132

AUSTRAL observing program serves as a testbed for the future VLBI Global Observing System (VGOS; Petrachenko et al. 2009).

In this paper we give an overview of the AUSTRAL observing program so far and in the future, summarise the results, and show that in a certain aspect they are better than standard IVS products. Finally, plans for system upgrades to full VGOS compatibility for the AuScope antennas are presented.

2 The AUSTRAL Observing Program

Since commencement of full operations in 2011, the AuScope VLBI antennas in Hobart (Hb), Katherine (Ke), and Yarragadee (Yg) have been steadily increasing their participation in geodetic observations. In 2012, the array contributed to 68 IVS sessions, then 101 in 2013, and 178 in 2014. Besides supporting almost every IVS-R1/R4 session,[1] an independent VLBI observing program is performed within AuScope. In the AUSTRAL 24 h sessions the three AuScope telescopes observe together with the 12 m telescope in Warkworth (Ww), New Zealand, and the 15 m dish in Hartebeesthoek (Ht), South Africa. The 26 m telescopes at Hartebeesthoek (Hh) and Hobart (Ho) also participate on occasion. Recently, the rate of these AUSTRAL campaigns has been increased tremendously, including continuous AUST campaigns (AUST Cont) over 15 days each in December 2013 and September 2014, as well as 48 h weekend sessions additional to the traditional 24 h AUSTRAL experiments. For the year 2015, the AuScope antennas are scheduled for 241 observing days, supporting about 120 AUSTRAL sessions (including two 15 days continuous campaigns in February and June), 89 IVS R1/R4 sessions, and 25 other IVS experiments.

2.1 Pre-VGOS Scheduling

As a prototype for the VGOS system, the AUSTRAL antennas are small but fast (5°/s in azimuth, 1.5°/s in elevation) antennas. The reduced sensitivity due to the small dish size is compensated by an increased recording rate of 1 Gbps, which is four times more than the 256 Mbps standardly used in IVS sessions. In general, the sensitivity of the antennas (usually measured as the system equivalent flux

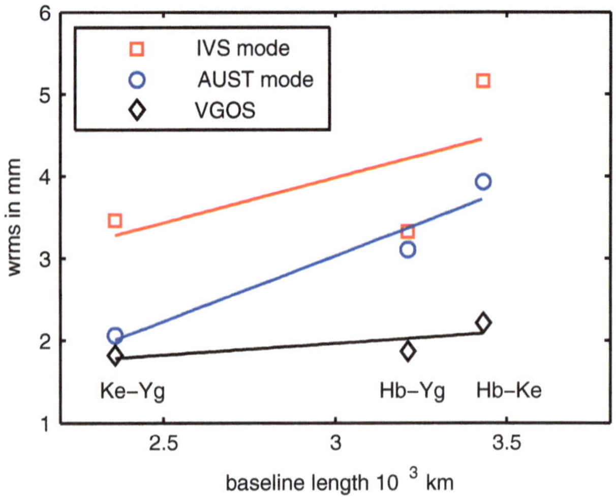

Fig. 1 Simulated wrms in baseline lengths for the AuScope baselines, when three different observing modes were used in the scheduling. The specific parameters of each mode are those of Table 1

Table 1 The three observing modes and their parameters (measurement noise σ_τ, antenna sensitivity in SEFD and the assumed data rate) used in the scheduling and simulation

Mode	σ_τ	SEFD	Data rate
IVS mode	50 ps	4000	256 Mbps
AUST mode	50 ps	4000	1 Gbps
VGOS	4 ps	2500	16 Gbps

density – SEFD), together with the strength of an observed radio source determine the duration of an observation, that is the integration time necessary to collect enough signal from the observed source. On the other hand, an increased data rate means more information and hence shortens the necessary observation time. Petrachenko et al. (2009) showed that the overall accuracy of geodetic products can be improved by using small telescopes and an increased recording rate, allowing for much more individual observations than the systems today. This is the concept of the future VGOS system. In Fig. 1 we show simulations comparing three different observing modes, the IVS mode, the mode of the AUSTRALs and the future VGOS mode. Using the Vienna VLBI Software (VieVS; Böhm et al. 2012), we first generated the schedules for these three observing modes, simulated them following the procedure described by Pany et al. (2010) and analysed the simulated observations. The parameters of the schedule and the simulation are given in Table 1. Due to the cooled receivers of the VGOS system, we expect an improved antenna sensitivity (SEFD = 2500 versus 4000) as well as lower measurement noise (σ_τ = 4 ps versus 50 ps) than in the present system. We find that the 1 Gbps AUSTRAL mode lets us expect much better results than the traditional IVS mode. Also, a further positive effect

[1]Rapid (R-) sessions are performed twice a week by the IVS. Using a fairly stable network of about ten radio telescopes, these R1 and R4 sessions provide the standard geodetic VLBI measurements of the Earth orientation parameters and data for the terrestrial and celestial reference frames.

of the planned system upgrade to the VGOS system is clearly visible.

2.2 Improved Scheduling

The scheduling of the AUSTRAL sessions is done at the Vienna University of Technology using the scheduling module of the VieVS software (Sun et al. 2014). Scheduling of a five station network with only limited common visibility between the stations on the eastern and western edges (Ht, Ww) is not trivial and subject to constant optimisation efforts (Mayer et al. 2015). In the beginning (before AUST30 on July 03, 2014), we had about 10–20 scans per hour and scan lengths of 100–200 s on average. A change of the a priori source list, from a few selected strong sources to the complete list as commonly used, as well as other minor improvements in the scheduling algorithm (see Mayer et al. 2015) led to a maximum of 30 scans per hour per station and average scan lengths of 50 s from AUST30 onwards. This is one reason to expect the results of the AUSTRALs to further improve, once more sessions were observed and analysed (cf. Sect. 3.1 and Fig. 3). A the moment we use the standard source list for IVS observations[2] and then set a limit of 0.5 Jansky for the minimal strength of sources that should be observed in the AUSTRAL sessions.

3 Results

We present results in terms of baseline length residuals, which are the estimated differences between the a priori (positions and velocities were taken from the VieTRF13; Krásná et al. 2014) and the observed baseline length per 24 h session. Hereby, changes in baseline lengths due to plate tectonics and geophysically induced station motions are accounted for through adequate models (Petit et al. 2010). The analysis was done using the VieVS software, estimating one set of station coordinates, source coordinates and Earth orientation parameters per session. The datum for the terrestrial reference frame was set by a no-net-rotation and no-net translation condition on all stations, the one for the celestial frame by a no-net-rotation condition on the defining sources of the International Celestial Reference Frame (Ma et al. 2009). Tropospheric and clock parameters were estimated every 60 min. The observations were weighted by the inverse of their formal uncertainties.

In Fig. 2 the residuals with their error bars are shown for 81 AUSTRAL sessions, including AUST Cont13 (November 28–December 15, 2013) and AUST Cont14 (September 4–21, 2014). For comparison, also results of the IVS R1/R4 sessions from 2011 to the end of 2014 as well as from the Cont14 campaign (May 6–20, 2014) are shown. The latter was a 15 day continuous VLBI campaign in May 2014 observing with a global network of 17 stations, including all AUSTRAL antennas as well as the 26 m legacy antenna in Hobart (Ho). Due to the large and global network and hence high number of observations, but also due to the fact that in Cont14 a recording rate of 512 Mbps was used, Cont14 results are expected to be superior to the standard IVS results. For the subsequent statistics, we removed results where the estimated offset to the a priori baseline length was more than ±5 cm or where the estimated error was larger than 3 cm.

Overall we find that the residuals of all baselines and session types are in good agreement. However, the dense time series also reveals some systematics for certain baselines, whose origins have not yet been identified. There is for example a long term periodic signal in the Hb-Ke baseline and a slight trend in the Ke-Yg baseline for the latest sessions. Also, all baselines with Ht show large offsets, even within a short period of continuous observations. This needs further investigation.

In Table 2 the repeatabilities in terms of baseline length wrms are shown for the three types of sessions, the AUSTRALs, the IVS R1/R4 sessions, and for Cont14. We find that the AUSTRALs give better baseline length repeatabilities than the standard IVS R1/R4 sessions. An exception is the Ht-Ww baseline, which was only observed in three IVS rapid sessions. On the other hand, the results of Cont14 are superior. One reason for this could be the fact that the AUSTRAL sessions generally show larger residuals than the IVS sessions. Without having identified the causes yet, while IVS sessions typically show a wrms of the post-fit residuals (session fit) of about 30 ps, we find about 50 ps for the AUSTRALs.

3.1 Steady Improvements

In the small network of the AUSTRAL sessions, problems at one station can have a significant effect on the performance of the whole session. When for example the data of one station is lost, this can drop the overall percentage of successfully observed scans down to 50%. Within the AUSTRAL campaigns, we typically have to deal with short-notice stations fall-outs, wind stows, problems with the back ends and the recording, or even disk module failures. The high cadence observing program, as such unique in the IVS community, also demands for changes to the usual work flow and organisation at each station. In order to minimize the number of casual operational failures, in AuScope we are working on more automated procedures. For example, all three AuScope telescopes are remotely operated from the AuScope observing room at the University campus in

[2] Available at ftp://gemini.gsfc.nasa.gov/pub/sked/catalogs.

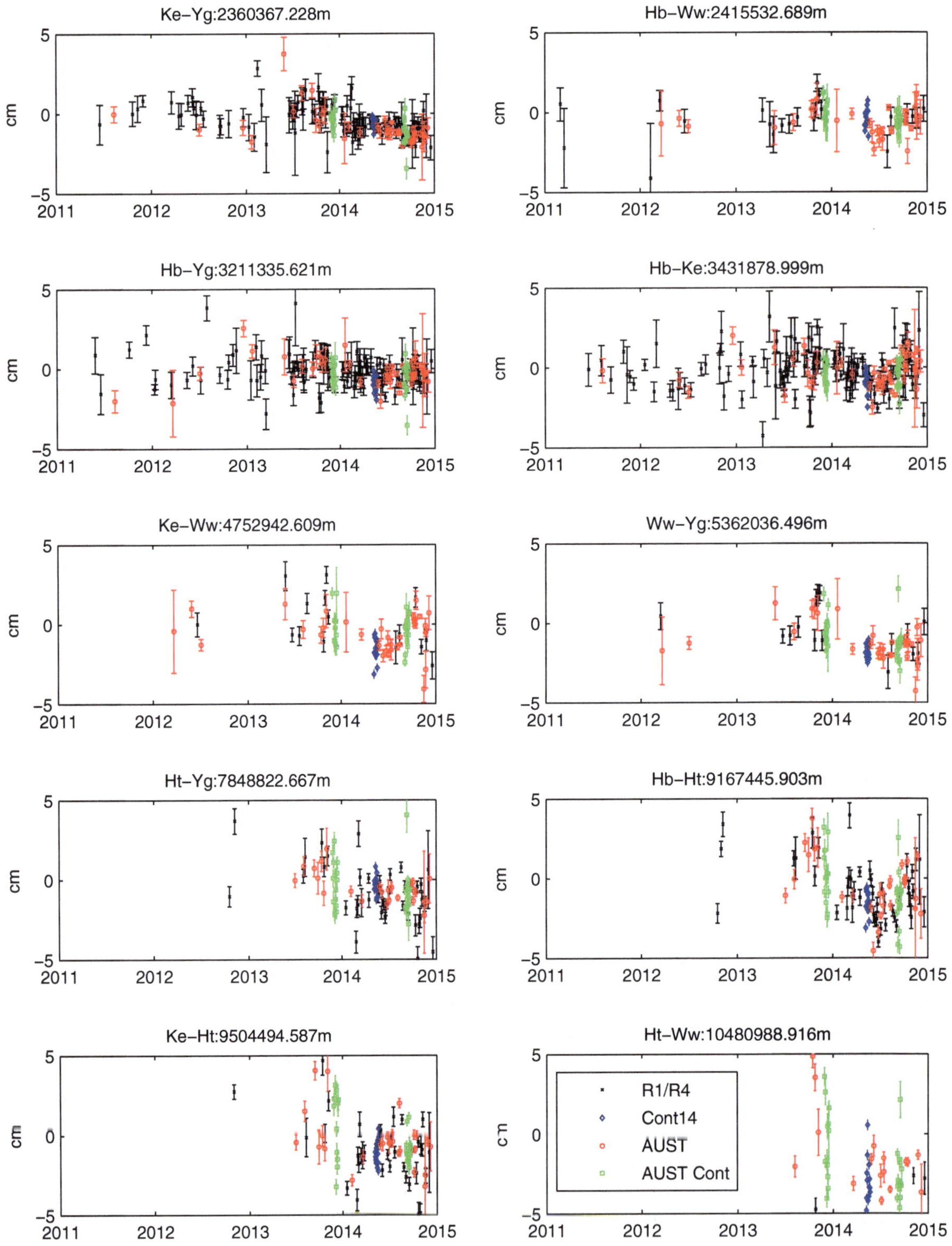

Fig. 2 Baseline length residuals (observed minus a priori) and error bars of the AUSTRAL baselines. AUSTRAL sessions are shown in *red*, the AUSTCont13 and 14 campaigns in *green*, results of the IVS R1/R4 sessions in *black* and Cont14 in *blue*

Table 2 Baseline length repeatabilities (wrms) as determined in the AUSTRALs, the R1/R4 sessions from 2011 to 2014, and Cont14

Baseline	(Length)	AUST	(#)	R1/R4	(#)	Cont14	(#)
Hb-Ho	(296 m)	–		9.5 mm	(31)	1.9 mm	(15)
Ke-Yg	(2360 km)	7.0 mm	(74)	8.5 mm	(107)	3.1 mm	(15)
Hb-Ww	(2416 km)	6.9 mm	(60)	9.4 mm	(22)	5.8 mm	(15)
Hb-Yg	(3211 km)	7.1 mm	(77)	8.5 mm	(113)	4.8 mm	(15)
Hb-Ke	(3432 km)	8.8 mm	(78)	11.2 mm	(121)	6.6 mm	(15)
Ke-Ww	(4753 km)	9.5 mm	(58)	15.7 mm	(15)	3.7 mm	(15)
Ww-Yg	(5362 km)	8.7 mm	(58)	13.8 mm	(15)	4.3 mm	(15)
Ht-Yg	(7849 km)	9.9 mm	(57)	15.2 mm	(40)	5.5 mm	(15)
Hb-Ht	(9167 km)	15.9 mm	(58)	17.2 mm	(44)	6.9 mm	(15)
Ht-Ke	(9504 km)	14.4 mm	(55)	18.5 mm	(34)	7.2 mm	(15)
Ht-Ww	(10,481 km)	19.0 mm	(43)	9.6 mm	(3)	13.2 mm	(13)

Results are shown for the baselines of the AUSTRALs, including the antennas in Hobart (Hb – 12 m, Ho – 26 m), Katherine (Ke), Yarragadee (Yg), Hartebeesthoek (Ht – 15 m), and Warkworth (Ww). In brackets, the number of sessions that a baseline was observed is given

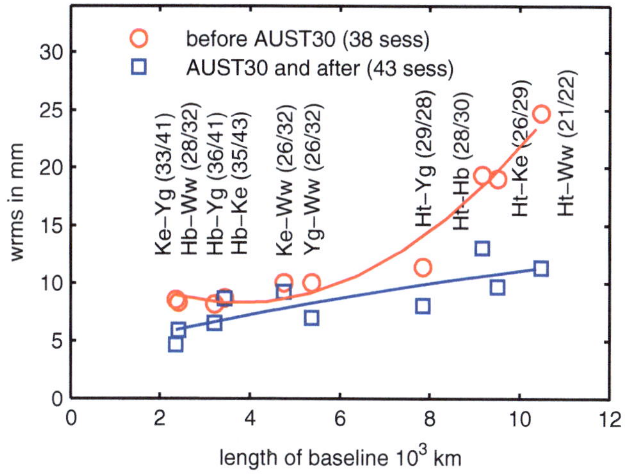

Fig. 3 Baseline length repeatabilities (wrms) of the AUSTRAL sessions before AUST30 (*red*) and after AUST30 (*blue*). In brackets, the number of used sessions for each baseline (before AUST30/AUST 30 and after) is given. For better visibility, we fitted quadratic trends to all baselines of the two solutions

Hobart. In Fig. 3 AUSTRAL results in terms of baseline length repeatabilities are shown for the initial AUSTRAL sessions (before AUST30 on July 3 2014) and the more recent ones. We find a clear improvement after AUST30, particularly for the longer baselines. We think that this is predominantly due to the improved scheduling mentioned in Sect. 2.2, but also a result of better observing and improved operations.

3.2 Astrometry

Within the AUSTRAL program, special astrometry sessions are performed once per month. The goal is astrometry of radio sources of the ICRF2 (Ma et al. 2009) with limited

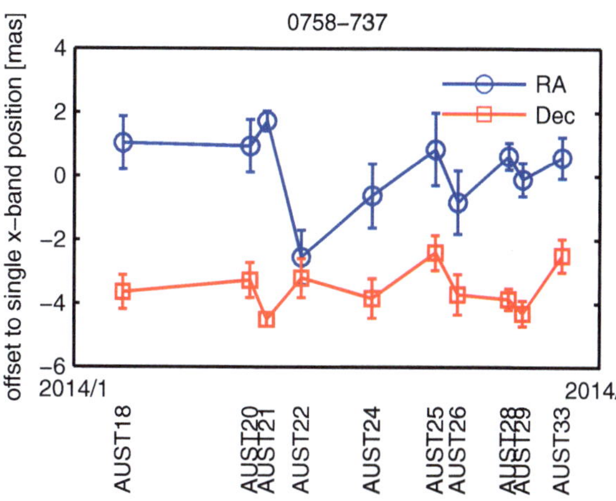

Fig. 4 Estimated source position offsets in right ascension (RA) and declination (Dec), as observed with the AUSTRAL S/X band observations. The a priori position was determined of single X band observations (Petrov et al. 2009)

number of observations. For the AUST-Astro sessions, the lower flux limit is at about 100 mJy. As a result of the effort for improved homogeneity for the defining sources of the ICRF2, some southern defining sources have a limited number of observations as in the ICRF2 official document (Ma et al. 2009) and therefore with the AUSTRAL observations we wish to improve their statistics. We also look for new sources, which are not usually observed with geodetic VLBI and whose positions are known from single X band observations (Petrov et al. 2009). Having the results of the first AUST-Astro experiments, we can report successful detections of three new sources (0743-673, 0758-737, 1511-360). In Fig. 4 the estimated source position for 0758-737 is shown, as observed in ten AUSTRAL sessions in 2014. We find a clear offset in declination of −3 mas between the reported X band position (RA $= 7^h 57^m 14^s.077$, Dec $= -73° 53' 9''.369$) and our S/X measurements. As soon as more observations are available to other new sources, it can be investigated whether this offset is a common feature due to the switching from X to S/X band. This will be topic of future investigations. Besides that, several AUSTRAL experiments are used for intense observations to sources experiencing special constellations, for example, close appearances of the Sun or other large gravitating bodies in the line of sight to a source. These observations will then be used for studying effects of General relativity.

4 Summary and Outlook

The AUSTRAL observing program greatly improves the number of observations of both southern baselines and southern radio sources. Geodetic results in terms of wrms in baseline lengths are consistent with (and slightly better than) standard IVS products for the observed baselines. Improved scheduling and continuation of the high cadence observing program are expected to yield further improved results, with 241 sessions including the AUSTRAL radio telescopes to be analysed and observed in 2015.

Recently approved funding from the Australian Research Council is expected to enable the upgrade of all three AuScope sites with a broadband receiver system and 16 Gbps sampler/recording system by the end of 2015.

The increased exposure of the AUSTRAL sessions to observing failures at a single station is a great motivator for us to further improve the reliability of our observations. Under the header of *Dynamic observing* we perform research in the areas of short-term and adaptive scheduling as well as real time quality control during observations e.g. through real time correlation of certain scans. Other areas of research that will be supported by dedicated observations within the AUSTRAL experiments are the operation of sibling telescopes, i.e. the co-location of a small VGOS-style dish and a legacy antenna, in Hobart and Hartebeesthoek as well as studying the effects of source structure in classical and VGOS geodetic VLBI observations.

Acknowledgements The AuScope Initiative is funded under the National Collaborative Research Infrastructure Strategy (NCRIS), an Australian Commonwealth Government Programme. LP, JM, and SS would like to thank the Australian Research Council for Fellowships FS1000100037, FS110200045, and DE130101399. We also acknowledge the International VLBI Service for Geodesy and Astrometry (IVS) for managing and providing VLBI data. The excellent suggestions by three reviewers were highly appreciated and helped to improve the paper.

References

Böhm J, Böhm S, Nilsson T, Pany A, Plank L, Spicakova H, Teke K, Schuh H (2012) The new Vienna VLBI software VieVS. In: Proceedings of the 2009 IAG symposium, Buenos Aires, Argentina, International Association of Geodesy Symposia, vol 136, 31 August–4 September 2009

Krásná H, Böhm J, Plank L, Nilsson T, Schuh H (2014) Atmospheric effects on VLBI-derived terrestrial and celestial reference frames. In: Rizos C, Willis P (eds) Earth in the edge: science for a sustainable planet. IAG Symposia, vol 139, pp 203–207. Springer, Berlin

Lovell J, McCallum J, Reid P, McCulloch P, Baynes B, Dickey J, Shabala S, Watson C, Titov O, Ruddick R, Twilley R, Reynolds C, Tingay S, Shield P, Adada R, Ellingsen S, Morgan J, Bignall H (2013) The AuScope geodetic VLBI array. J Geodesy 87:527–538

Ma C et al. (2009) The second realization of the international celestial reference frame by very long baseline interferometry. In: IERS Technical Note 35. Frankfurt am Main: Verlag des Bundesamts für Kartographie und Geodäsie. Presented on behalf of the IERS / IVS Working Group

Mayer D, Böhm J, Lovell J, Plank L, Sun J, Titov O (2015) Scheduling strategies for the AuScope VLBI network. In: VGI - Österreichische Zeitschrift für Vermessung und Geoinformation. Schwerpunktheft zur IUGG 2015. www.ovg.at

Pany A, Böhm J, MacMillan DS, Schuh H, Nilsson T, Wresnik J (2010) Monte Carlo simulations of the impact of troposphere, clock and measurement errors on the repeatability of VLBI positions. J Geodesy 85(1):39–50

Petit G, Luzum B (eds) (2010) IERS conventions 2010. Frankfurt am Main: Verlag des Bundesamtes für Kartographie und Geodäsie. IERS Technical Note No. 36

Petrachenko B et al. (2009) Progress report of the IVS VLBI2010 committee: design aspects of the VLBI2010 system. NASA/TM-2009-214180 ftp://ivscc.gsfc.nasa.gov/pub/misc/V2C/TM-2009-214180.pdf

Petrov L, Phillips C, Bertarini A, Deller A, Pogrebenko S, Mujunen A (2009) Use of the long baseline array in Australia for precise geodesy and absolute astrometry. Publ Astron Soc Aust 26:75–84

Schuh H, Behrend D (2012) VLBI: a fascinating technique for geodesy and astrometry. J Geodyn 61:68–80

Sun J, Böhm J, Nilsson T, Krásná H, Böhm S, Schuh H (2014) New VLBI2010 scheduling strategies and implications on the terrestrial reference frames. J Geodesy 88:449–461

Implementation of the ETRS89 in Europe: Current Status and Challenges

Carine Bruyninx, Zuheir Altamimi, Elmar Brockmann, Alessandro Caporali, Rolf Dach, Jan Dousa, Rui Fernandes, Michail Gianniou, Heinz Habrich, Johannes Ihde, Lotti Jivall, Ambrus Kenyeres, Martin Lidberg, Rosa Pacione, Markku Poutanen, Karolina Szafranek, Wolfgang Söhne, Günter Stangl, João Torres, and Christof Völksen

Abstract

The EUREF (Reference Frame Sub-Commission for Europe) Permanent GNSS Network (EPN) serves as the backbone for the realization of, and access to, the European Terrestrial Reference System (ETRS89). The cumulative site positions and velocities for the EPN stations are used for national ETRS89 densifications and geo-information applications. EUREF has developed specific guidelines through which European countries ask validation of their national ETRS89 densification campaigns. Today, the majority of the European countries has passed this process and a large part of European National Mapping and Cartographic Agencies have officially adopted ETRS89. In addition, ETRS89 plays a fundamental role in INSPIRE (Infrastructure for Spatial Information in the European Community).

In the meantime, some geographical zones are experiencing considerable deformation, degrading the lifetime of their national ETRS89 realizations. In order to derive a cross-boundary European velocity model that can be applied by different countries on top of their validated ETRS89 realization (while keeping consistency with their neighboring countries) the EUREF Technical Working Group has set up two initiatives: the EPN densification project and the EUREF Working Group on Deformation Models.

Keywords

Deformations • ETRS89 • EUREF • GNSS

C. Bruyninx (✉)
Royal Observatory of Belgium, Av. Circulaire 3, B-1180 Brussels, Belgium
e-mail: C.Bruyninx@oma.be

Z. Altamimi
IGN/LAREG, Université Paris Diderot, 35 rue Hélène Brion, 75205 Paris Cedex 13, France

E. Brockmann
Bundesamt für Landestopografie swisstopo, Seftigenstrasse 264, 3084 Wabern, Switzerland

A. Caporali
University of Padua, via Gradenigo 6, 35131 Padova, Italy

R. Dach
University of Bern, Astronomical Insitute, Sidlerstrasse 5, CH-3012 Bern, Switzerland

J. Dousa
Geodetic Observatory Pecny, RIGTC/NTIS, Ustecka 98, Zdiby 250 66, Czech Republic

R. Fernandes
University of Beira Interior, Instituto D. Luiz, R. Marquês d'Ávila e Bolama, 6201-001 Covilhã, Portugal

M. Gianniou
National Cadastre and Mapping Agency S.A, Mesogion Ave. 288, 15562 Athens, Greece

H. Habrich • J. Ihde • W. Söhne
Bundesamt für Kartographie und Geodäsie (BKG), Richard-Strauss-Allee 11, 60598 Frankfurt am Main, Germany

L. Jivall • M. Lidberg
Lantmäteriet, Lantmäterigatan 2C, SE-80182 Gävle, Sweden

T. van Dam (ed.), *REFAG 2014*, International Association of Geodesy Symposia 146,
DOI 10.1007/1345_2015_130

1 Background

The IAG (International Association of Geodesy) sub-commission 1.3a Europe, EUREF, is a joint effort of European research agencies and National Mapping and Cartographic Agencies (NMCA) with the goal to define, realize, maintain and promote the adoption of the European Reference System (ETRS89) and the European Vertical Reference System (EVRS). The motivations for creating the ETRS89 were threefold: scientific (for geodesy and geophysics), technical (as a reference for NMCAs) and practical (for positioning and navigation). Consequently, EUREF has to reconcile the needs of the scientists, requiring the most precise coordinates, updated each time when a new improved computation is available, with the needs of the NMCAs, which also require precise coordinates but request minimal updates in order to not affect national ETRS89 realizations. For that purpose the members of the EUREF Technical Working Group (TWG, which acts as the EUREF steering committee) are representing both NMCAs (eight members in 2014) and research agencies (seven members).

Following a resolution adopted in Firenze (Italy) in 1990, EUREF defined the European Terrestrial Reference System 1989 (ETRS89) as coincident with ITRS (International Terrestrial Reference System) at the epoch 1989.0 and fixed to the stable part of the Eurasian Plate. The motivation and direct consequence of such a definition is that coordinates of stations located in the stable part of Europe have minimal time dependency. It follows that the relationship between the two systems is entirely defined by a similarity transformation formula (Boucher and Altamimi 1992). Therefore for any new release of the International Terrestrial Reference Frame (ITRF), say ITRFyy, a corresponding ETRFyy would be defined, through a determined set of 14-parameter transformation formula (Altamimi and Boucher 2002; Altamimi 2009). After the release of ITRF2005, and noticing frame shifts between ETRF2005 and ETRF2000 (inherited from the corresponding ITRF solutions), the TWG recommended the usage of ETRF2000 which was kept as the standard ETRS89 frame. This recommendation has the advantage to homogenize and ensure compatibility of future and/or updated realizations of the ETRS89 throughout Europe. The transformation from ITRFyy to ETRF2000 contains the transformation from ITRFyy to ITRF2000 (see http://itrf.ensg.ign.fr/trans_para.php), followed by the transformation from ITRF2000 to ETRF2000 (Boucher and Altamimi 2011). The INSPIRE (INfrastructure for SPatial InfoRmation in Europe) Directive, which entered into force on the 15th May 2007 (see http://inspire.jrc.it/directive/l_10820070425en00010014.pdf), addresses 34 spatial data themes needed for environmental applications subdivided in the three annexes of the directive. Annex 1 deals with Coordinate Reference Systems and specifies the datum for three-dimensional and two-dimensional coordinate reference systems as follows:

- datum of the ETRS89 in areas within its geographical scope;
- datum of the ITRS or other geodetic coordinate reference systems compliant with ITRS in areas that are outside the geographical scope of ETRS89.

The practical consequence of this directive is that the ETRS89 has been adopted in the majority of European countries, but there is a big variety of map projections in use and the geo-information needs to be converted by providing on-line tools (transformation services) which requires the involvement of other communities.

2 Implementation of the ETRS89

2.1 EUREF Permanent Network

The maintenance and access to the ETRS89 is guaranteed by the EUREF Permanent GNSS Network (EPN). It contains (October 2014) 264 active permanent GNSS stations (see Fig. 1) running in a well-organized environment. The weekly EPN site positions estimated by the EPN analysis centers are used by the EPN Reference Frame Coordinator to update the multi-year estimates of the EPN site positions and velocities every 15 weeks. The coordinates are made available in both the ETRS89 (realization ETRF2000) and the ITRS (currently IGb08, Rebischung et al. 2012; Rebischung 2013) on the web site of the EPN Central Bureau (http://www.epncb.oma.be).

A. Kenyeres
FOMI Satellite Geodetic Observatory, 1592 Budapest, P.O. Box 585, Hungary

R. Pacione
E-GEOS Centro di Geodesia Spaziale (ASI/CGS), Contrada Terlecchia, 75100 Matera, Italy

M. Poutanen
FGI, Geodeetinrinne 2, 02430 Masala, Finland

K. Szafranek
Military University of Technology, Gen. Sylwestra Kaliskiego 2, 00-908 Warsaw, Poland

G. Stangl
Institute for Space Research, Schmiedlstr. 6, 8042 Graz, Austria

J. Torres
Artop, Av. Marconi 14-A, 1000-205 Lisbon, Portugal

C. Völksen
Bavarian Academy of Sciences and Humanities, Alfons-Goppel-Str. 11, 80539 Munich, Germany

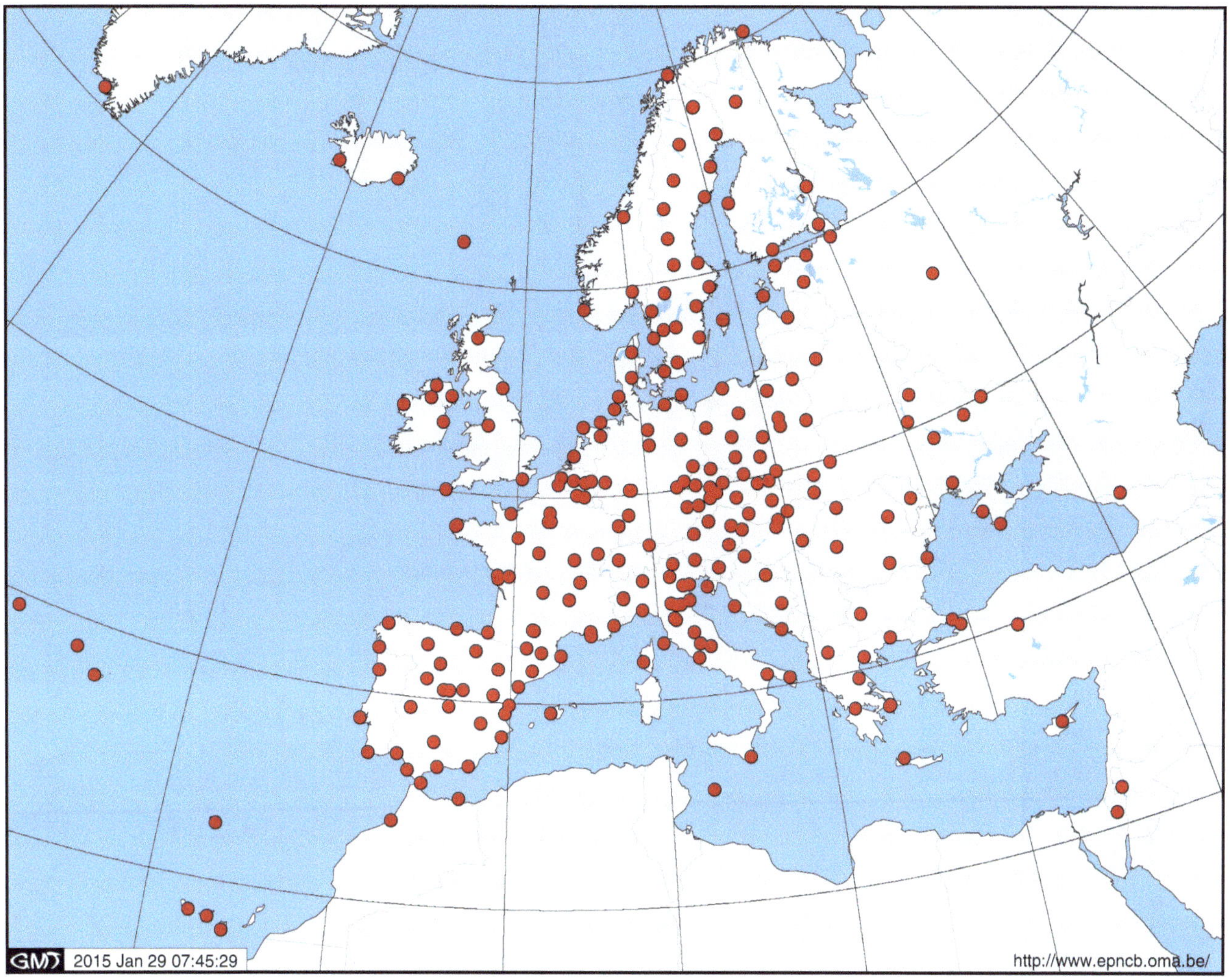

Fig. 1 Locations of the EPN tracking stations (status October 2014)

2.2 Site Categorization

The EUREF TWG provides "Guidelines for EUREF Densifications" (Bruyninx et al. 2013) for organizations wishing to ask EUREF the validation of their (mostly national) GNSS ETRS89 densification campaign. The guidelines require that each densification of the ETRS89 is first computed in the ITRS and then converted in ETRS89. Consequently, the guidelines include not only GNSS data analysis instructions, but also explain that the densification has to be tied to the ITRS using minimal constraints on the EUREF's multi-year coordinates of the so-called EPN Class A stations at the mean epoch of the observation of the campaign. The position and velocity of these EPN Class A stations (205 stations by October 2014) are known with 1-cm accuracy at all epochs of the station's lifetime and their velocity repeatability (determined by comparing the velocities obtained from successive multi-year solutions) is below 0.5 mm/year.

Finally, the guidelines provide the detailed list of deliverables and the content of the report that has to be delivered to the EUREF TWG at least 2 weeks before one of its triannual meetings. During the TWG meeting, the agency requesting validation presents its project to the TWG members. In practice, the TWG often requests corrections and the final densification report is presented and validated at the following TWG meeting. The whole process is finalized at the annual EUREF symposium where the densification project is presented to the plenary followed by a symposium resolution indicating the classification of the campaign:

- Class A (1 cm accuracy independent of epoch); typically densifications covering more than 2 years of observations on permanent GNSS markers
- Class B (1 cm accuracy at the epoch of observations); typically all GNSS campaigns since 1993
- Class C (5 cm accuracy at the epoch of observations).

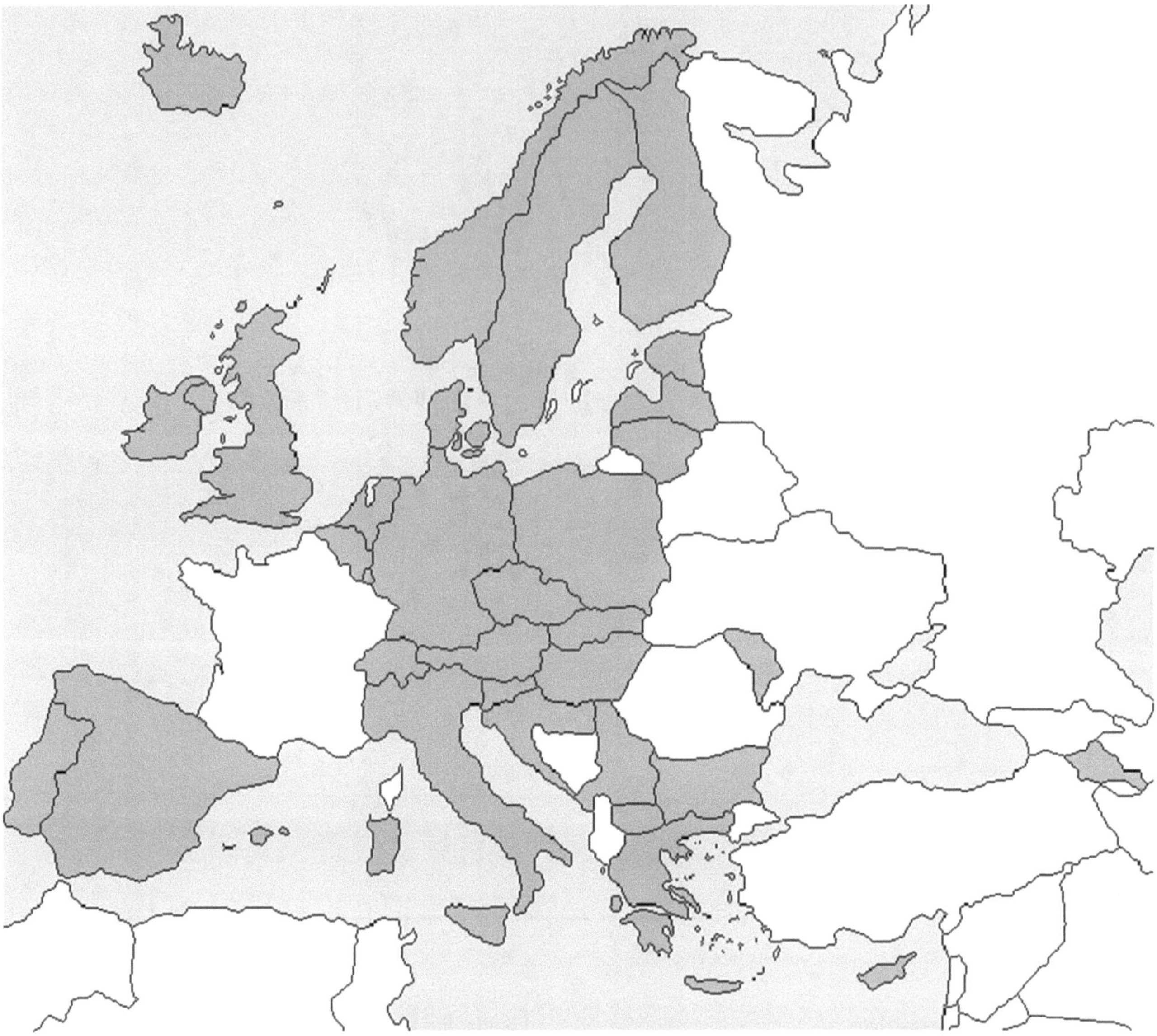

Fig. 2 Countries (in *grey*) whose national ETRS89 densification has been validated by the EUREF Technical Working Group

This process has been followed by the majority of the European countries (see Fig. 2). Also other countries are using the ETRS89, but did not ask a validation from the TWG (e.g. France).

The official national ETRS89 coordinates used in the different countries agree at the few cm level (see Fig. 3) with the most recent multi-year coordinates computed by EUREF for the EPN Class A stations. This is perfectly within the expectations as the EUREF's multi-year solution is in the ETRF2000 at epoch 2005.0 and based on the latest EPN antenna calibration model (individual antenna calibrations complemented with type-mean antenna calibrations given in igs08.atx, http://igscb.jpl.nasa.gov/igscb/station/general/igs08.atx). The coordinates of the national ETRS89 densifications are however provided in different ETRS89 realizations at the epoch of the densification campaign. In addition, they are computed using the antenna calibration models available at that time.

3 Challenges

3.1 Introduction

In many countries users have access to the ETRS89 using Virtual Reference Stations (VRS). VRS software uses the data from a set of GNSS reference stations with known ETRS89 coordinates (at a specific epoch t_0). These data are then processed in a central facility to generate VRS correction data. This central data analysis is normally performed

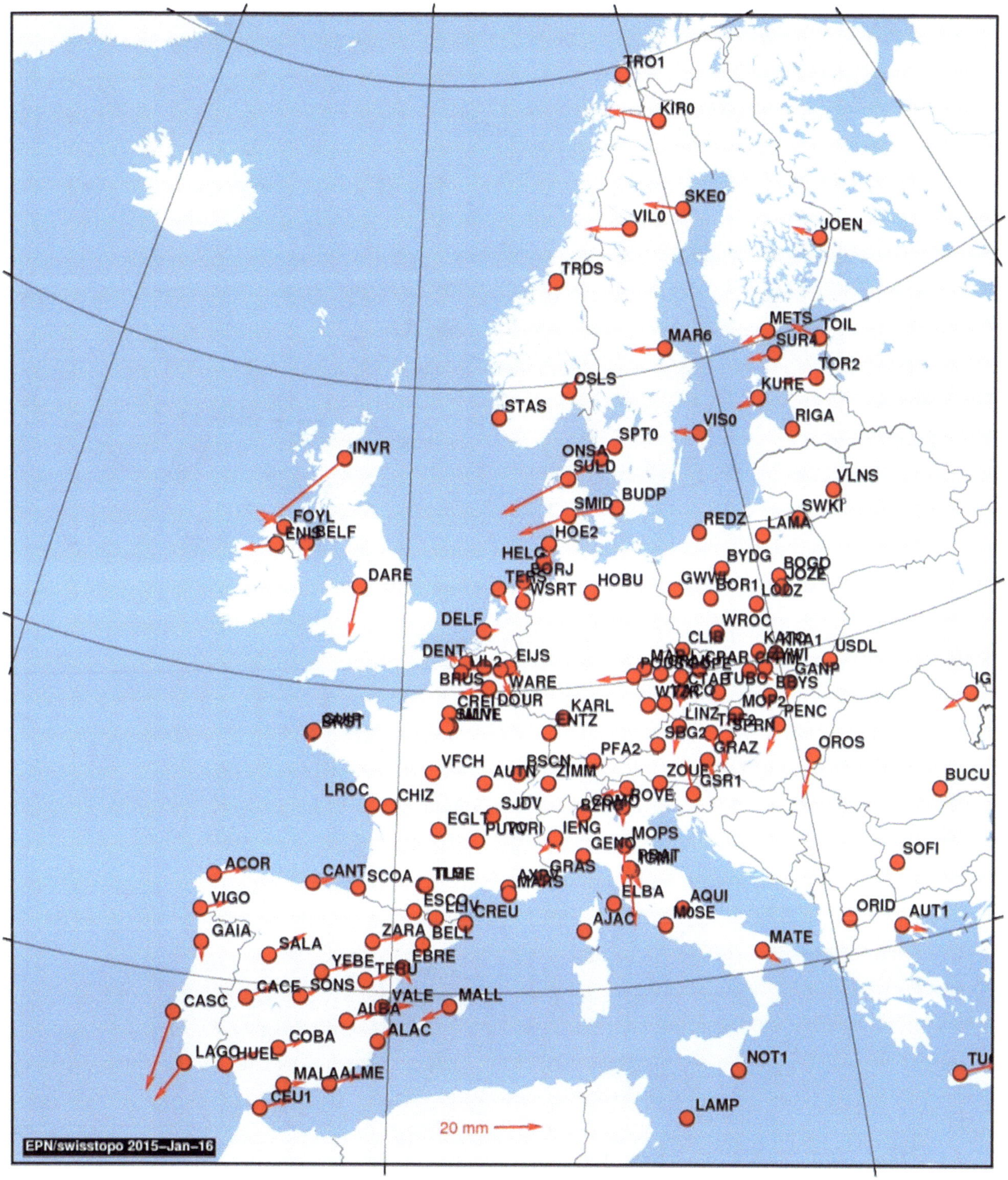

Fig. 3 Horizontal differences between the official national ETRS89 coordinates (at their epoch of definition) with the most recent multi-year coordinates (EPN class A only, week 1,800; status January 16, 2015)

in the ITRS at the epoch of observation and consequently also the coordinates of the reference stations need to be computed in the ITRS at the epoch of observations. The transformation from ETRS89 (epoch t_0) to ITRS (epoch of obs.) consists of two steps: a transformation of ETRS89 from epoch t_0 to the epoch of observations, followed by a transformation from ETRS89 to ITRS (both at the epoch of obs.).

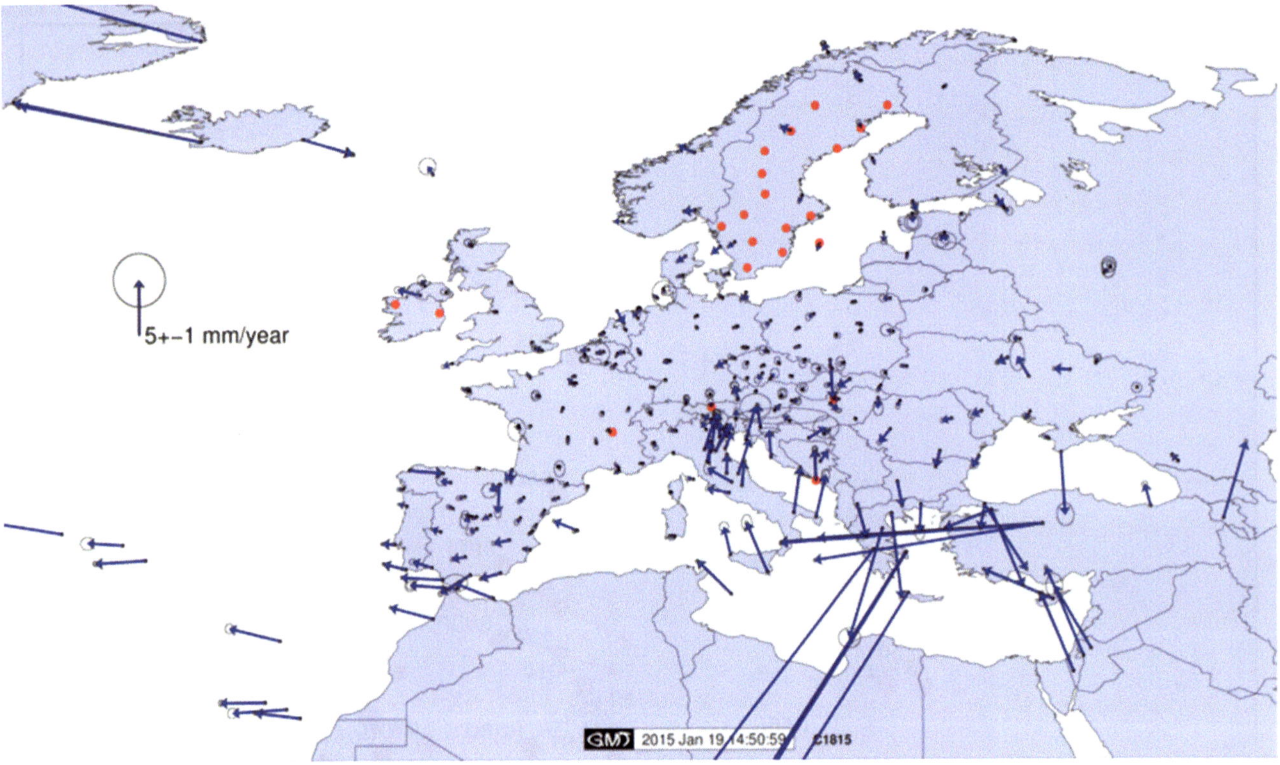

Fig. 4 Horizontal ETRS89 velocities of the EPN Class A stations (*red dots* indicate Class B stations)

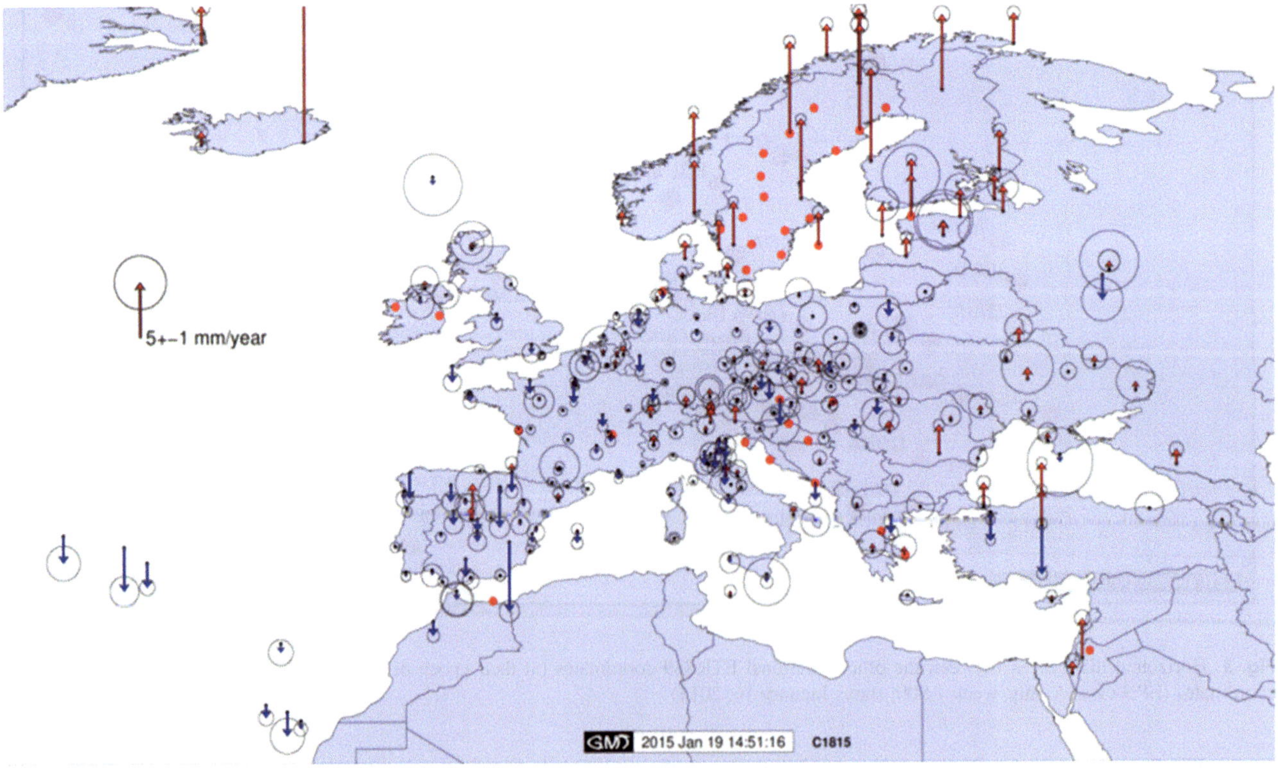

Fig. 5 Vertical ETRS89 velocities of the EPN Class A stations (*red dots* indicate Class B stations)

The ETRS89 has been designed to have minimal site velocities in the 'stable part' of Europe and therefore the first step of the transformation can be neglected in many countries, while the second step is performed using the standard ETRS89/ITRS transformation issued by EUREF. However, ETRS89 velocities can reach 1 cm/year in the up-component in Scandinavia (Lidberg et al. 2006) and 3 cm/year in the horizontal in Southern Europe (Caporali et al. 2011) as can be seen from Figs. 4 and 5. Consequently, when using ETRS89 as a national reference system, the affected countries need to implement mechanisms to deal with the changing ETRS89 coordinates. Possible countermeasures are more frequent updates of the national ETRS89 densifications or the explicit introduction of the first step of the transformation described above. This last option requires the knowledge of local velocities or a model for intraplate velocities. The next sections will give the examples of Sweden and Greece.

3.2 Sweden

The Swedish national ETRS89 realization, SWEREF99 (Jivall and Lidberg 2000) was validated by the EUREF TWG in 2000. Users can get access to SWEREF99 through SWEPOS (Fig. 6), the Swedish Network of Permanent GNSS Reference Stations (Lilje et al. 2014) that provides VRS corrections. SWEREF99 coordinates refer to the epoch 1999.5 and due to post-glacial rebound (PGR), SWEPOS stations are affected by relative coordinate changes (with respect to 2014) of 9, 7 and 11 cm in the up-component for the northern, central and southern part of the network, respectively.

To take this deformation into account, Sweden attributed in the VRS software a velocity to the SWEPOS reference stations corresponding to the 3-dimensional NKG_RF03vel model (Lidberg et al. 2006). This model allows for the extrapolation of the original SWEREF coordinates from the epoch 1999.5 to the epoch of observation (first step of the transformation). The obtained ETRS89 coordinates at the epoch of observation are then transformed to the ITRS by the VRS software itself, which has the standard transformation from ETRS89 to ITRS hardcoded in its software (see scheme in Fig. 7).

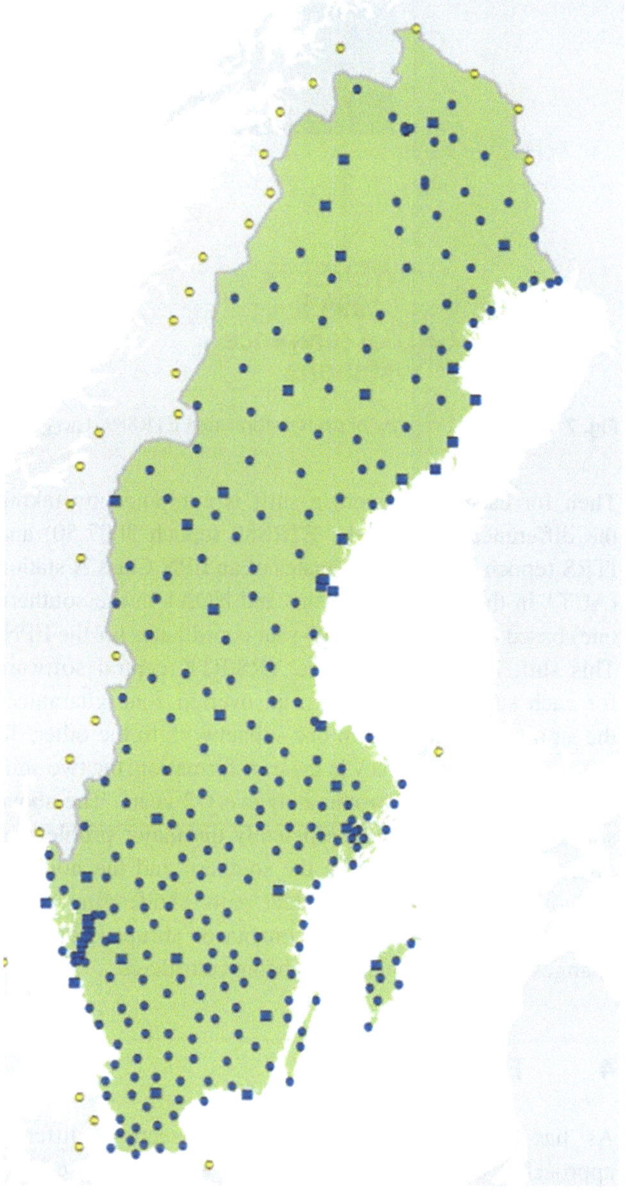

Fig. 6 SWEPOS network of reference stations

3.3 Greece

For historical reasons, Greece uses the HTRS07 (Katsampalos et al. 2010) as national ETRS89 realization (epoch 2007.5). The HTRS07 coordinates of the HEPOS stations have a 2-cm level agreement with the ETRS89 station positions included in the EUREF GR 2007 campaign which was validated by EUREF as a Class B densification of the ETRS89 (epoch 2007.83, Gianniou 2010a). Greece is located on the boundary region between the Eurasian and Nubian plates while the southern part of Greece lies on the Aegean plate, a smaller plate which is moving southwest. This causes large differences in the observed HEPOS station velocities (see Fig. 8).

Greece distributes the ETRS89 using VRS/RTK services provided by the HEPOS network. The HEPOS VRS software handles the transformation ETRS89 (epoch 2007.50) $\rightarrow$ ETRS89 (epoch of obs.) $\rightarrow$ ITRS (epoch of obs.) in a different way than the Swedish approach. Greece is using a single shift to account for the two-step transformation. For that purpose HEPOS is split in two subnetworks (see Fig. 8), based on the tectonic characteristics of the Greek region.

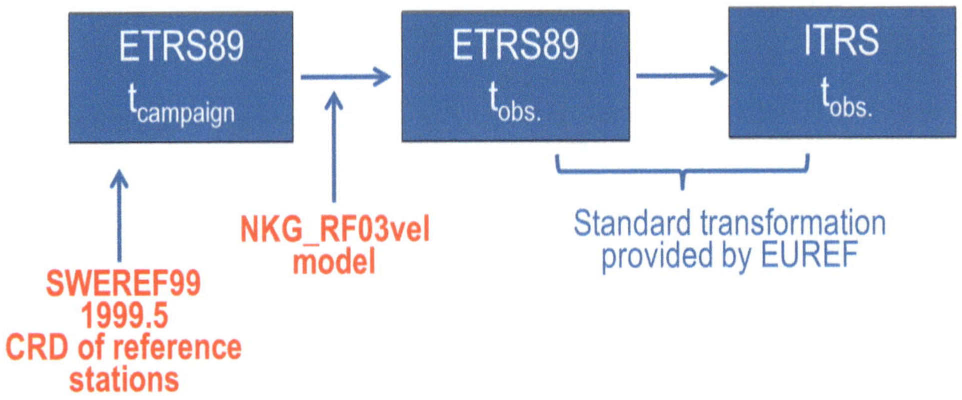

Fig. 7 Schematic overview of the transformation ETRS89 ($t_{SWEREF99}$) to ITRS (epoch of obs.) implemented in SWEPOS

Then for each subnetwork a shift is computed by taking the difference between the ETRS89 (epoch 2007.50) and ITRS (epoch of obs.) coordinates of an EPN Class A station (AUT1 in the northern network and NOA1 in the southern one) based on EUREF's multi-year coordinates for the EPN. This shift is then used in the VRS/RTK central software for each subnetwork (Fig. 9); an overlap zone guarantees the smooth transition from one subnetwork to the other. To ensure sufficient accuracy in the transformation, the two shift values are updated approximately every 2 years. The above solution proved to handle efficiently the major problem of different velocities between the southern and the northern part of Greece. Methods to treat more challenging issues such as local geological phenomena or abrupt coordinate changes due to earthquakes are being considered.

4 EUREF Strategy

As has been shown in the previous section, different approaches can be used to take into account the residual ETRS89 velocities when maintaining a national ETRS89 densification. Common to these approaches is the need to know the ETRS89 velocities (in some reference points or through a geophysical model). The availability of a homogeneous dense European velocity model could help the countries to achieve more cross-boundary consistency. For that purpose EUREF initiated two activities: the EPN densification and the EUREF Working Group "Deformation Models".

4.1 EPN Densification

To take advantage of the many national agencies that routinely analyze the data from their GNSS networks consisting of stations not all included in the EPN (their RINEX data are not necessarily freely available), EUREF decided in 2010 to initiate a densification of the EPN. It is done by the EPN Reference Frame Coordinator who collects and combines the weekly SINEX solutions provided by European countries for their dense national active GNSS networks with the weekly EPN SINEX solutions. Then all available weekly combined solutions are stacked to obtain a consistent cumulative position/velocity solution. Both the weekly combinations as well as the stacking are done using the CATREF software (Altamimi et al. 2007) applying the same approach and parameters as for the generation of the EPN densification of the IGb08, ensuring full consistency from the global to local level. The total number of stations included in the EPN densification exceeded 2,500 as of December 2014, see Fig. 10. In the near future, EUREF will focus on collecting and maintaining meta-data from the EPN densification stations in order to increase the reliability of the analysis of their residual time series.

4.2 EUREF Working Group on "Deformation Models"

The EPN densification products are a valuable contribution to several groups and projects such as the IAG Working Group on "The Integration of Dense Velocity Fields in the ITRF", EPOS (European Plate Observatory System) and EUPOS (European Positioning System). But, they will be of crucial interest for the EUREF working group on Deformation Models (created in 2012) which relies on having access to dense and accurate 3D-crustal velocities of GNSS reference stations to obtain velocity models and significantly improve the prediction of the time evolution of coordinates and overcome the limitations in the use of the ETRS89. Activities of the working group comprise the investigation of methods for applying geophysical deformation models to the ETRS89 and the work on the interpolation methodologies

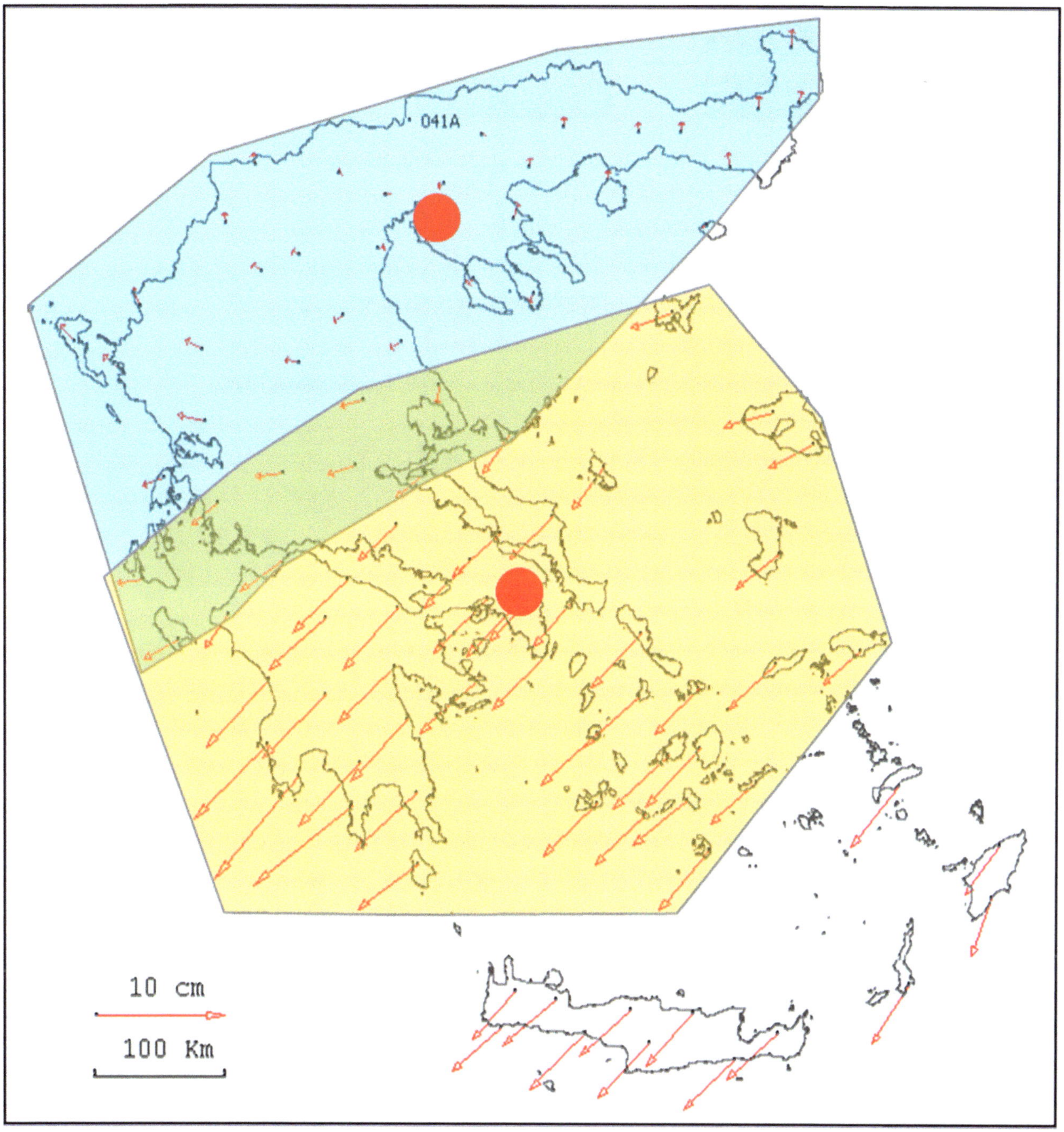

Fig. 8 Displacements in the HEPOS network: period 11/2007 to 11/2009 relative to the station 041A (Gianniou 2010b). The *red dots* indicate the location of the EPN Class A stations that were used to compute the shift ETRS89 (2007.5) to ITRS (epoch of obs.) for each subnetwork

to develop methods to create a gridded representation of the deformation models.

5 Summary

The European Terrestrial Reference System (ETRS89) is the regional reference system to be used within Europe. It is tied to the stable part of the European plate and it is derived from the global International Terrestrial Reference System (ITRS). It is part of the legislation of the majority of the European countries. The EUREF sub-commission has developed guidelines on how a country should generate a national densification of the ETRS89. EUREF is also keeping track of the differences between the official national ETRS89 coordinates and the ones estimated by EUREF. They agree at the few cm level. Using the ETRS89 in geographical zones like Scandinavia and Southern Europe is challenging because

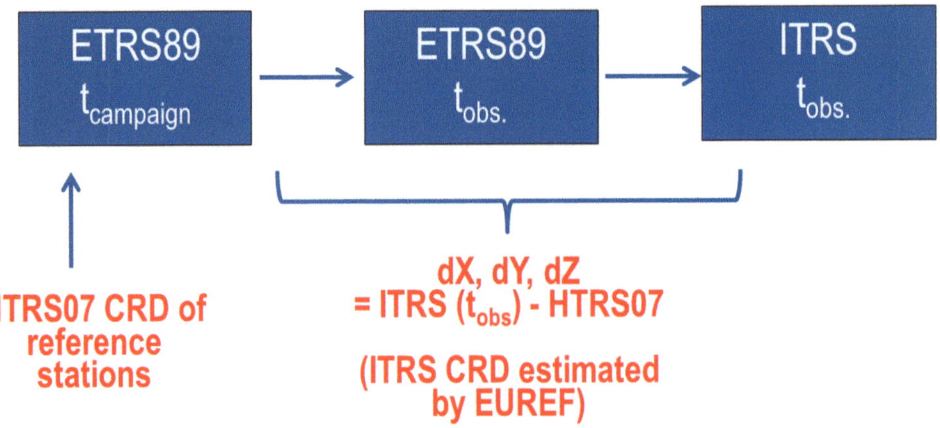

Fig. 9 Schematic overview of the transformation ETRS89 (2007.5) to ITRS (epoch of obs.) implemented in HEPOS

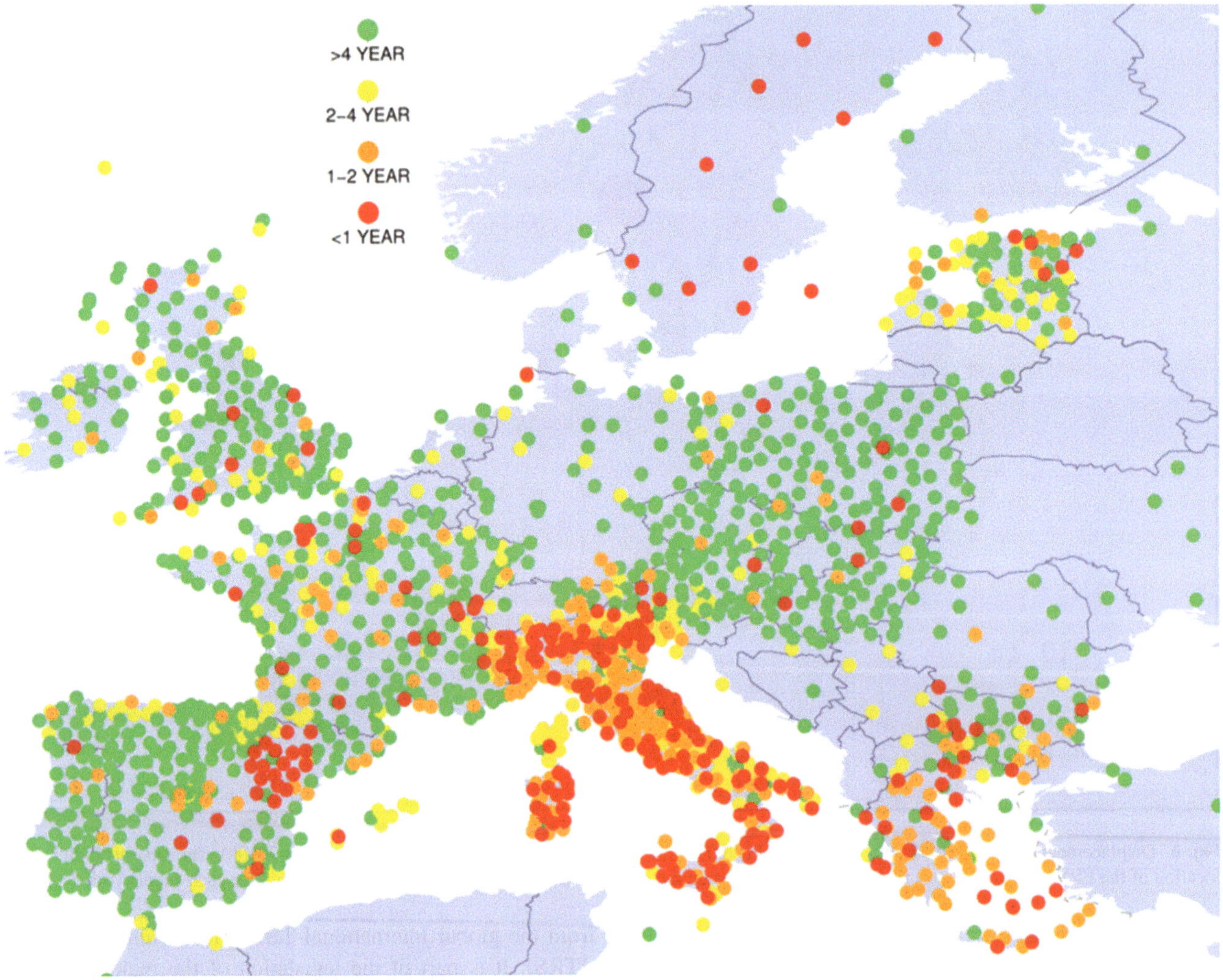

Fig. 10 EPN densification stations; the *color codes* indicate the length of the available weekly SINEX solutions

ground deformations in these areas cause residual motions in the ETRS89 which could reach 1–3 cm a year. Greece and Sweden have each developed their proper way of dealing with this challenge. Common is the need for known ETRS89 velocities. The working group on "Deformation models" and the "EPN densification" are EUREF's first steps toward a long-term approach to this problem by aiming at developing a European dense velocity model.

References

Altamimi Z (2009) ETRS89 realization: current status, ETRF2005 and future development. Bull Geod Geomat LXVIII(3):255–267

Altamimi Z, Boucher C (2002) The ITRS and ETRS89 relationship: new results from ITRF2000. Verlag des Bundesamts für Kartographie und Geodäsie, Frankfurt Am Main (23):49–52

Altamimi Z, Sillard P, Boucher C (2007) CATREF software: combination and analysis of terrestrial reference frames. LAREG, Technical Report, Institut Géographique National, Paris, France

Boucher C, Altamimi Z (1992) The EUREF terrestrial reference system and its first realizations. Veröffentlichungen der Bayerischen Kommission für die Internationale Erdmessung, ed. Bayerische Akadmey der Wissenschaften (52):205–208

Boucher C, Altamimi Z (2011) Memo: specifications for reference frame fixing in the analysis of a EUREF GPS campaign. http://etrs89.ensg.ign.fr/memo-V8.pdf. Accessed 20 Jan 2015

Bruyninx C, Altamimi Z, Caporali A, Kenyeres A, Lidberg M, Stangl G, Torres JA (2013) Guidelines for EUREF densifications. ftp://epncb.oma.be/pub/general/Guidelines_for_EUREF_Densifications.pdf. Accessed 20 Jan 2015

Caporali A, Lidberg M, Stangl G (2011) Lifetime of ETRS89 coordinates (2011). EUREF symposium, Chisinau, Moldova, 25–28 May 2011. http://www.euref.eu/symposia/2011Chisinau/01-03-p-Caporali.pdf. Accessed 20 Jan 2015

Gianniou M (2010a) EUREF GR 2007 campaign – final report. ftp://olggps.oeaw.ac.at/pub/EUREF_camp/EUREF_GR_2007_Final_Report.pdf. Accessed 20 Jan 2015

Gianniou M (2010b)Tectonic deformations in Greece and the operation of HEPOS network. EUREF 2010 symposium, June 2–4 2010, Gävle, Sweden. http://users.teiath.gr/mgianniou/index_en_files/Gianniou_2010_EUREF.pdf. Accessed 20 Jan 2015

Jivall L, Lidberg M (2000) SWEREF-99 an updated EUREF realization for Sweden. Veröffentlichungen der Bayerischen Kommission für die Internationale Erdmessung, ed. Bayerische Akadmie der Wissenschaften (61):167–175

Katsampalos K, Kotsakis C, Gianniou M (2010) Hellenic terrestrial reference system 2007 (HTRS07): a regional realization of ETRS89 over Greece in support of HEPOS. Bull Geod Geomat LXIX(2–3):329–348

Lidberg M, Johansson JM, Scherneck HG (2006) Geodetic reference frames in the presence of crustal deformation – with focus on Nordic conditions. Mitteilungen des Bundesamtes für Kartographie un Geodäsie 40:189–198

Lilje M, Wiklund P, Hedling G (2014) The use of GNSS in Sweden and the national CORS network SWEPOS. FIG XXV International Congress, June 16–21 2014, Kuala Lumpur, Malaysia, p 11

Rebischung P (2013) [IGSMAIL-6663] IGb08: an update on IGS08. http://igscb.jpl.nasa.gov/pipermail/igsmail/2012/007853.html. Accessed 20 Jan 2015

Rebischung P, Griffiths J, Ray J, Schmid R, Collilieux X, Garayt B (2012) IGS08: the IGS realization of ITRF2008. GPS Sol 16(4):483–494. doi:10.1007/s10291-011-0248-2

Defining a Local Reference Frame Using a Plate Motion Model and Deformation Model

Richard Stanaway, Craig Roberts, Chris Rizos, Nic Donnelly, Chris Crook, and Joel Haasdyk

Abstract

As GNSS point-positioning becomes more precise and accessible to a wider spectrum of users, the issue of misalignment between GNSS positioning reference frames and spatial data reference frames used in GIS will become more apparent. Positions of plate-fixed features within GNSS reference frames are kinematic in nature due to global plate motions and other geophysical phenomena including seismic deformation and post-glacial rebound. Coordinates within GIS and applications such as Google Earth on the other hand, are typically fixed to the Earth's surface and tectonic plate and may be misaligned with global reference frames unless a kinematic model is applied to the data.

The problem becomes more apparent when data acquired at different epochs are combined in the absence of a kinematic model. Should a GNSS point-position or baseline vector solution be transformed to the epoch of existing spatial data, or should the spatial data be transformed to the epoch of the point-position? In either case, data acquired at different epochs within a GNSS frame will need to be transformed to a common epoch for the purpose of combination, interpretation and analysis. Furthermore, localised deformation analysis studies using remote sensing techniques such as InSAR and Lidar require removal of any secular plate motion signal prior to meaningful analysis. Presently, it is more computationally efficient to transform GNSS observations to a formalised reference epoch for spatial data.

A logical approach to the problem is to develop a Local Reference Frame (LRF) which is fixed to the crust within a defined polygon, and which is also directly traceable to GNSS reference frames such as the International Terrestrial Reference Frame (ITRF) by means of a Plate-Motion Model (PMM) and residual Deformation Model (DM). In plate boundary zones where crustal deformation is significant such as New Zealand, a PMM is of limited application and an "Absolute" Deformation Model (ADM) can be used to describe the full transformation between reference frames. PMMs are specified by an Euler Pole which can also be defined by the rotation rates of the three Cartesian axes. The Euler Pole is estimated by inversion of a selection of station ITRF site velocities. A residual DM can be estimated by kriging or least-squares collocation of site-velocity residuals within the PMM and application of a fault locking model where elastic strain or seismic deformation is evident.

R. Stanaway (✉) • C. Roberts • C. Rizos
School of Civil and Environmental Engineering, University of New South Wales, Sydney, NSW, Australia
e-mail: richard.stanaway@student.unsw.edu.au; c.roberts@unsw.edu.au

N. Donnelly • C. Crook
National Geodetic Office, Land Information New Zealand, Wellington, New Zealand

J. Haasdyk
Survey Infrastructure and Geodesy, NSW Land and Property Information, Bathurst, NSW, Australia

T. van Dam (ed.), *REFAG 2014*, International Association of Geodesy Symposia 146,
DOI 10.1007/1345_2015_147

Use of a PMM and associated DM enables ITRF positions and vectors (e.g. from GNSS observations) to be transformed to a local frame to support GIS data integration and combination of data acquired using terrestrial positioning techniques such as Terrestrial Laser Scanning and conventional Total Station surveys. A case-study for the development of a new Australian Terrestrial Reference Frame is presented.

Keywords
Deformation model • Plate motion model • Reference frame

1 Introduction

The rapid improvement in mass-market positioning precision is presenting new challenges to users and managers of spatial data. GNSS positioning is inherently undertaken within reference frames closely aligned to ITRF (e.g. WGS84 for GPS and PZ-90 for GLONASS). In the near future it is anticipated that centimetre-level positioning precision will be attainable by a wide spectrum of users of personal navigation devices. This increase in precision will result from improvements to GNSS orbit products, GNSS augmentation systems and software. Such precision does pose a dilemma for managers of geodetic infrastructure and spatial data: Spatial data infrastructure is currently referenced to ground-fixed reference frames where coordinates of stable features are not expected to change significantly as a function of time, especially in tectonically stable regions. The dilemma arises where GNSS derived precise positions in terms of a kinematic global reference frame are used erroneously in the context of existing spatial data defined in ground-fixed frames (e.g. static geodetic datums).

Presently there are two approaches to resolving the misalignment between positioning and spatial data reference frames. One approach is to transform existing spatial data sets defined in a ground-fixed frame to the epoch of a GNSS precise position. Another approach is to transform GNSS precise positions to a local ground-fixed frame. Both approaches make use of a conformal transformation (e.g. a 14 parameter transformation) or time-dependent block shift derived from an ITRF site velocity model e.g. Stanaway et al. (2014). The first approach is not yet widely implemented or tested in GIS software and kinematic transformation algorithms are still in development. Furthermore, the computational overhead of transforming large volumes of spatial data "on-the-fly" can be a limitation with this approach. The second approach is presently in more widespread use (e.g. in geodetic analysis software) as it is well suited to current generation geodetic datums which are inherently fixed to the Earth's crust at a defined reference epoch. Either approach must be used to combine and analyse spatial data acquired at different epochs of ITRF. In the absence of metadata, the difference between a GNSS precise position and a precise spatial database can be significant.

Local Reference Frames (LRF) fixed to stable portions of the Earth's crust are ideally suited to support spatial data integration over longer periods of time as site velocities are minimised with respect to the local frame. However, there remains the issue of how GNSS precise positions relate to spatial data defined in a ground-fixed frame and the 14 parameter transformation and gridded deformation model approaches each have their limitations. 14 parameter transformations include scale and scale-rate parameters which, if non-zero, implicitly define uniformly distributed deformation of the local frame. Gridded deformation models can better accommodate localised and variable deformation, however they maybe inefficient over large areas of stable tectonic plates. For tectonically stable regions, a plate motion model (PMM) (Altamimi et al. 2011, 2012) can be used to transform GNSS point positions to a local frame. Where higher precision is required a residual deformation model can also be applied if intraplate deformation is significant. In the USA, residual deformation and block-rotation models have been used in Horizontal Time-Dependent Positioning (HTDP) software (Snay 1999; Pearson and Snay 2012) since 2000.

The advantage of a PMM is that it is inherently distortion-free as it is defined by the rotation of a stable portion of the tectonic plate. Localised or intraplate deformation is more clearly visualised where the rigid plate rotation component is removed.

Regional Reference Frames (RRF) fixed to tectonic plates such as ETRF89 (Boucher and Altamimi 1992), NAD83 (Schwarz 1983) and GDA94 (Steed 1995) have been defined from earlier realisations of ITRF, and 14 parameter transformations are required to transform positions within these RRF to ITRF at a specified epoch. Furthermore, residual deformation within these RRF are evident as non-zero station velocities for stations in deforming zones within the RRF.

This paper shows how a PMM and residual DM can be used to define an LRF. The simplest realisation of a LRF is by a four parameter transformation from ITRF to the local frame: three rotation rate parameters of the Cartesian axes of the local frame within ITRF and an epoch-difference parameter. A case-study in support of a new Australian Terrestrial Reference Frame is presented.

2 Plate Motion Models

The Earth's surface comprises of a series of stable tectonic plates rotating slowly over the mantle and deforming zones generally located near plate boundaries. Recent studies e.g. Bird (2003), DeMets et al. (2010), Altamimi et al. (2011), Argus et al. (2011) and Kreemer et al. (2014) have better defined the extent of smaller tectonic plates (microplates) and stable crustal blocks within these deforming zones. Euler Poles estimated for each of these plates and crustal blocks can be adapted to define a stable LRF to support land surveying and GIS activities. Where deformation of a plate or block is significant a residual DM can be applied for higher precision applications. Estimation of Euler Poles of tectonic plates using space geodetic and geophysical observations is well documented e.g. DeMets et al. (1990). An Euler Pole can be defined using space geodetic techniques by least-squares inversion of n sites with ITRF site velocities estimated from analysis of the ITRF site time-series (Eqs. 1–9) adapted from Goudarzi et al. (2014). Site velocities are typically defined during the interseismic period, so any known coseismic and postseismic deformation should be isolated from the time-series analysis. Elastic strain accumulation arising from locked faults near a site should also be modelled using elastic half-space models in order to estimate interseismic back-slip (McCaffrey 2002).

$$\mathbf{\Omega}^{plate} = \left(\mathbf{A}^T\mathbf{W}\mathbf{A}\right)^{-1}\left(\mathbf{A}^T\mathbf{W}\mathbf{L}\right) \tag{1}$$

where,

$$\mathbf{\Omega}^{plate} = \begin{bmatrix} \omega_x \\ \omega_y \\ \omega_z \end{bmatrix} \quad \mathbf{A} = \begin{bmatrix} 0 & z_1 & -y_1 \\ -z_1 & 0 & x_1 \\ y_1 & -x_1 & 0 \\ \vdots & \vdots & \vdots \\ 0 & z_n & -y_n \\ -z_n & 0 & x_n \\ y_n & -x_n & 0 \end{bmatrix}_{ITRF} \quad \mathbf{L} = \begin{bmatrix} Vx_1 \\ Vy_1 \\ Vz_1 \\ \vdots \\ Vx_n \\ Vy_n \\ Vz_n \end{bmatrix}_{ITRF}$$

and,

$\mathbf{\Omega}^{plate}$ is the Euler Pole (rotation rate of axes ω_x ω_y ω_z in Rad/year)
A is the design matrix of ITRF cartesian site coordinates (m) x_1 y_1 z_1 to x_n y_n z_n
W is the weight matrix (if applicable)
L is the observation matrix of ITRF site velocities (m/year) Vx_1 Vy_1 Vz_1 to Vx_n Vy_n Vz_n

The Euler Pole can also be expressed using Eqs. (2–4):

$$\text{Pole rotation rate (in Rad/yr)}\ \omega_{plate} = \sqrt{\omega_x{}^2 + \omega_y{}^2 + \omega_z{}^2} \\ + \text{is anti} - \text{clockwise about pole)} \tag{2}$$

$$\text{Pole latitude}\ \phi_{plate} = \tan^{-1}\frac{\omega_z}{\sqrt{\omega_x{}^2 + \omega_y{}^2}} \tag{3}$$

$$\text{Pole longitude}\ \lambda_{plate} = \frac{\omega_y}{\omega_x} \tag{4}$$

Site velocities (m/year) for any specific point can be computed directly from the Euler Pole model using (5):

$$\begin{bmatrix} Vx \\ Vy \\ Vz \end{bmatrix} = \begin{bmatrix} \omega_Y z - \omega_Z y \\ \omega_Z x - \omega_X z \\ \omega_X y - \omega_Y x \end{bmatrix} \tag{5}$$

The Euler Pole can be expressed as a three parameter conformal transformation as follows:
(rotation rates in radians per year)

$$\dot{r}_x = -\omega_x \quad \dot{r}_y = -\omega_y \quad \dot{r}_z = -\omega_z \tag{6}$$

(Note: Equation (6) uses the coordinate frame rotation convention. If the position vector notation convention is used, the signs of rotation rates and derived rotations are reversed)

The rotation rates in Eq. (6) can be expressed conventionally as arcseconds per year using (7).

$$r_{(\text{arc sec}/yr)} = \frac{648000\,\omega_{(Rad/yr)}}{\pi} \tag{7}$$

Velocity residuals are computed using Eq. (8)

$$\mathbf{v} = \mathbf{A}\mathbf{\Omega}^{plate} - \mathbf{L} \tag{8}$$

The reference standard deviation for the Euler Pole inversion is computed using Eqn. (9)

$$\text{S}_\text{o} = \sqrt{\frac{\mathbf{v}^\mathbf{T}\mathbf{W}\mathbf{v}}{r}} \tag{9}$$

where r is the degree of freedom ($r = 2n - 3$) where n is the number of stations used in the inversion.

The standard deviation of each of the rotation parameters is derived by scaling the square-root of diagonal components of the variance-covariance matrix or inverted normal matrix $(\mathbf{A}^T\mathbf{W}\mathbf{A})^{-1}$ by S_o

3 Residual Deformation Models

Analysis of site velocity residuals with respect to plate rotation can be used to visualise interseismic deformation and to develop a residual DM. Such a model can be developed by kriging or least-squares collocation (LSC). Alternatively, a fault locking model (McCaffrey 2002) can be used to estimate back-slip on known locked faults to form an apriori deformation model. Site velocity residuals can then be correlated with estimated back-slip from the apriori model to further refine the deformation model. This approach has been used in HTDP software used in the USA (Snay 1999). The deformation model can be presented as a grid model of velocity residuals with respect to the stable plate. An Absolute Deformation Model (ADM) can be formed by combination of the residual DM and a PMM in grid format to represent the velocity field in terms of ITRF. The ADM approach has been used in the development of the New Zealand Deformation Model 2000 (NZDM2000) (LINZ 2015).

4 Australian Case-Study: Stable Australian Plate Reference Frame (SAPRF)

To illustrate how the PMM and residual DM approach for defining a LRF can be applied in practice, a case-study is presented showing development of a Stable Australian Plate Reference Frame (SAPRF2014). Geoscience Australia have published the latest IGb08 (GPS realisation of ITRF2008) (Rebischung et al. 2012) set-of-station coordinate (SSC) solution (Geoscience 2014a) and associated SINEX file for the Asia-Pacific Reference Frame (APREF) encompassing the extent of the Australian continent (Geoscience 2014b). GPS data for all continuous GPS (CORS) sites forming the APREF network were processed using the Bernese GPS software Version 5.0 (Dach et al. 2007) and the ITRF site velocities for all stations in the network were estimated using the CATREF software (Altamimi et al. 2004). Known coseismic and equipment change offsets were isolated from the velocity estimation and a power-law noise model applied to estimate more realistic station velocity uncertainties from the APREF GPS time-series (John Dawson, personal communication).

46 AuScope and Australian Regional GNSS Network (ARGN) stations (Fig. 1) were used for the inversion of the Euler Pole of the Australian plate that fitted the following criteria:

1. Station located within the Australian continental landmass including Tasmania
2. Antenna mounts and reinforced concrete pillars anchored to cratonic bedrock
3. ITRF site velocity (horizontal component) uncertainty <0.45 mm/year (rooftop, tower, jetties or clay soil locations are excluded from analysis – e.g. MOBS, ADE1, PERT, BUR2)
4. Well distributed selection of stations over the Australian continental landmass

The mean horizontal velocity uncertainty of the 46 stations is 0.4 mm/year with a standard deviation of 0.04 mm/year, hence no weighting was applied to the inversion. Figure 2 shows the ITRF site velocities of the selected network.

The ITRF2008 Euler Pole (Rad/year) for the Australian plate was estimated by inversion of the 46 site velocities using Eq. (1)

$$\mathbf{\Omega}^{AustPlate} = \begin{bmatrix} 7.2905\text{E}^{-9} \\ 5.7479\text{E}^{-9} \\ 5.8807\text{E}^{-9} \end{bmatrix}^{p}$$

The standard deviation of the rotation rates using (Eqns. 8 and 9) are:

$$\sigma_{\omega_x} = 4.512\text{E}^{-11}\text{Rad/yr} \quad \sigma_{\omega_y} = 4.147\text{E}^{-11}\text{Rad/yr}$$
$$\sigma_{\omega_z} = 3.652\text{E}^{-11}\text{Rad/yr}$$

The equivalent Euler Pole rates using (Eqns. 2, 3 and 4) are:

$$\omega_p = 0.630°/\text{Ma} \quad \phi_p = 32.35° \quad \lambda_p = 45.17°$$

The SAPRF2014 Euler Pole is closely aligned with the published ITRF2008 Euler Pole for the Australian Plate which was estimated from a sparser network of 19 sites forming a subset of the 46 sites used in this paper (Altamimi et al. 2012). Site velocities estimated using the ITRF2008 PMM differ by 0.3 mm/year from velocities estimated from SAPRF2014.

The equivalent SAPRF2014 to ITRF2008 transformation parameters and uncertainties were computed using (Eqns. 6 and 7):

$$\dot{r}_x = -1.5038\text{E}^{-3}\text{arcsec/yr} \quad \dot{r}_y = -1.1856\text{E}^{-3}\text{arcsec/yr}$$
$$\dot{r}_z = -1.2130\text{E}^{-3}\text{arcsec/yr}$$
$$\sigma_{\dot{r}_x} = 9.31\text{E}^{-6}\text{arcsec/yr} \quad \sigma_{\dot{r}_y} = 8.55\text{E}^{-6}\text{arcsec/yr}$$
$$\sigma_{\dot{r}_z} = 7.53\text{E}^{-6}\text{arcsec/yr}$$

The rotation rates are multiplied by an epoch-difference (Δt) to compute the rotation parameters between ITRF2008 and SAPRF2014 at different epochs.

The rotation rate parameters can be used in a 14 parameter transformation model with zeros for all other parameters.

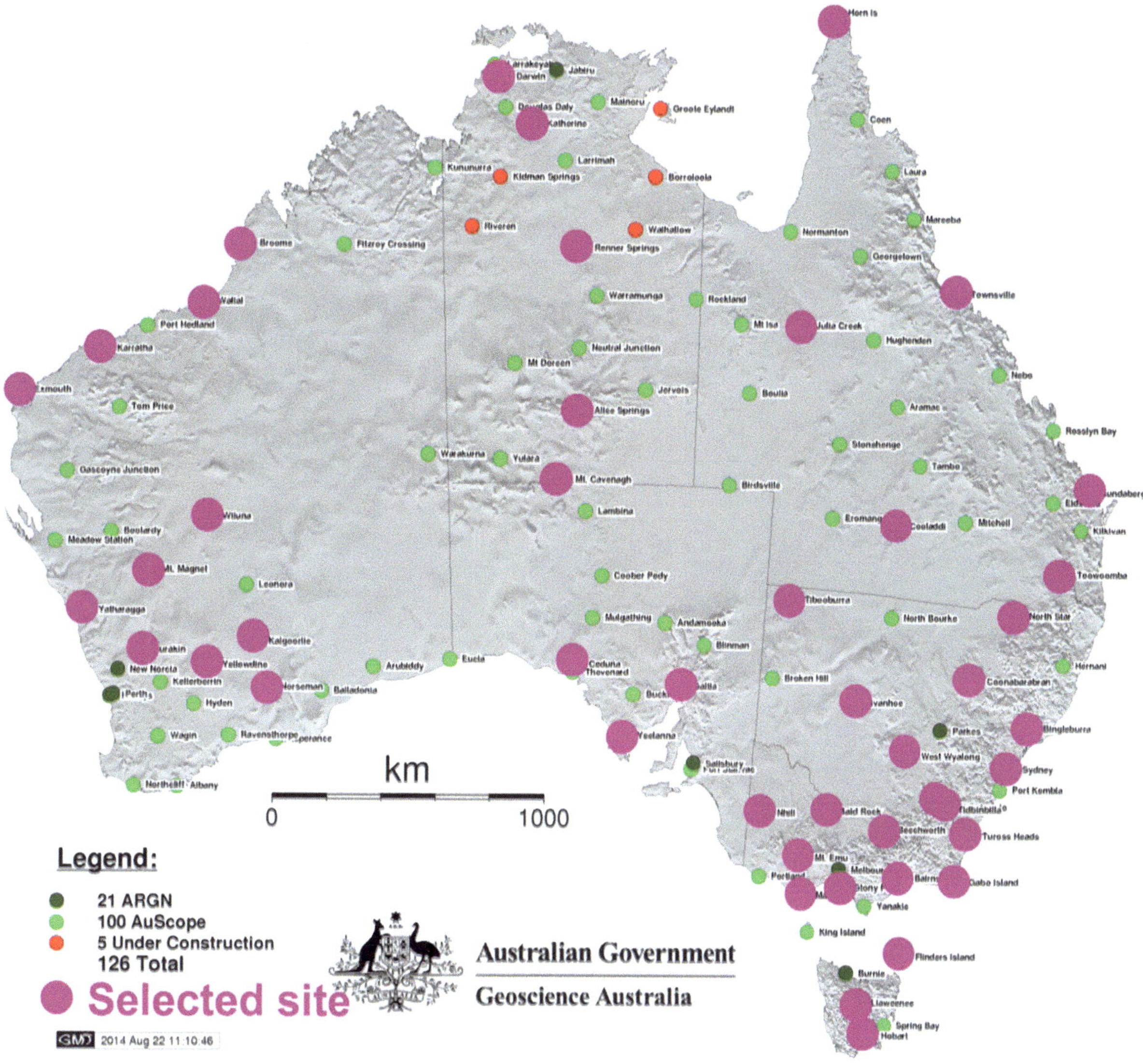

Fig. 1 CORS selection used to estimate pole of stable Australian plate

The velocity residuals were then computed using Eq. (8) and are shown in Fig. 3. The velocity residuals are largely within the uncertainty of the site velocities used for the inversion and indicates that the Australian continent is stable at the level of uncertainty of the observations during periods of interseismic stability. Tregoning et al. (2013) show that large regional plate boundary earthquakes result in observable deformation within the Australian continent at the 10 mm level. Their study shows agreement between observed seismic deformation (both coseismic and postseismic) and modelling. As the uncertainties currently exceed any interseismic deformation signal no residual DM has been developed for SAPRF2014. By 2016 many of the AuScope stations (Fig. 1, green circles) used for the inversion of a refined SAPRF will have a sufficiently long time-series to improve the uncertainties of the site velocities and better quantify the magnitude of any intraplate deformation.

5 Application of SAPRF2014 in Practice

SAPRF2014 can be used as a basis for representation of spatial data in Australia as coordinates of stable features (e.g. bedrock) within the SAPRF2014 will change by less than 0.4 mm/year in the absence of any seismic deformation (local or large regional earthquakes). Kinematic ITRF2008 coordinates can be transformed to SAPRF2014 coordinates by a four parameter transformation (three rotation rates and a difference in epoch). The reference epoch for SAPRF2014 can be arbitrary, however in Australia GDA94 (ITRF92 realised at epoch 1994.0) has been the mandated national geodetic datum since 2000 (Intergovernmental Committee on Surveying and Mapping (ICSM) 2014), and so a reference epoch of 1994.0 would be beneficial to support data integration and surveying until spatial software improvements

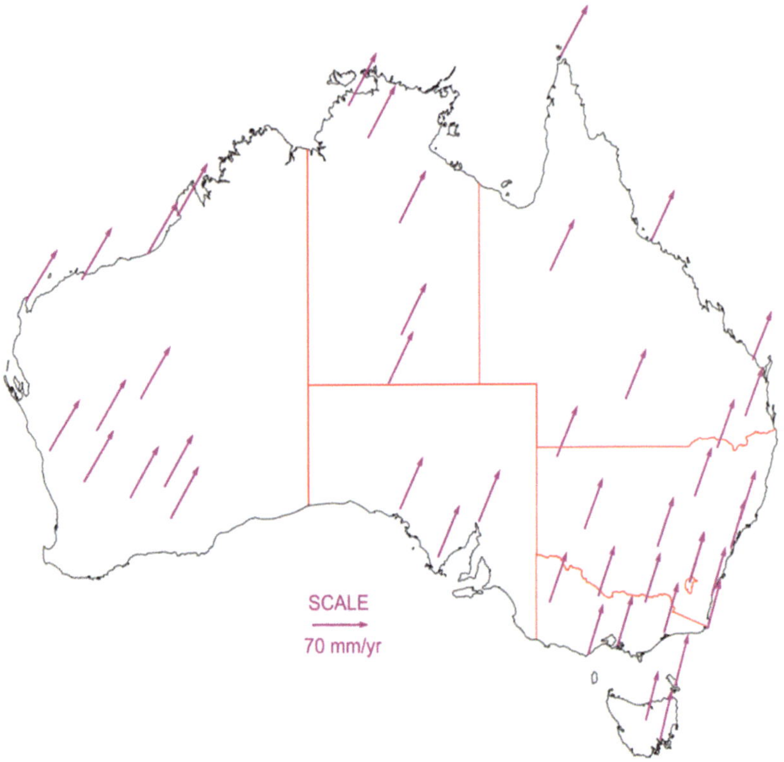

Fig. 2 ITRF site velocities for selected ARGN and AuScope stations

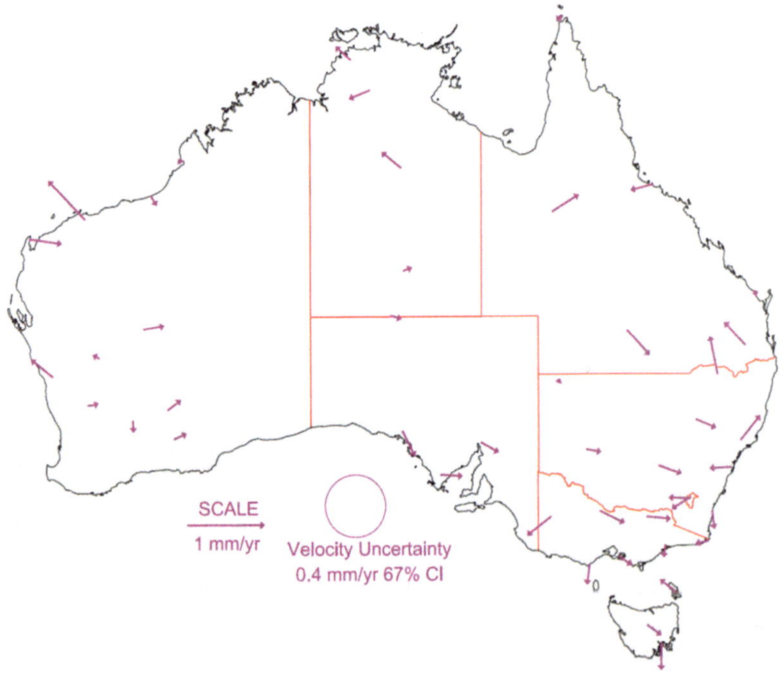

Fig. 3 Velocity residuals for stable Australian plate. The uncertainty of the velocities (0.4 mm/year) are generally greater than the velocity residuals which indicates that intraplate deformation is currently insignificant during interseismic periods

can handle kinematic ITRF coordinates in a robust and assured fashion. Adoption of a 1994.0 epoch for SAPRF2014 in order to maintain consistency with the existing datum would at present result in an increase of uncertainties of up to 5 mm for most locations on the Australian continent (from 3 mm at the current epoch). This uncertainty would be expected to decrease once intraplate deformation rates are better defined. SAPRF2014 at epoch 1994.0 could be described as SAPRF2014(1994.0) in order to clearly show the reference epoch for frame coordinates and velocities. GDA94 currently has significant distortions of up to 300 mm (Haasdyk et al. 2013) and a datum update or readjustment is warranted to minimise these existing distortions. SAPRF ellipsoid heights would be fully consistent with ITRF and

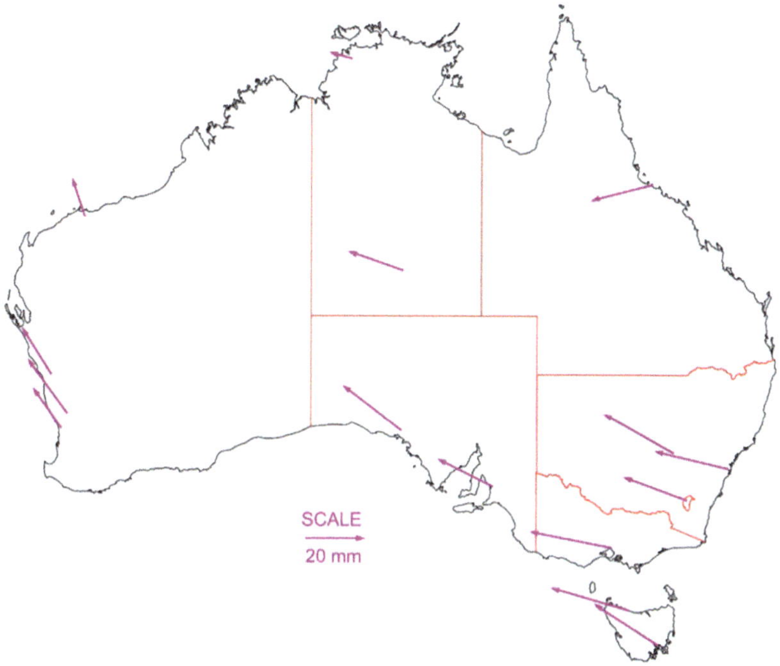

Fig. 4 SAPRF2014 and GDA94 coordinate difference (GDA94 minus SAPRF2014 at epoch 1994.0) at ARGN stations that realise GDA94 as published in 2012

the difference between the GDA94 and ITRF2008 ellipsoid is between 70 and 120 mm (Stanaway and Roberts 2015). Figure 4 shows the differences between GDA94 (as gazetted in 2012) (Commonwealth of Australia 2012) and SAPRF2014 at epoch 1994.0 for the stations which define GDA94 on the Australian continent. The mean distortion is 20 mm in the East component, 10 mm in the North component. The difference is predominantly due to imprecision of the original ITRF92 realisation, coseismic deformation and postseismic relaxation arising from large plate boundary earthquakes. The distortion between GDA94 and ITRF2008 at epoch 1994.0 could be estimated by kriging of observed residuals between the two realisations and presented in a grid format for high precision transformations between GDA94 and SAPRF2014 (1994.0).

6 Conclusion

This paper shows that the inherent stability of many tectonic plates can be used to provide a temporally stable local reference frame to support integration, analysis and management of spatial data. A Plate Motion Model (PMM), which is distortion free, provides a simple four parameter transformation (three rotation rates and epoch difference) allowing reversible transformations between local and global reference frames such as ITRF2008. Using a PMM to describe the uniform movement of a tectonic plate also allows any localised and intraplate residual deformation to be better visualised with the option of higher precision deformation modelling to facilitate higher precision applications.

The Australian case-study describing the development of a Stable Australian Plate Reference Frame shows that the described approach is an improvement on the current mandated geodetic datum in Australia, but still features coordinates that can be considered static for all but the most precise applications.

Acknowledgements John Dawson and Guorong Hu from Geoscience Australia provided the APREF 2014 combination solution and associated SINEX file used in this paper. Their assistance and support is greatly appreciated.

References

Altamimi Z, Sillard P, Boucher C (2004) CATREF Software : combination and analysis of terrestrial reference frames. Publication LAREG SP08, Institut Géographique National

Altamimi Z, Collilieux X, Métivier L (2011) ITRF2008: an improved solution of the international terrestrial reference frame. J Geod 85(8):457–473

Altamimi Z, Métivier L, Collilieux X (2012) ITRF2008 plate motion model. J Geophys Res Solid Earth 117(B7), B07402

Argus D, Gordon R, DeMets C (2011) Geologically current motion of 56 plates relative to the no-net-rotation reference frame. Geochem Geophys Geosyst 12:Q11001. doi:10.1029/2011GC003751

Bird P (2003) An updated digital model of plate boundaries. Geochem Geophys Geosyst 4(3):1027. doi:10.1029/2001GC000252

Boucher C, Altamimi Z (1992) The EUREF terrestrial reference system and its first realizations. EUREF Meeting, Bern, March 4–6

Commonwealth of Australia (2012) ComLaw, National Measurement Act 1960 – Recognized-value standard of measurement of position 2012 (No. 1)

Dach R, Hugentobler U, Fridez P, Meindl M (eds). (2007) Bernese GPS Software version 5.0. user manual, Astronomical Institute, University of Bern, Bern

DeMets C, Gordon R, Argus D, Stein S (1990) Current plate motions. Geophys J Int 101:425–478
DeMets C, Gordon R, Argus D (2010) Geologically current plate motions. Geophys J Int 181(1):1–80. doi:10.1111/j.1365-246X.2009.04491.x
Geoscience Australia (2014a) APREF Network Solution ftp://ftp.ga.gov.au//geodesy-outgoing/gnss/solutions/misc/combination.xyz.vel.2014.0. Accessed 30 Oct 2014
Geoscience Australia (2014b) APREF SINEX file ftp://ftp.ga.gov.au//geodesy-outgoing/gnss/solutions/misc/XVSOLFIN.SNX. Accessed 30 Oct 2014
Goudarzi M, Cocard M, Santerre R (2014) EPC: Matlab software to estimate Euler pole parameters. GPS Solutions 1:153–162. doi:10.1007/s10291-013-0354-4
Haasdyk J, Roberts CA (2013) Monitoring station movement using a state-wide simultaneous 'adjustment of everything' – implications for a next-generation Australian datum. In: Proceedings of the IGNSS symposium 2013 (IGNSS2013), 16–18 July 2013, Gold Coast, 15 pp
Intergovernmental Committee on Surveying and Mapping (ICSM) (2014) Geocentric Datum of Australia 1994 (GDA94) http://www.icsm.gov.au/gda/. Accessed 30 Oct 2014
Kreemer C, Blewitt G, Klein E (2014) A geodetic plate motion and Global Strain Rate Model. Geochem Geophys Geosyst 15:3849–3889. doi:10.1002/2014GC005407
LINZ (2015) New Zealand Deformation Model 2000. http://www.linz.govt.nz/data/geodetic-system/datums-projections-and-heights/geodetic-datums/new-zealand-geodetic-datum-2000. Accessed 25 May 2015
McCaffrey R (2002) Crustal block rotations and plate coupling. In: Stein S, Freymueller J (eds) Plate boundary zones, vol 30. AGU Geodynamics Series, pp 101–122
Pearson C, Snay R (2012) Introducing HTDP 3.1 to transform coordinates across time and spatial reference frames. GPS Solutions. doi:10.1007/s10291-012-0255-y
Rebischung P, Griffiths J, Ray J, Schmid R, Collilieux X, Garayt B (2012) IGS08: the IGS realization of ITRF2008. GPS Solutions 16(4):483–494
Schwarz C (ed) (1983) North American Datum of 1983, NOAA Professional Paper NOS2, National Geodetic Survey, National Oceanic and Atmospheric Administration, US. Department of Commerce
Snay R (1999) Using the HTDP software to transform spatial coordinates across time and between reference frames. Surv Land Inf Syst 59(1):15–25
Stanaway R, Roberts CA (2015) A high-precision deformation model to support geodetic datum modernisation in Australia. In: Rizos C; Willis P (eds) Proceedings of the IAG Scientific Assembly, Potsdam, 1–6 September 2013, International Association of Geodesy Symposia, vol 143
Stanaway R, Roberts CA, Blick G (2014) realisation of a geodetic datum using a gridded absolute deformation model (ADM). In: Rizos C, Willis P (eds) Earth on the edge: science for a sustainable planet proceedings of the IAG general assembly, Melbourne, Australia, June 28–July 2, 2011, International Association of Geodesy Symposia, vol 139. pp 259–256
Steed J (1995) The Geocentric Datum of Australia – a co-ordinate system for the 21st century. Second Australian National GIS in the Geosciences Forum, Canberra Australia, pp 29–31
Tregoning P, Burgette R, McClusky SC, Lejeune S, McQueen H, Watson CS (2013) A decade of horizontal deformation from great earthquakes. J Geophys Res. doi:10.1002/jgrb.50154

A Development of the Russian Geodetic Reference Network

Suriya Tatevian and Sergei Kuzin

Abstract

Two geodetic systems are now used on the Russian territory: Regional Geodetic Coordinate System (GSC-11) and geocentric coordinate system (PZ-90.11) for the GLONASS system and for different orbital calculations. The GSC network consists of 35 (50 in future) permanent fundamental sites, equipped with dual GPS/GLONASS receivers within 700–800 km distance from each other, and of about 300 first class geodetic points. Several sites of this network are already co-located with SLR, VLBI and DORIS stations, established on Russian territory with some additional co-locations to improve the accuracy and stability of the reference network in the near future. This network will fix the national coordinate system all over the country with mean square errors at the level of 1–5 cm for absolute coordinates and for relative positioning within errors of several mm. It is assumed to integrate this network with the ITRF and to realize the unified East European – North Asian Reference Frame. The improved global geocentric system PZ-90.11, which is now used for the GLONASS system, is fixed at the 2010 epoch. This system was obtained by the processing of a large amount of GPS, GLONASS, SLR and DORIS data, obtained at the Russian and international sites. The orientation of the coordinate axes, linear scale and origin of the system coincide with the ITRF system at the 0.5 cm-level. Transformation parameters between PZ-90.11, ITRF2008 and GSC-11 are considered.

GLONASS and GPS data, obtained at the 15 new sites of the Russian fundamental geodetic network during the 2013.5–2014.5 year, were processed separately with the use of PPP strategy. The differences between coordinates of the new Russian sites, estimated by only GPS or GLONASS measurements, are in the limits of 1–10 mm.

Keywords

Geodetic reference system • GLONASS • GPS

1 Introduction

Two geodetic systems are now used on Russian territory: Regional Geodetic Coordinate System (GSC-11) and geocentric global coordinate system (PZ-90.11) for the GLONASS system and for different orbital and ballistic calculations (Boucher and Altamimi 2001). The GSC network consists of 35 (more than 50 in the nearest future) permanent core sites, equipped with dual GPS/GLONASS receivers. It is assumed to integrate this network with the ITRF and to realize the unified East European – North Asian Reference Frame. The positioning accuracy estimated by the use of GLONASS and GPS measurements, obtained at the new sites of the GSC-11, was analyzed and compared. The improved global geocentric reference system PZ-90.11, created by the State Geodetic Service, is now used for the GLONASS system. This system is fixed at the 2010.0 epoch

Dr. S. Tatevian passed away on 16 Jul 2015

S. Tatevian (✉) • S. Kuzin
Institute of Astronomy Russian Academy of Science (INASAN), 48, Pyatnitskaya St., 119017 Moscow, Russia
e-mail: statev@inasan.ru

T. van Dam (ed.), *REFAG 2014*, International Association of Geodesy Symposia 146,
DOI 10.1007/1345_2015_181

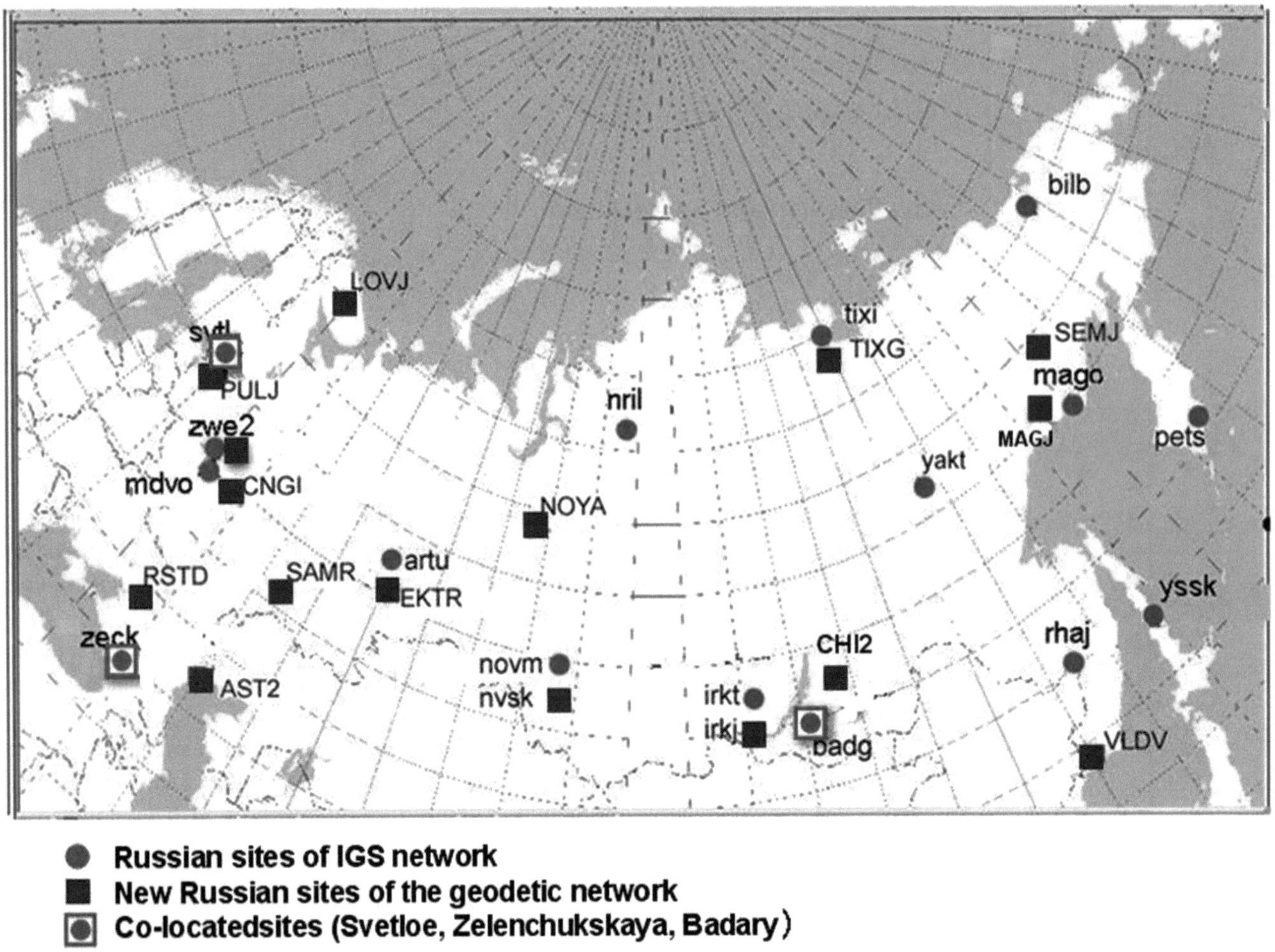

Fig. 1 Permanent sites of the GSC network

and was obtained by the processing of a large amount of GPS, SLR and DORIS data (Noll 2010), obtained at the Russian and international co-located sites. Transformation parameters between PZ-90.11, ITRF2008 and GSC-11 are presented.

2 Geodetic Reference System GSC-11

The main aims of the terrestrial geodetic system (GSC-11), which is developed now in Russia with the view of replace the systems CK-42 and CK-95 (Demyanov and Tatevian 2000), are the definition, realization and maintenance of the unified East European – North Asian Reference Frame, integrated with ITRF (Altamimi et al. 2011). Z-axis of this system is oriented to the conditional Pole (Petit and Luzum 2010) and X-axis is directed to the point of intersection of equatorial plane and zero meridian. The origin of GSC-11 system is a geometrical center of the Earth's ellipsoid (Table 2). This geocentric coordinate system will fix the national geodetic coordinate system all over the country with mean square errors at the level of 1–5 cm for absolute coordinates and for relative positioning within errors of several mm. At present, the network consists of 35 (50 in the nearest future) fundamental permanent sites, equipped with dual GPS/GLONASS receivers, within 700–800 km distance from each other, and of about 300 first class geodetic points, augmented by hundreds of lower class points (Demyanov and Tatevian 2000). The fundamental sites are equally distributed with a special account of the local geodynamic activity (Fig. 1).

Several sites of the network are already co-located with SLR, VLBI and DORIS stations, established on Russian territory, and some additional co-locations to improve the accuracy and stability of the reference network are expected in the nearest future. Normal heights, determined by precise leveling and gravimetric data at the core sites, are used for the level control. At the core sites of the network gravimetric measurements with the use of ballistic gravimeters will be repeated every 5 years. According to the Molodensky (1948) theory the value of a geodetic height above the general Earth ellipsoid, estimated by

Table 1 Parameters of the Earth's ellipsoid and constants used for the PZ-90.11 and GSC-11

Coordinate system	Geocentric gravity constant (atmosphere included) **fm** (km^3/s^2)	Angular velocity of the Earth $\dot{\omega}$ (rad/s)	Earth oblateness α	Semi major axis **a** (m)
PZ90-11 (2010)	398,600.4418	7.292115×10^{-5}	1/298.25784	6,378,136
GSC – 2011.0	398,600.4415	7.292115×10^{-5}	1/298.2564151	63,788,136.5

GNSS measurements, should strictly coincides with the sum of the normal height and quasi-geoid height. These data provide an independent control of the geocentric reference system.

The quasi geoid heights are estimated with the use of the global gravity field model. The last version of the global model has been developed at the Central Research Institute of Geodesy and Cartography (Moscow) in 2008 (Demyanov and Sermyagin 2009). Now the new version, estimated with use of GOCE data, is in preparation. The ellipsoid with a semi major axis equal to 6,378,136.5 ± 2–3 mm has been adopted for the GSC reference system.

Table 2 Transformation parameters between PZ-90.2, PZ-90.11 and ITRF 2008 systems

ΔX	ΔY	ΔZ	σx	σy	σz	*m* scale
mm			s·10^{-3}			
PZ90.02 – PZ90.11 (GLONASS)						
−9.7	−8.6	−191.5	0.295	−0.046	18.0	−1.7e-11
PZ 90.02 – ITRF						
3.3	22.4	−194.1	0.564	0.1370	17.3	−3.4e-11
PZ 90.11 – ITRF						
9.7	8.9	21.2	0.068	0.156	0.212	7.6e-11

Table 3 Dual GPS/GLONASS receivers at the GSC sites

Station	Receiver	Antenna	Radome
AST2	TPS PREGO	TPSCR3_GGD	CONE
CHI2	TPS E_GGD	TPSCR3_GGD	CONE
CNG1	TRE_G3TH_4	JAV_RINGANT_G3T	JAVC
EKTR	TPS LEGACY	JAV_RINGANT_G3T	JAVC
LOVJ	TPS PREGO	JNSMARANT_GGD	NONE
MAGJ	TPS LEGACY_E	JPSREGANT_SD_E	NONE
MOBJ	JPS LEGACY	JPSREGANT_SD_E	NONE
NOYA	JPS SIGMA	JAV_RINGANT_G3T	JAVC
NSKL	TPS ODYSSEY_E	TPSCR3_GGD	CONE
PULJ	TPS LEGACY	JAV_RINGANT_G3T	JAVC
RSTD	TPS PREGO	TPSCR3_GGD	CONE
SAMR	JPS PREGO	TPSCR3_GGD	CONE
SEMJ	TRE_G3TH_4	JAV_RINGANT_G3T	JAVC
TIXG	TPS ODYSSEY_E	TPSCR3_GGD	NONE
VLDV	TPS HGGDT	TPSCR3_GGD	NONE

3 Geocentric Global Coordinate System PZ-90.11

The geocentric coordinate system, as implemented in PZ-90.11, is a practical realization of the terrestrial reference frame at the epoch 2010.0. The PZ-90.11 is an improved version of the PZ-90.2 system. This frame is fixed by the sites of the global network, which includes all Russian GPS/GLONASS, SLR and VLBI sites, as well as a large number of IGS and DORIS sites. The accuracy of the PZ-90.11 geocentric coordinates is at cm level with a mean square error 1–5 cm. Relative positions are estimated at 0.5–1.9 cm level. The orientation of coordinate axes, linear scale and origin of the system agrees with the ITRF system at the 0.5 cm level. Gravity field of the Earth in the PZ-90.11 is presented by the planetary models with spherical harmonics up to 70° (PZ-2002/70) and up to 360° (PZ-2002/360). The model PZ-2002/70 was obtained with the use of dynamic method of space geodesy through joined adjustment of a priori gravity model parameters and geocentric coordinates of the tracking sites. This model is recommended for orbital calculations. For the development of the (PZ-2002/360) model data from the available global catalogues of the mean gravity anomalies by 30′ × 30′ trapeziums have been analyzed together with tracking data. Parameters of the Earth's ellipsoid and constants used for the PZ-90.11 and GSC-11 are shown in Table 1 and transformation parameters between PZ-90.2, PZ-90.11 and ITRF 2008 systems are shown in Table 2.

4 Processing of the GPS and GLONASS Data

GLONASS and GPS data, obtained at the 15 new sites of the Russian fundamental geodetic network during the 2013.5–2014.5 year, were analyzed at the Institute of astronomy (Moscow) (Kuzin et al. 2007; Tatevyan and Kuzin 2010). The stations are equally distributed and equipped with dual GPS/GLONASS receivers of different types (Table 3).

At the time of calculations there were 29 GLONASS satellites in constellation, including 23 operational satellites,

3 satellites in maintenance, 2 spares and one spacecraft in flight tests phase (www.glonass-ianc.rsa.ru). The spacecrafts are mostly of the GLONASS-M type. These satellites have a 7 year life span, power supply of 1,450 W, and clock stability at the level of 10^{-14}. Second civil signal at L2 frequency band is added and the attitude control accuracy has been improved to 0.5°. An improved solar panel pointing (2°) provides the better dynamical model of satellite orbits and less level of unpredicted accelerations. An estimation of the GLONASS-M standards, comparable with those of GPS-II/IIA satellites, shows an optimistic trend to the achievement of GLONASS and GPS compatibility and interoperability in terms of stability and accuracy.

Daily measurements with 30 s recording of the data from every station have been processed with the use of GIPSY-OASIS II (Linux version 6.1.2), developed at the JPL (NASA) (Webb and Zumberge 1995). The strategy of the PPP (precise point positioning) (Zumberge et al. 1997) has been applied for the analyses of GPS and GLONASS measurements separately, with the use of utility gd2p.pl (GNSS data to position) of the software.

The models common for the analyses of the GPS and GLONASS data have been applied according to the IERS standards 2010 (Petit and Luzum 2010):

- Epoch-2008;
- Ocean tidal models
- Solid Earth's tides;
- Troposphere model GPT2 (Lagler et al. 2013);
- Ocean load pole tide model for station position;
- Use second order ionosphere correction (IRI2021 model);
- Troposphere horizontal gradients estimated stochastically with the use of "random walk" strategy;
- Elevation angle for ground site −7°;
- de-weight low elevation data (weight ={1/sqrt(sin(e)})
- Receiver's phase center variations were accounted with use of IGS "antex" file (http://igscb.jpl.nasa.gov/igscb/station/general/antex14.txt);
- The coordinates are referred to the IGS08 coordinate frame, which is very close to the ITRF2008;
- The GPS Block II solar radiation pressure model was used for GLONASS satellites, with the assumption that the attitude control scheme is almost the same for GLONASS and GPS satellites.

However there are some differences in the process of GPS and GLONASS data analysis, namely:

- For GPS solutions the transmitter clock parameters and GPS orbit were fixed to the precise ephemeris and clock corrections, produced at the JPL with the use of IGS/FLINN (Fiducial Laboratorial International Natural-Science Network) data.
- GPS solutions account ambiguities of the phase measurements, but for GLONASS – these corrections not accounted for.

Table 4 Differences of coordinates, estimated separately by GPS and GLONASS data at the new sites of GSC-11 for 1 year period (in mm)

Site	(glo-gps) X mm	Rms	(glo-gps) Y mm	Rms	(glo-gps) Z mm	Rms
ast2	−2.95	9.76	0.87	9.60	−3.67	9.11
chit	−4.00	6.14	−2.23	6.82	−14.56	7.92
cngl	5.98	2.17	3.47	1.95	−14.48	3.36
ektr	4.35	10.18	−3.31	10.78	−12.19	12.13
lovj	11.39	4.56	4.42	4.24	−1.96	7.77
magj	3.68	2.29	−0.22	1.86	−9.22	3.74
mobj	−9.25	2.90	−6.45	2.63	−8.94	4.59
noya	−11.89	1.87	5.54	2.69	−13.27	4.96
nsk1	−18.15	3.58	−11.18	4.85	−16.33	6.24
pulj	−13.45	3.79	−2.65	2.40	−6.49	5.94
rstd	−5.10	5.01	4.21	4.87	−0.61	5.25
samr	−4.58	2.90	−2.28	3.15	−6.36	4.33
semj	−5.61	5.05	−4.40	5.21	−4.23	7.21
tixg	−0.74	1.82	−2.42	1.90	−12.58	5.16
vldv	3.93	10.63	1.54	10.50	−6.04	9.98

- Until recently the IGS data analysis center at the JPL didn't produce clock corrections for GLONASS solutions.
- For GLONASS solutions we have used clock corrections and orbits, provided by the GFZ/Potsdam Analytical IGS center.

Outcome of the performed computations shows that mean square residuals (*rms*) of the new sites of the Russian geodetic network were determined with the precision (*rms*) 1–10 mm by the use of measurements during one year time period. The differences between coordinates of the new Russian sites, estimated by only GPS or GLONASS measurements, are in the same limits, but they are strongly dependant on the amount and time period of measurements (Table 4).

5 Summary

The regional terrestrial reference frame, covering the large territory from the European part of Russia to the North-Eastern border of the Eurasian plate, will provide a representative covering of the largest stable areas (the Siberian and the East European) of the Eurasian plate. The precise data, obtained at the permanent sites, will provide more information for monitoring and studies of the tectonic movements in Eastern Siberia and along the Alaska-Aleutian-Kurils ark. This geodynamical active region requires attention to disaster risk reduction by all countries of the North Pacific. The Russian fundamental geodetic network is considered as a component of the multinational Circum-Pacific geodetic monitoring network that could operate in real time.

Acknowledgments Authors express gratitude to the colleagues from the Central Research Institute of Geodesy and Cartography (Moscow) for the information on the GSC and PZ-90.11 systems and measurement data from the GSC sites.

References

Altamimi Z, Collilieux X, Metivier L (2011) ITRF2008 an improved solution of the International terrestrial reference frame. J Geod 85:457–473. doi:10.1007/s00190-011-0444-4

Boucher C, Altamimi Z (2001) ITRS, PZ-90 and WGS 84: current realizations and the related transformation parameters. J Geod 75(11):613–619

Demyanov GV, Sermyagin R (2009) Planetary models of the Earth's gravity field. Geod Cartogr 10:8–12 (in Russian)

Demyanov GV, Tatevian SK (2000) Integrated geodynamical network in Russia. (Scientific objectives and realization). Phys Chem Earth (A) 25(12):819–822

Kuzin S, Revnivykh S, Tatevian S (2007) GLONASS as a key element of the Russian positioning service. Adv Space Res 39(10):1531–1538

Lagler K et al (2013) GPT2: empirical slant delay model for radio space geodetic techniques. Geophys Res Lett 40:1069–1073. doi:10.1002/grl.50288

Molodensky MS (1948) Outer gravity field and figure of the physical Earth's surface. Izvestiya RAS 12(3):193–211 (in Russian)

Noll CE (2010) The crustal dynamic data information system: a resource to support scientific analysis using space geodesy. Adv Space Res 45:1421–1440

Petit G, Luzum B (eds) (2010) IERS conventions, IERS Technical Note 36, Frankfurt am Main: Verlag des Bundesamts fur Kartographie und Geodasie, 179 pp, ISBN 3-89888-989-6

Tatevyan S, Kuzin S (2010) On the combined use of GPS/GLONASS techniques for the development of the Russian geodetic reference network. In: Kenji S (ed) Advances in geosciences, vol 26. WSPC, pp 23–32

Webb F, Zumberge J (eds) (1995) An introduction to GIPSY-OASIS II. Report JPLM D-11088, Jet Propulsion Laboratory, Pasadena

Zumberge JF, Heflin MB, Jefferson DC, Watkins MM, Webb FH (1997) Precise point positioning for the efficient and robust analysis of GPS data from large networks. J Geophys Res 102(B3):5005–5016

Weighted vs. Unweighted MCs for the Datum Definition in Regional Networks

M. Chatzinikos and C. Kotsakis

Abstract

Minimum constraints (MCs) is a standard tool for the datum definition in geodetic network adjustment and they are regularly used for the alignment of regional GNSS networks to successive realizations of the International Terrestrial Reference System. Their implementation has been restricted to an unweighted setting without using any weighting for the reference stations that participate in the datum definition process. The aim of this paper is to discuss the optimal choice problem for the weight matrix of the reference stations within the context of MC network adjustment, and to expose its relevance for practical applications having particularly in mind network densification problems. Numerical examples are presented from the alignment of two regional GNSS networks to ITRF2008 using weighted and unweighted MCs. The results obtained from our analysis offer a preliminary view of the expected improvement in the estimation accuracy of MC solutions due to the optimal weighting of the used reference stations.

Keywords

Datum definition • Minimum constraints • Network adjustment • Reference station weighting

1 Introduction

The use of minimum constraints (MCs) in geodetic network adjustment is a well known tool of fundamental importance for the realization and densification of reference coordinate frames. Although the formal meaning of the term "minimum constraints" is associated with any set of sufficient and non-distorting datum conditions whose number is equal to the rank defect of the network's observational model (e.g. Sillard and Boucher 2001), herein we adapt to its usage that is currently followed in most geodetic studies. Therefore, we refer to MCs in the sense described by Altamimi et al. (2002) which closely corresponds to the so-called inner constraints that were introduced by Meissl (1969) and Blaha (1971) for the optimum datum definition in geodetic networks. These datum constraints are applied over a set of reference stations that are included in the network adjustment, and they result in the nullification of the (non-estimable) frame transformation parameters between the adjusted coordinates and the prior coordinates of those stations. Their implementation for the integration of regional GNSS networks to the International Terrestrial Reference Frame (ITRF) was suggested by Altamimi (2003) and since then several IAG regional frame sub-commissions have been promoting this strategy for: (1) the alignment of their weekly combined solutions to successive ITRF realizations and (2) the densification of their permanent reference networks by the national mapping agencies in different countries (e.g. Bruyninx et al. 2013).

To the authors' knowledge the use of MCs is always utilized without the aid of a weight matrix for the reference stations, although this option has been sporadically mentioned in some studies (e.g. Angermann et al. 2004;

M. Chatzinikos (✉) • C. Kotsakis
Department of Geodesy and Surveying, Aristotle University of Thessaloniki, Thessaloniki 54124, Greece
e-mail: mchatzin@topo.auth.gr; kotsaki@topo.auth.gr

T. van Dam (ed.), *REFAG 2014*, International Association of Geodesy Symposia 146,
DOI 10.1007/1345_2015_184

Heinkelmann et al. 2007). Recently, a more general MC framework was presented by Kotsakis (2013, 2015) which incorporates a weighting scheme for the reference stations based on certain optimality criteria for the adjusted network coordinates. It is well known that a network solution by the classic (unweighted) MCs is already optimal in the sense of minimizing the propagated data noise to the estimated coordinates of the reference stations. Such a solution, however, does not provide optimal control (of the propagated data noise) on the estimated coordinates of other network stations, and it also ignores the random errors of the reference stations coordinates and their propagated effect to the final network solution – the latter represents what we shall briefly refer to as datum noise effect. In Kotsakis (2013) it was shown that both of these limitations can be handled through a weight matrix for the reference stations which is computed by straightforward closed-form analytic formulae. Generally speaking, this weighting tool allows us to obtain a minimally constrained solution in the same frame of the reference stations, with optimal accuracy over the entire network by minimizing the total effect of the data and datum noise on all estimated coordinates.

The aim of this paper is to present, for the first time, numerical results from the implementation of the weighted MCs in the alignment of regional GNSS networks to the ITRF2008 frame. Our tests are performed to weekly solutions of different networks with continental and national size, namely the SIRGAS reference network and a Hellenic reference network. The presented comparisons between the unweighted and weighted MC solutions refer to the weekly rms of their coordinate differences and the average station accuracy that is obtained in each case for every week.

2 Weighted MCs in Network Adjustment

In this section we give an overview of the optimally weighted MCs in geodetic network adjustment. For the sake of simplicity the notation herein is slightly different from the one used in the original presentation by Kotsakis (2013) which should be consulted for a more detailed description and related theoretical proofs.

We start with a *singular system* of normal equations (NEQ)

$$\underbrace{\begin{bmatrix} \mathbf{N_x} & \mathbf{N_{x\bar{x}}} \\ \mathbf{N_{\bar{x}x}} & \mathbf{N_{\bar{x}}} \end{bmatrix}}_{\mathbf{N}} \underbrace{\begin{bmatrix} \mathbf{x}-\mathbf{x}^o \\ \bar{\mathbf{x}}-\bar{\mathbf{x}}^o \end{bmatrix}}_{\mathbf{X}-\mathbf{X}^o} = \underbrace{\begin{bmatrix} \mathbf{u_x} \\ \mathbf{u_{\bar{x}}} \end{bmatrix}}_{\mathbf{u}} \quad (1)$$

which is deduced from the least-squares adjustment of a geodetic network in the usual linearized context. The partitioning of the above system indicates the network separation into two parts: (1) a set of reference stations (with unknown coordinates $\mathbf{x}$ and known approximate coordinates $\mathbf{x}^o$) that will be used in the datum definition process and (2) a set of non-reference stations (with unknown coordinates $\bar{\mathbf{x}}$ and known approximate coordinates $\bar{\mathbf{x}}^o$). It is considered that the rank defect of Eq. (1) is caused by the datum deficiency of the observables and that any nuisance parameters have been eliminated beforehand without causing any additional singularity problems.

Based on Eq. (1) we seek a network solution to be expressed in the same frame of the reference stations using the minimum required information for the datum definition. This requires an auxiliary (and consistent) set of linear constraints that do not distort the geometrical information of the original observations. Such a set can be generally expressed as

$$\underbrace{\begin{bmatrix} \mathbf{Q} & \mathbf{0} \end{bmatrix}}_{\mathbf{H}} \underbrace{\begin{bmatrix} \mathbf{x}-\mathbf{x}^o \\ \bar{\mathbf{x}}-\bar{\mathbf{x}}^o \end{bmatrix}}_{\mathbf{X}-\mathbf{X}^o} = \underbrace{\mathbf{Q}\left(\mathbf{x}^{ref}-\mathbf{x}^o\right)}_{\mathbf{c}} \quad (2)$$

or equivalently

$$\mathbf{Q}\left(\mathbf{x}-\mathbf{x}^{ref}\right) = \mathbf{0} \quad (3)$$

where $\mathbf{x}^{ref}$ is the known coordinate vector of the reference stations in the desired frame, which does not necessarily coincide with their approximate coordinates that were used in the NEQ formation stage. The constraint matrix $\mathbf{Q}$ should satisfy some general algebraic properties that are imposed by the MC theory in rank-deficient linear models (see e.g. Koch 1999) but it remains otherwise unspecified. In fact there are infinite options for the selection of this matrix – herein it will be uniquely determined on the basis of an optimal criterion for the estimated coordinates in the desired frame of the adjusted network.

Considering Eq. (2), the minimally constrained solution of Eq. (1) has the general form

$$\widehat{\mathbf{X}} = \mathbf{X}^o + \left(\mathbf{N}+\mathbf{H}^T\mathbf{H}\right)^{-1}\left(\mathbf{u}+\mathbf{H}^T\mathbf{c}\right) \quad (4)$$

while its covariance (CV) matrix consists of two independent components

$$\boldsymbol{\Sigma}_{\widehat{\mathbf{X}}} = \boldsymbol{\Sigma}_{\widehat{\mathbf{X}}}^{obs} + \boldsymbol{\Sigma}_{\widehat{\mathbf{X}}}^{mc} \quad (5)$$

The first component quantifies the data noise effect in the estimated coordinates and the second component represents the datum noise effect due to random errors in the known coordinates $\mathbf{x}^{ref}$ of the reference stations. Detailed analytic expressions for both of these components can be found in Kotsakis (2013, 2015).

The sought constraint matrix $\mathbf{Q}$ is hidden in the general matrix $\mathbf{H}$ (see Eq. (2)) and it imposes the datum definition based on the prior information of the reference stations. It is reasonable to choose this matrix such that the error variances of the estimated coordinates over the entire network are minimized. In this way, the frame information of the reference stations is "transferred" in an optimal way through Eq. (3) into the adjusted network or, conversely, the adjusted network is optimally "aligned" to the frame of the reference stations.

The aforementioned requirement can be expressed in terms of the minimization principle

$$\min_{\mathbf{Q}} tr\left(\boldsymbol{\Sigma}_{\widehat{\mathbf{x}}}^{obs} + \boldsymbol{\Sigma}_{\widehat{\mathbf{x}}}^{mc}\right) \tag{6}$$

which leads to the following optimal form of the constraint matrix (for the proof see Kotsakis 2013)

$$\mathbf{Q} = \mathbf{E}_{\mathbf{x}}\left(\boldsymbol{\Sigma} + \boldsymbol{\Sigma}_{\mathbf{x}}^{ref}\right)^{-1} \tag{7}$$

where $\mathbf{E}_{\mathbf{x}}$ is the usual MC matrix whose columns and rows refer to the reference stations and the non-estimable frame parameters, respectively, of the underlying network (i.e. the classic inner-constraint matrix).

The matrix $\boldsymbol{\Sigma}_{\mathbf{x}}^{ref}$ is the prior CV matrix of the known coordinates of the reference stations, and the matrix $\boldsymbol{\Sigma}$ is given by the equation (ibid.)

$$\left(\mathbf{N}+\mathbf{E}^{T}\mathbf{E}\right)^{-1} = \left[\begin{array}{c|c} \boldsymbol{\Sigma} & \# \\ \hline \# & \# \end{array}\right] \tag{8}$$

where $\mathbf{E}$ is the usual MC matrix for the entire network, i.e. $\mathbf{E} = \left[\mathbf{E}_{\mathbf{x}}\ \mathbf{E}_{\overline{\mathbf{x}}}\right]$. Note that the matrix partitioning in the last equation conforms to the NEQ partition of Eq. (1).

The fundamental result of Eq. (7) shows that a weighted type of MCs should be applied to the reference stations in order to obtain an optimal solution in the desired frame. Consequently, the weighted MCs do not represent just an additional option for the datum definition in network adjustment problems but they are, in fact, the optimum scheme under the criterion of Eq. (6) among any other choice of minimum constraints for the underlying network. The weight matrix depends on the components $\boldsymbol{\Sigma}$ and $\boldsymbol{\Sigma}_{\mathbf{x}}^{ref}$, each of which controls the influence of the reference stations into the datum definition process with regard to the data and datum noise effect, respectively. The first component is dictated by the network's own characteristics as per Eq. (8), whereas the second component depends on the prior accuracy of the reference coordinates in the desired frame; for more details see Kotsakis (2013).

3 Numerical Tests

Two case studies are presented to compare the performance of weighted and unweighted MCs for the constrained adjustment of regional GNSS networks. The first example refers to the alignment of 51 weekly solutions of the SIRGAS reference network (Sánchez et al. 2013) to the ITRF2008 frame (Altamimi et al. 2011). The second example employs a Hellenic reference network that was recently used for studying the Greek horizontal velocity field (Chatzinikos 2013) and it examines its alignment to the ITRF2008 frame over a sample of 300 weekly solutions. The weighted/unweighted MCs are applied to the weekly unconstrained NEQs of each network according to the formulation of the previous section. The prior coordinates of the reference stations ($\mathbf{x}^{ref}$) and their full CV matrix ($\boldsymbol{\Sigma}_{\mathbf{x}}^{ref}$) were extracted from the official ITRF2008 sinex file by reducing them to the mean epoch of every week with the use of the reference stations velocities and their associated CV matrix.

It is noted that the unconstrained NEQs originating from the GNSS data processing in the context of the weekly network adjustments are not strictly singular. In order to conform with the theoretical setting of MCs, a preliminary correction was applied to remove the (weak) frame origin information from the original weekly NEQs of each network. This step was implemented according to the covariance projection formulae and the pseudoinverse computation algorithm given in Pope (1973); relevant algorithms are also given in Sillard and Boucher (2001). In this way we are able to work with truly rank-deficient NEQs, and thus exploit the genuine properties of MCs without distorting the geometrical content of the network observations.

3.1 SIRGAS Network

The entire network consists of 274 stations, 18 of which were used as reference stations for its weekly alignment to ITRF2008 based on the no-net-translation (NNT) condition (this is similar to the frame alignment methodology used by the SIRGAS analysis center). The geographical distribution of all network stations is shown in Fig. 1. Both weighted and unweighted MC solutions were computed on a weekly basis for a one-year period (2013.5–2014.5) starting from GPS week 1750.

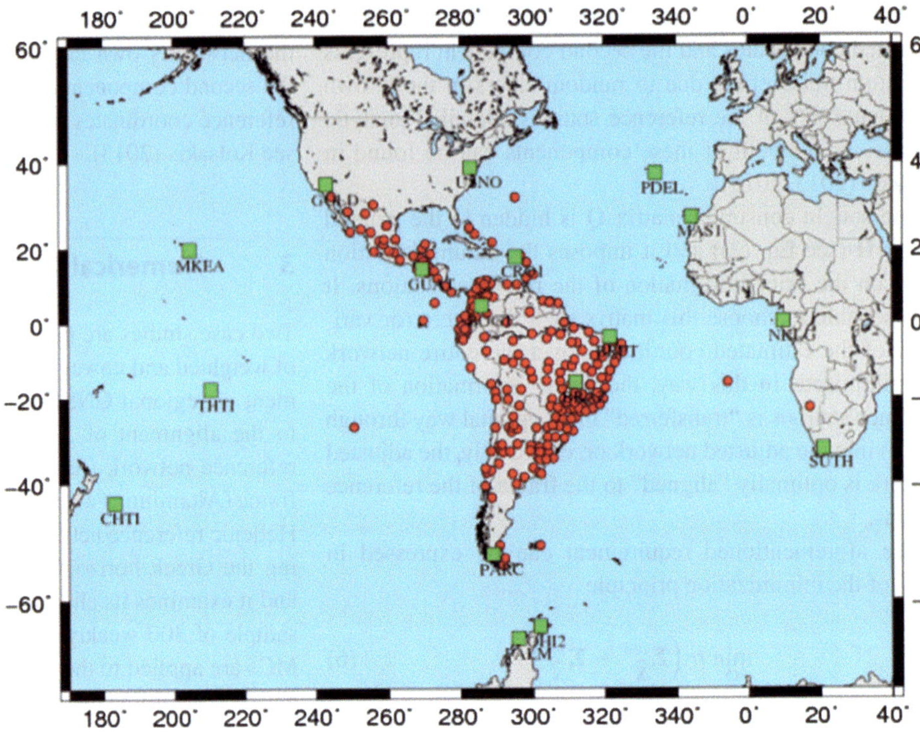

Fig. 1 The SIRGAS network (*green squares*: reference stations, *red circles*: non-reference stations)

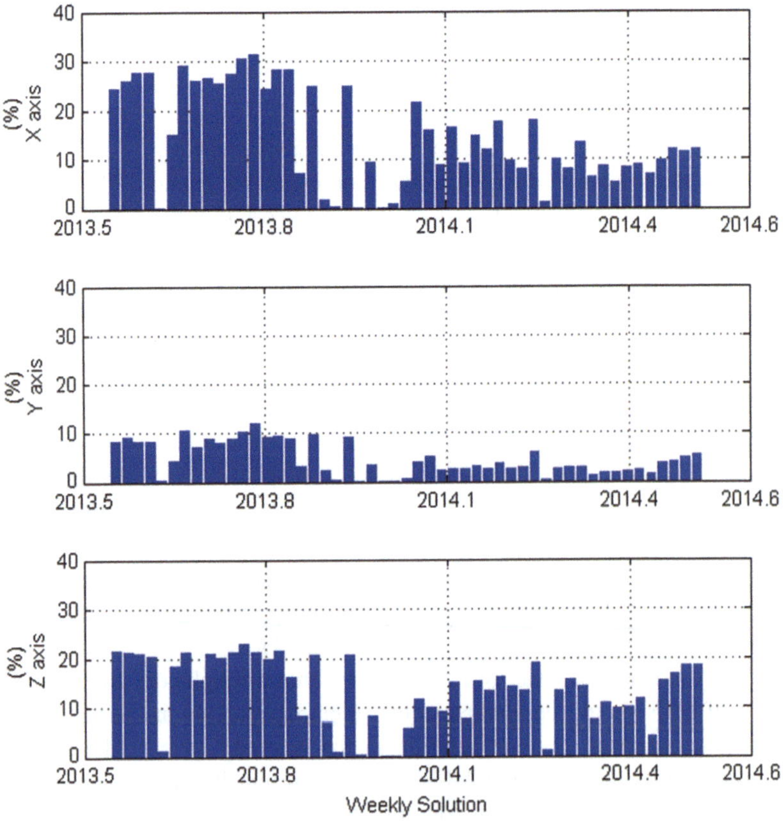

Fig. 2 Average percentage improvement of the estimation accuracy in the ITRF2008 weekly coordinates by the optimally weighted MCs in the SIRGAS network

As expected, our results revealed that the traces of the weekly CV matrices $\Sigma_{\hat{X}}$ are always smaller in the weighted MC solutions thus implying better accuracy for the estimated coordinates over the entire network. The average improvement of the estimation accuracy in each Cartesian coordinate component for every week is depicted in Fig. 2. The most significant improvement occurs in the X and Z components whose mean error variances

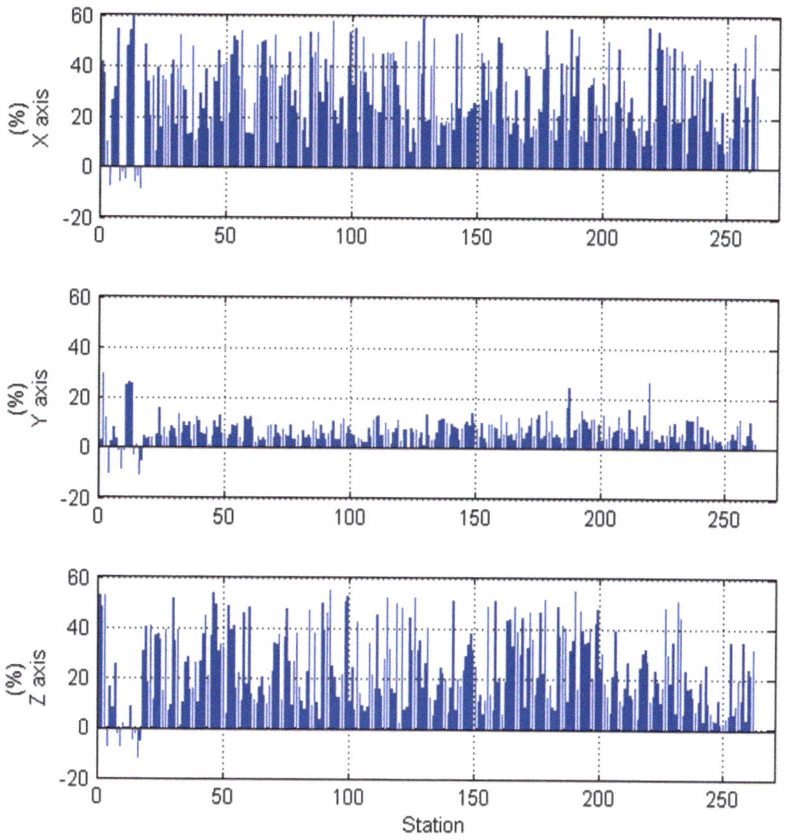

Fig. 3 Percentage improvement of the estimation accuracy in the ITRF2008 weekly coordinates *for each station* by the optimally weighted MCs in the SIRGAS network (GPS week: 1752)

(over all network stations) are reduced by up to 31% and 23%, respectively. Note that more than half of the weekly solutions showed an average decrease by more than 10% in the error variances of the estimated X and Z coordinates.

A representative example of the accuracy improvement in the estimated coordinates at each network station, for a particular GPS week, is shown in Fig. 3. Overall, the weighted MC solution leads to coordinate accuracies that are better (in average sense over all stations) by 30%, 7% and 24% in the X, Y and Z component, respectively. In a significant number of stations the accuracy improvement exceeds the 50% level for their estimated X and Z coordinates. However, seven out of the eighteen reference stations showed worse accuracy by almost 10% in their estimated coordinates that were determined by the weighted MC solution (see left side of the plots in Fig. 3).

The coordinate differences between the unweighted and weighted MC solutions, in terms of their weekly rms for each Cartesian component, are depicted in Fig. 4. Since both solutions are minimally constrained based on the NNT condition (with or without weighting for the reference stations), this rms reflects the apparent shifts between the unweighted/weighted MC solutions of the SIRGAS network along each coordinate axis. Their values range from 0 to 4 mm throughout the considered period (see Fig. 4).

3.2 Hellenic Network

The second example refers to a GNSS network consisting of 68 Greek permanent stations that belong to the Leica SmartNet network and also 16 EUREF reference stations which were used for the alignment to ITRF2008 based on the

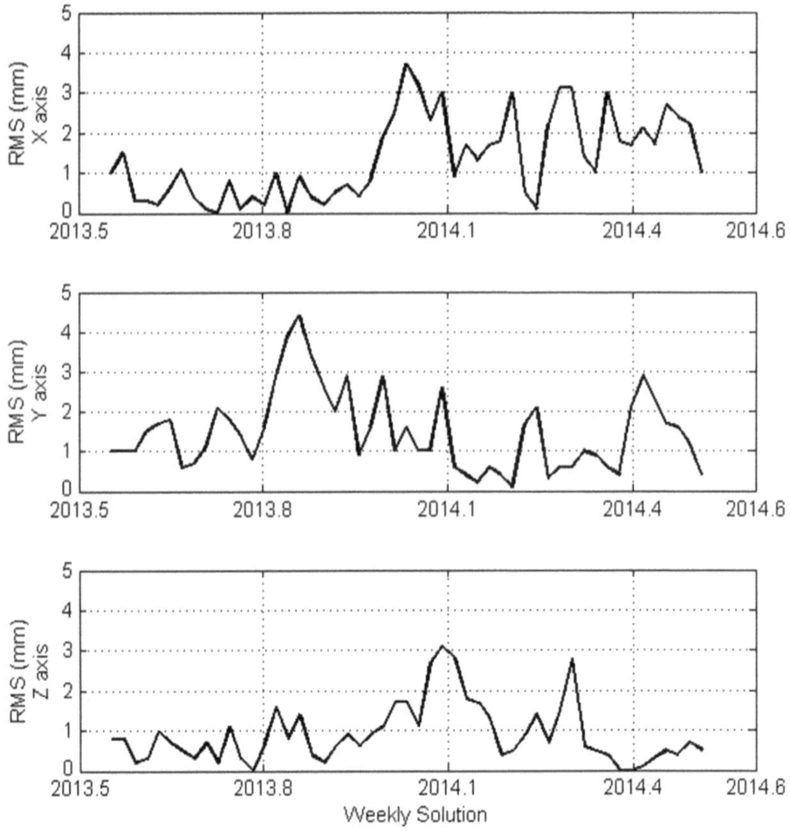

Fig. 4 Weekly rms of the coordinate differences between the weighted and unweighted MC solutions in the SIRGAS network

NNT condition (see Fig. 5). Three hundred weekly solutions, using unweighted and weighted MCs, were computed in this network for a 6-year period (2007–2013) starting from GPS week 1408.

Similarly to the previous example, the weighted MCs always improve the average accuracy (mean error variance over all network stations) of the weekly estimated coordinates. The percentage levels of this accuracy improvement for each week are shown in Fig. 6. For the Y and Z coordinates, the average reduction of their error variances throughout the six-year period is 11% and 5%, reaching up to 22% and 10% in some weeks. On the other hand, the accuracy of the X coordinates seems to be unaffected by the use of the weighted MCs since their mean error variances are not reduced by more than 2–3% in the considered period.

A representative example of the accuracy improvement in the estimated coordinates at each network station, for a particular GPS week, is shown in Fig. 7. As in the previous test with the SIRGAS network, we see again that some of the 16 reference stations have worse accuracy in their weekly estimated coordinates (especially in the Y component) obtained by the weighted MC solution.

The weekly rms of the coordinate differences between the weighted and unweighted MC solutions in the Hellenic network is shown in Fig. 8. The largest differences occur mainly in the Z component and they reach up to 4 mm, whereas for the other two components they remain below 2 mm in most weekly solutions.

4 Conclusions

The use of a weight matrix for the optimal handling of reference stations in MC network adjustment has been studied in this paper. The theoretical background of the weighting methodology was briefly presented along with some first numerical results from its implementation with real data. Our tests were performed in two regional GNSS networks using weighted and unweighted MCs for their weekly alignment to the ITRF2008 frame. A significant improvement in the

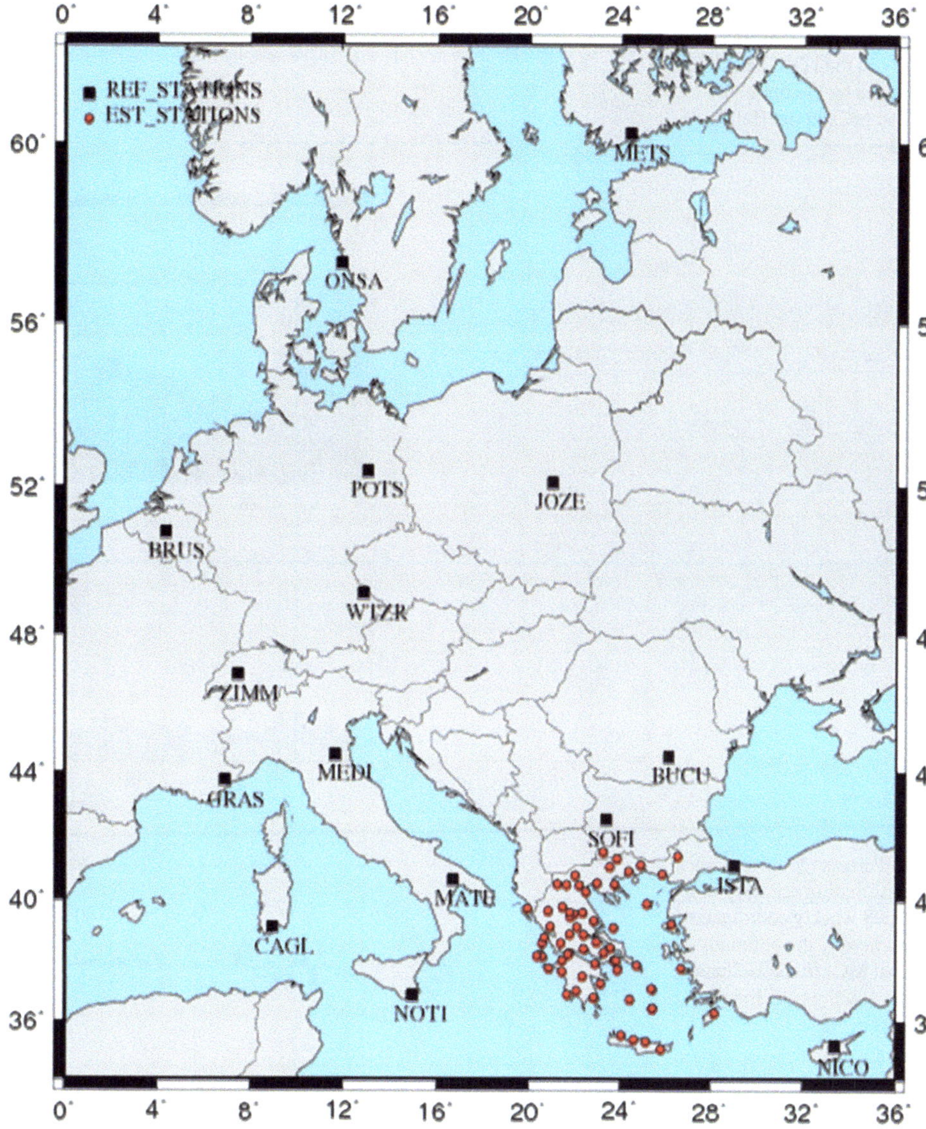

Fig. 5 The Hellenic network (*black squares*: reference stations, *red circles*: non-reference stations)

estimation accuracy of the constrained weekly solutions was achieved in both networks by the optimally weighted MCs. The percentage reduction of the error variances of the estimated coordinates varied both spatially (over different stations) and temporally (over different weeks), especially in the SIRGAS network, and it was in most cases well above the 15% level with maximum up to 50%.

The differences of the estimated weekly coordinates in ITRF2008 by the two adjustment approaches remain the same over all network stations – this is expected since the datum definition in both cases is based on the NNT condition either with or without weighting for the reference stations – but they vary among different weeks. In general, the weekly shifts between the unweighted and weighted MC solutions range within 0–4 mm for all coordinate components as shown in Figs. 4 and 8.

It is noted that, in current geodetic practice, the known coordinates of the reference stations refer to a secular frame, such as the ITRF, without accounting for nonlinear effects in their temporal evolution. The use of MCs is a valuable tool to align a series of network solutions to such a secular frame in order to investigate unmodeled seasonal signals of geophysical interest in the resulting coordinate time series. The weighting scheme presented in this paper is "blind" to such unmodeled signals while its rationale is to guarantee that the adjusted network is integrated into the secular frame without any geometrical distortion and in a way that the estimated coordinates of all stations have the best accu-

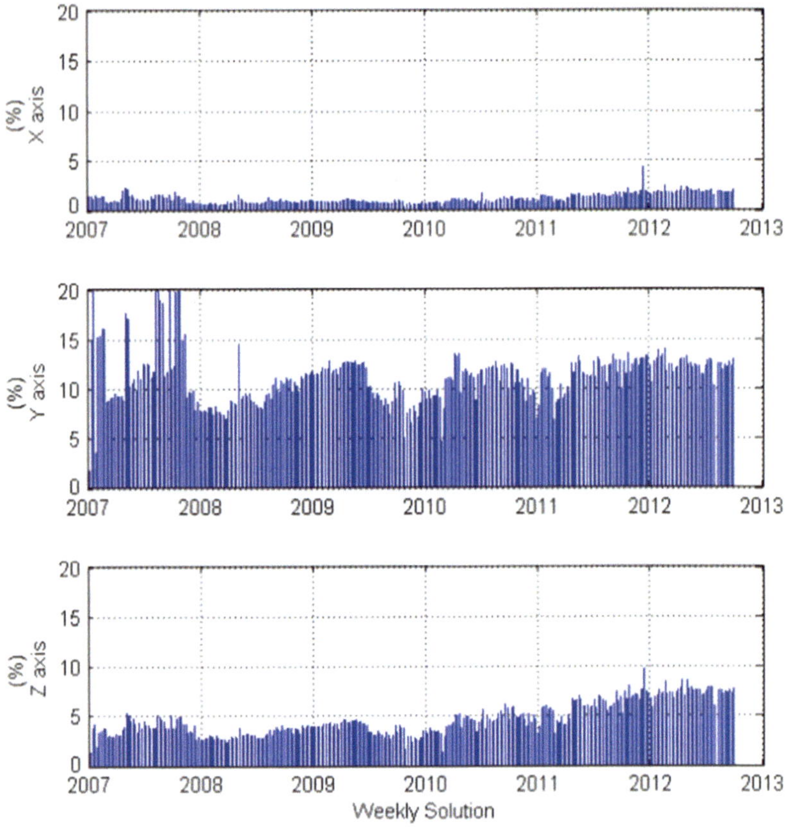

Fig. 6 Average percentage improvement of the estimation accuracy in the ITRF2008 weekly coordinates by the optimally weighted MCs in the Hellenic network

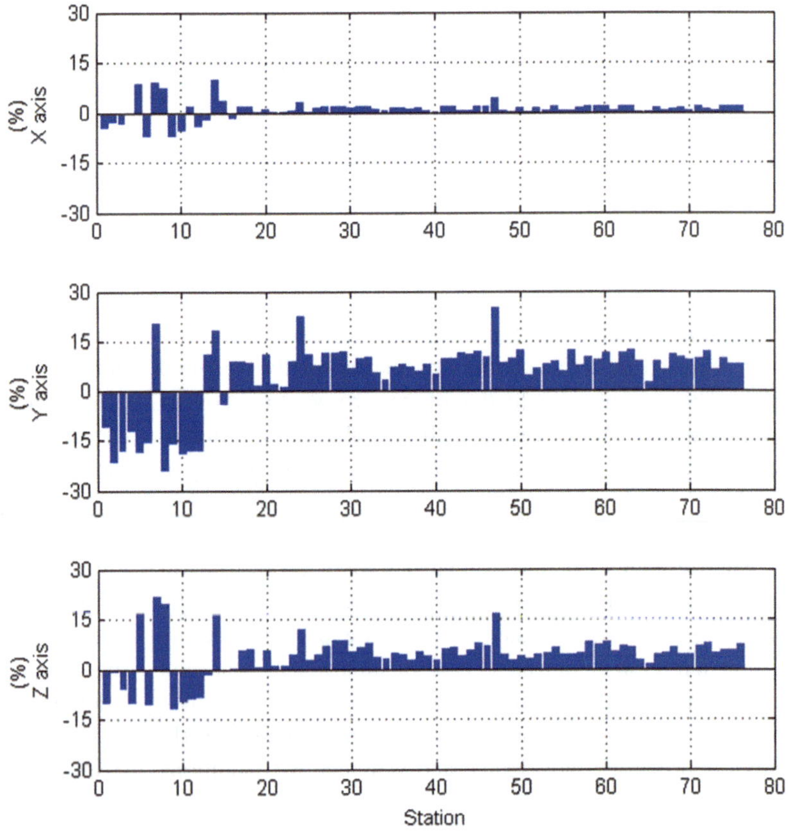

Fig. 7 Percentage improvement of the estimation accuracy in the ITRF2008 weekly coordinates *for each station* by the optimally weighted MCs in the Hellenic network (GPS week: 1669)

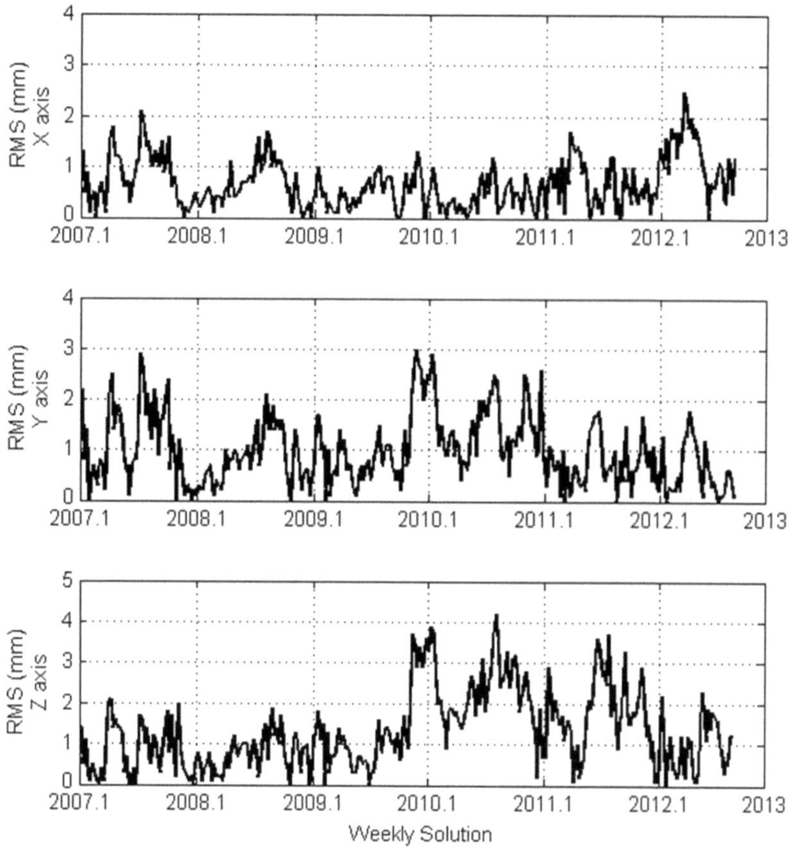

Fig. 8 Weekly rms of the coordinate differences between the weighted and unweighted MC solutions in the Hellenic network

racy (with respect to that secular frame). Whether or not our weighting methodology provides any actual advantages to better identify unmodeled effects within the estimated coordinate time series of ITRF-aligned networks is of course an interesting question that requires separate investigation.

References

Altamimi Z (2003) Discussion on how to express a regional GPS solution in the ITRF. EUREF Publication No. 12, Verlag des Bundesamtes für Kartographie und Geodäsie, Frankfurt, pp 162–167

Altamimi Z, Boucher C, Sillard P (2002) New trends for the realization of the International Terrestrial Reference System. Adv Spac Res 30:175–184

Altamimi Z, Collilieux X, Métivier L (2011) ITRF2008: an improved solution of the international terrestrial reference frame. J Geod 85:457–473

Angermann D et al. (2004) ITRS Combination Center at DGFI: a terrestrial reference frame realization 2003. DGK, Reihe B, Heft Nr 313

Blaha G (1971) Inner adjustment constraints with emphasis on range observations. Department of Geodetic Science, The Ohio State University, OSU Report No. 148, Columbus

Bruyninx C, Altamimi Z, Caporali A, Kenyeres A, Lidberg M, Stangl G, Torres JA (2013) Guidelines for EUREF Densifications. IAG Sub-commission for the European Reference Frame - EUREF. ftp://epncb.oma.be/pub/general/Guidelines_for_EUREF_Densifications.pdf

Chatzinikos M (2013) Study of the Earth's crust displacements in the area of Greece by analyzing GNSS data. PhD Thesis, School of Rural and Surveying Engineering, Aristotle University of Thessaloniki

Heinkelmann R, Boehm J, Schuh H (2007) Effects of geodetic datum definition on the celestial and terrestrial reference frames determined by VLBI. In: Boehm et al (eds) Proceedings of the 18th European VLBI for geodesy and astrometry working meeting, Technische Universität Wien, Heft Nr. 79, pp 200–205

Koch K-R (1999) Parameter estimation and hypothesis testing in linear models, 2nd edn. Springer-Verlag, Berlin Heidelberg

Kotsakis C (2013) Generalized inner constraints for geodetic network densification problems. J Geod 87:661–673

Kotsakis C (2015) Reference station weighting and frame optimality in minimally constrained networks. IAG Symposia, vol. 142, Springer Berlin Heidelberg, in press. doi:10.1007/1345_2015_139

Meissl P (1969) Zusammengfassung und Ausbau der inneren Fehlertheoric eines Punkthaufens. Deutsche Geodätische Kommission, Reihe A 61:8–21

Pope AJ (1973) The use of the 'solution space' in the analysis of geodetic network adjustments. Presented at the IAG symposium on computational methods in geometric geodesy, Oxford, September 2–8, 1973

Sánchez L, Seemüller W, Drewes H, Mateo L, González G, da Silva A, Pampillon J, Martinez W, Cioce V, Cisneros D, Cimbaro S (2013) Long-term stability of the SIRGAS reference frame and episodic station movements caused by the seismic activity in the SIRGAS region. In: IAG Symposia, vol 138. Springer Berlin Heidelberg, pp 153–161

Sillard P, Boucher C (2001) A review of algebraic constraints in terrestrial reference frame datum definition. J Geod 75:63–73

Is Nubia Plate Rigid? A Geodetic Study of the Relative Motion of Different Cratonic Areas Within Africa

Mary Njoroge, Rocco Malservisi, Denis Voytenko, and Matthias Hackl

Abstract

The Nubia plate is considered to be a rigid plate and, as such, is used in the realization of International Terrestrial Reference Frame (ITRF). Geophysical and geological observations suggest that there is intraplate deformation within the Nubia plate along the Cameroon volcanic line and the Okavango rift. To test this hypothesis and to evaluate rigid plate motion, we divide the plate into three regions and calculate six Euler vectors based on available long-term GPS data.

We process the data using GIPSY-OASIS 6.2 and analyze the resulting time series for long term, annual, and semiannual signals. We calculate uncertainties for secular velocity using the Allan variance of the rate technique. We also analyze the color of the noise of each time series as a function of latitude and climatic region, and show that it is not latitude-dependent.

Although geological and geophysical studies indicate the possibility of intraplate deformation, the current Global Positioning System (GPS) network cannot identify deformation within the Nubia plate, suggesting that it is behaving as a rigid plate within uncertainty.

Keywords

Euler vectors • Nubia plate • Plate rigidity • Reference frame

1 Introduction

Plate rigidity is one of the main paradigms of plate tectonics and is a fundamental assumption in the definition of a global reference frame like the ITRF (e.g. Altamimi et al. 2011). Although still far from optimal, the recent increase in GPS instrumentation within the African region allows us to better understand the applicability of the rigidity assumption to the Nubia plate.

The Nubia plate corresponds to the western and largest part of Africa. It formed from the early Miocene division of the African region along the continental East African Rift system (EARs) (Roberts et al. 2012) and is bordered by four extensional boundaries on the east, northeast, west, and south, and one compressional boundary on the northwest (Chu and Gordon 1999; Bird 2003). The continental part of the Nubia plate is composed of three large Archean cratonic regions (West Africa, Congo, and South African Kalahari) with lithospheric mantle thickness greater than 300 km (Begg et al. 2009), indicating a low degree of recent tectonic activity. The cratons are separated by old suture zones of possibly weaker lithosphere (e.g., Begg et al. 2009; Tokam 2010). The Nubia plate and its counterpart, the Somalia plate on the East side of EARs, are linked together by three microplates: Victoria, Rovuma, and Lwandle (Fig. 1), which

The online version of this chapter (doi:10.1007/1345_2015_212) contains supplementary material, which is available to authorized users.

M. Njoroge (✉) • R. Malservisi • D. Voytenko
University of South Florida, Tampa, FL, USA
e-mail: mnjoroge@mail.usf.edu

M. Hackl
Allianz SE Reinsurance, Munich and Germany

T. van Dam (ed.), *REFAG 2014*, International Association of Geodesy Symposia 146,
DOI 10.1007/1345_2015_212

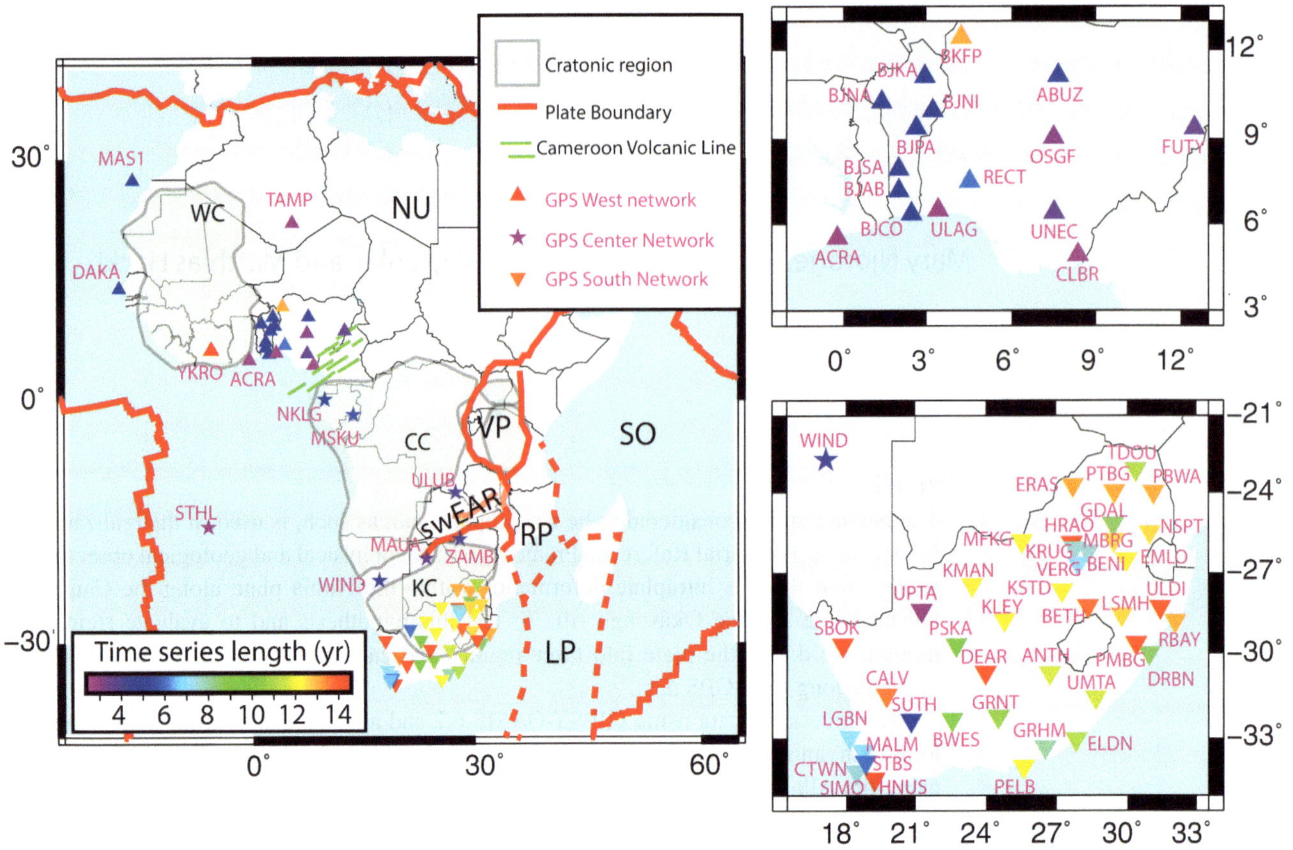

Fig. 1 Map of Africa showing the Nubia (NU) and Somalia (SO) plates and the three microplates: Victoria Plate (VP), Rovuma Plate (RP), and Lwandle Plate (LP). The swEAR indicates the South West continuation of the EARs. *Green lines* indicate the Cameroon volcanic line. *WC* indicates the West African craton, *CC* the Congo craton, and *KC* the Kalahari Craton (from Begg et al. 2009). *Red lines* show well-defined (*solid*) and assumed (*dashed*) plate boundaries (from Bird 2003 and Stamps et al. 2008). The three Nubia plate cratons (West Africa, Congo, and Kalahari) are labeled. GPS sites are color coded by time series length with a symbol indicating their regional network. *Right side* shows enlarged maps of the West (*top*) and South (*bottom*) networks

are separated by well-defined divergent boundaries (the different branches of EARs) (Hartnady 2002; Calais et al. 2006; Stamps et al. 2008; Saria et al. 2013). Plate reconstructions indicate that the Nubia plate underwent internal deformation along the suture zones during the breakup of Gondwana (e.g., Reeves and De Wit 2000; Eagles 2007; De Wit et al. 2008). Observed seismicity, geomorphology, and geophysical data suggest that the Cameroon volcanic line (CVL, the region separating West Africa and Congo cratons and a hot spot track) and the southwest propagation of the East African Rift System (swEAR) are tectonically active (e.g., Midzi et al. 1999; Modisi 2000; Hartnady 2002; Shemang and Molwalefhe 2011; Yu et al. 2015). Previous geodetic studies show that internal tectonic deformation within the Nubia plate is ≤0.6 mm/year and may be located along the swEAR (Malservisi et al. 2013; Saria et al. 2013). The uncertainties are however larger than the value itself, and the location of deformation is not well-constrained.

2 Data Acquisition and Processing

We use all of the publically available continuous GPS (cGPS) data within stable Nubia from the following archives: TRIGNET (ftp://ftp.trignet.co.za), AFREF (ftp://ftp.afrefdata.org), NIGNET (http://server.nignet.net/data), UNAVCO (ftp://data-out.unavco.org/), and CDDIS (http://cddis.nasa.gov/). To obtain a reliable velocity field, we only use sites with at least 2.5 years of data (Blewitt and Lavallée 2002; Bennett et al. 2007; Malservisi et al. 2013). The sites have also been used in previous geodetic studies, however determination of stability of monumentation is beyond the scope of this study.

We divide the sites into three main regions: South, corresponding to the Kalahari craton and South Africa (46 sites); Central, corresponding to the area between the swEAR and the CVL (7 sites); and West, including all the sites northwest of the CVL (mainly NIGNET and AMMA sites, 21 sites) (Table S1). Although the amount of analyzed data is still far from ideal for a large plate like Nubia, the velocity field presented here shows a significant improvement in time series length (3 years longer) and plate coverage (with 84 stations having more than 2.5 years of observation) compared to previous publications by Malservisi et al. (2013) and Saria et al. (2013).

We obtain daily static positions for each site using at least 20 hours of dual frequency observations. We process the data using the GIPSY–OASIS 6.2 software (Lichten and Border 1987) and the precise point positioning (PPP) method described by Zumberge et al. (1997) using orbit and clock data provided by jet propulsion laboratory (JPL). We perform phase ambiguity resolution using the single receiver algorithm (Bertiger et al. 2010), correct for ocean loading using FES2004 (Lyard et al. 2006), and calculate tropospheric delay using Vienna Mapping Functions (Boehm et al. 2006). We then align the solutions with IGb08 (Rebischung et al. 2011) through daily seven-parameter transformations (x-files) provided by JPL.

We analyze the time series for long-term trends to compute secular velocities of each site. We also analyze each component independently and correct the time series for jumps due to known equipment replacement or co-seismic signals. We visually inspect each time series and use the MATLAB code PATV (Selesnick et al. 2012) to identify other unknown jumps. We then fit each time series component using the equation

$$x(t_i) = a + vt_i + b\cos(2\pi t_i + \phi_a) + c\cos(4\pi t_i + \phi_s) + \sum_{j=1}^{m} d_j H(t_i - t_j) \quad (1)$$

where a is the position at reference time, v is the long-term secular velocity, b and ϕ_a are the amplitude and phase of the annual signal, c and ϕ_s are the amplitude and phase of the semiannual signal, m is the number of jumps within the time series at time t_j, d_j is the unknown amplitude of the jump, and $H(t_i - t_j)$ is the Heaviside function.

Following the description of Njoroge (2015) and Malservisi et al. (2015), we remove daily positions that differ by more than five times the nominal uncertainty from the time series and re-fit the data in an iterative way, until no outliers remain (generally, a single iteration is enough). We detrend the resulting clean time series to compute uncertainties. Note that applying this method to time series with large and often almost periodic gaps could be problematic. Analyzing annual or semi-annual signals in such time series affects the long-term rate much more than any estimation of velocity uncertainty. In our case, we found that the effects on the station TAMP are such that does not allow a reliable velocity estimation thus the station TAMP is not used in our analysis.

We estimate velocity uncertainties using the Allan Variance of the rate (AVR) (Hackl et al. 2011, 2013) with a combination of white and power law noise (Malservisi et al. 2013). Although we detrend the time series by removing annual and semi-annual signals, the AVR analysis indicates the presence of a periodic signal with a period between 70 and 100 days. Although we do not conduct a full spectral analysis to identify such a period, the best fit of the AVR occurs with a periodic signal of 89 days (approximately a quarter of a year), which we add to the error model. It is possible that such period is part of higher harmonic components of the yearly seasonal signal that are not removed by the filtering of Eq. (1).

Table S1 in the supplemental material has detailed information about the GPS stations and the observed secular velocities.

3 Rigid Block Motion

We apply the Euler theorem (McKenzie and Parker 1967) to calculate the rigid motion of each region with respect to IGb08. We use methodology described by Plattner et al. (2007) and Malservisi et al. (2013) to identity stations producing the best-fitting Euler vector for each region. As described in Malservisi et al. (2013) stations with larger residuals do not move according to rigid plate motion. A part from tectonic motion, various factors such as bad monumentation or local effects (e.g. water extraction or mining) could affect the motion. Thus we decided not to use those sites for Euler vector calculation. This process identifies the subset of stations used to calculate Euler vectors describing the rigid motion of the South, Central, and West regions and their combinations (West + Central, South + Central, and full Nubia). The reduced χ^2 of the obtained Euler vectors varies from 2.58 to 7.95, and the average rate residuals range from 0.33 to 0.61 mm/year. The Euler vectors are described in Tables S2 and S3.

It is important to note that we can identify a subset of stations with a reduced $\chi^2 \sim 1$ for each region. These stations have relatively long time series (approx. 6 years for West and Central and 12 years for South), but are often unevenly distributed and may not be representative of the local rigid block motion. We also note that velocity residuals computed using the larger number of stations are randomly oriented (Figs. S1 and S2). We thus suggest that the large reduced χ^2 is probably related to underestimated uncertainties instead of departure from rigid plate behavior. For these reasons, we prefer solutions with more stations even if the reduced $\chi^2 \gg 1$.

3.1 West Region Euler Vector

Although there are 20 stations in the West region, we only use 12 to calculate the Euler vector WEST (Fig. S1). Of the two colocated stations DAKR and DAKA, we kept DAKA (Table S1). We have no physical explanation for the large residuals of BJKA, BJNA, BJNI, BKFP and ULAG. It is likely that ACRA is influenced by anthropogenic activity (oil and groundwater extraction) while high residuals at YKRO may be related to the nearby Lake Kossou (Malservisi et al. 2013) observed similar behavior at sites close to lakes in South Africa). The reduced χ^2 of the resulting Euler vector is 2.58 and the mean rate residual is 0.49 mm/year (Table S2).

3.2 Central Region Euler Vector

It's the least sampled region with only seven stations. Malservisi et al. (2013) and Saria et al. (2013) showed that MSKU and ULUB have large residuals and were not used for our calculations (Table S1). Using the remaining five stations (Fig. S1) results in a reduced χ^2 of 2.65 and rate residual of 0.33 mm/year (Euler vector CENTRAL in Table S2).

3.3 South Region Euler Vector

South Africa is the region with the densest GPS coverage. Of the co-located sites HARB, HRAC, and HRAO; SUTH, SUT1, and SUTM; and UPTA and UPTN, we kept HRAO, SUTH, and UPTA respectively (Table S1). We used 28 stations to fit the Euler vector (Fig. S1). The reduced χ^2 and average residual are 7.95 and 0.39 mm/year respectively (Euler vector SOUTH in Table S2). This region also had stations with the longest time series and hence low velocity uncertainties, explaining the higher reduced χ^2. We tested the possibility of reducing the χ^2 by using sites in the driest and most stable part of the network identified by Malservisi et al. (2013) but this resulted in no significant improvement. Note that the homogeneous velocity field in the Cape Town area that Malservisi et al. (2013) suggested was related to strain accumulation is no longer visible, indicating that it may have been an artifact associated with the length of the time series.

4 Combined Euler Vectors

4.1 West-Central Region Euler Vector

To obtain this Euler vector (WEST_C in Table S2) we eliminate three extra stations due to high residuals (FUTY, BJBA and BJPA) from those used for the WEST and CENTRAL Euler Vectors. Using the remaining 14 stations (Fig. S2), the resulting fit for the Euler vector has a reduced χ^2 of 4.89 and mean rate residual of 0.61 mm/year.

4.2 South-Central Region Euler Vector

This region consists of stations in the South and Central regions (SOUTH_C in Table S2). We used only 33 sites which were used in computation of SOUTH and CENTRAL Euler vector to calculate SOUTH_C Euler vector (Fig. S2). The reduced χ^2 and mean rate residual are 7.27 and 0.41 mm/year respectively.

4.3 Nubia Euler Vector

To calculate the NUBIA Euler vector, we rely on 42 sites used to calculate the WEST_C and SOUTH_C Euler vectors (Fig. S2). The reduced χ^2 and mean rate residual of this Euler vector calculation are 6.79 and 0.47 mm/year respectively (Table S2).

The large reduced χ^2 of some Euler vector fits may be related to underestimated uncertainties. This may be due to the error model used in the AVR interpolation (periodic, power-law, and white noise) or an underestimation of a higher time-correlated noise component (random walk). Another possibility is that flicker noise is much stronger than the time-correlated noise components, which can only be observed with much longer time series. This is particularly true using AVR, where only ¼ of the full time series length is used to calculate the variance. In both cases, the error model predicts smaller uncertainties resulting in the large reduced χ^2.

In all Euler vector calculations, most stations have residuals <0.5 mm/year, indicating a possible upper limit for the internal deformation of the Nubia plate. The few stations with residuals >1 mm/year appear to be stations with problematic behavior, short time series, or gaps. All other stations, with residuals between 0.5 and 1.0 mm/year, are more likely be affected by local phenomena (e.g., subsidence or anthropogenic effects) rather than tectonic motion.

5 Comparison of Euler Vectors

Traditionally, Euler vectors are compared separately by plotting the position of the Euler pole (with relative error ellipses) and its rate (with relative uncertainties). Since the three components of the Euler vector are

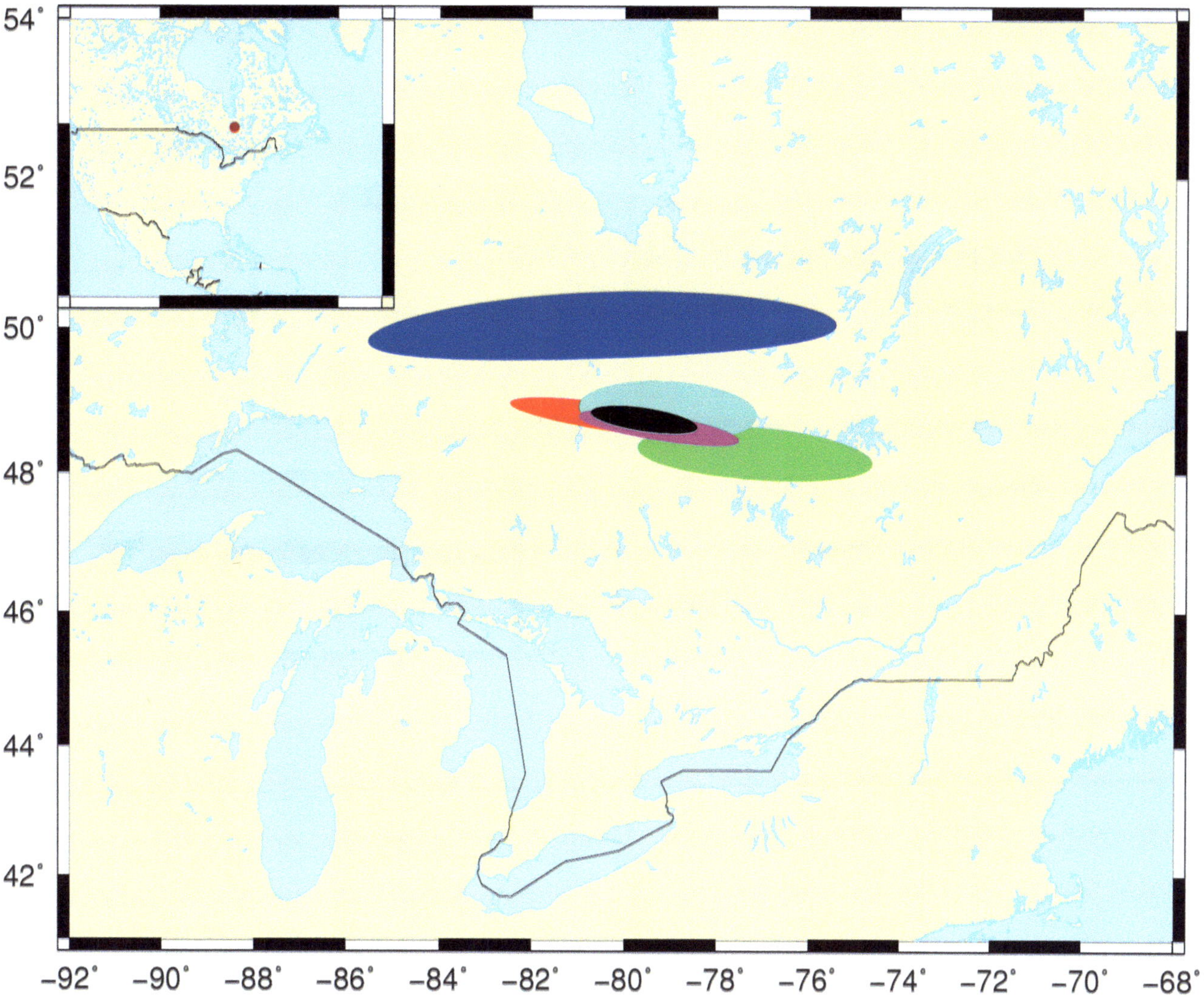

Fig. 2 Positions and 2σ error ellipses (95% confidence) of the six Euler poles: WEST (*blue*), CENTRAL (*green*), WEST_C (*cyan*), SOUTH_C (*magenta*), SOUTH (*red*), and NUBIA (*black*) in 2-dimensions, calculated with respect to the IGb08 reference frame. Within uncertainties, all of the poles except the WEST pole are compatible with each other. (The *brown circle* in the *inset* map indicates the location of the Euler poles.)

highly correlated, it is beneficial to compare the full vectors with error ellipsoids instead of ellipses. In order to make the error directly comparable, we decided to use for the 2D and 3D cases the same confidence intervals (68% and 95%) as described by Vanicek and Krakiwsky (1987). By comparing the six Euler poles calculated in this study using relative error we observe that five of them overlap at 95% confidence (Fig. 2) while four of them overlap at 68% confidence (Fig. S3). WEST Euler pole (Figs. 2 and S3) is significantly separated, suggesting that there may be relative motion between West Africa and the rest of Nubia. When comparing Euler vectors using the full covariance matrix (Fig. 3), we observe that all ellipsoids overlap at 95% confidence, indicating that at the current level of uncertainties, we cannot rule out rigid plate behavior. The ellipsoids are also partially overlapping (Fig. S4) at 68% confidence, meaning that the likelihood of rigid plate behavior for the full Nubia plate is significant, and that with current uncertainties and network geometry, the Nubia plate moves as a rigid block with respect to IGb08. Comparing our Nubia Euler vector with those of Altamimi et al. (2007), Nocquet et al. (2006) and Stamps et al. (2008) (Figs. S5 and S6) shows that the four vectors are compatible within uncertainties when using the full covariance (Fig. S6).

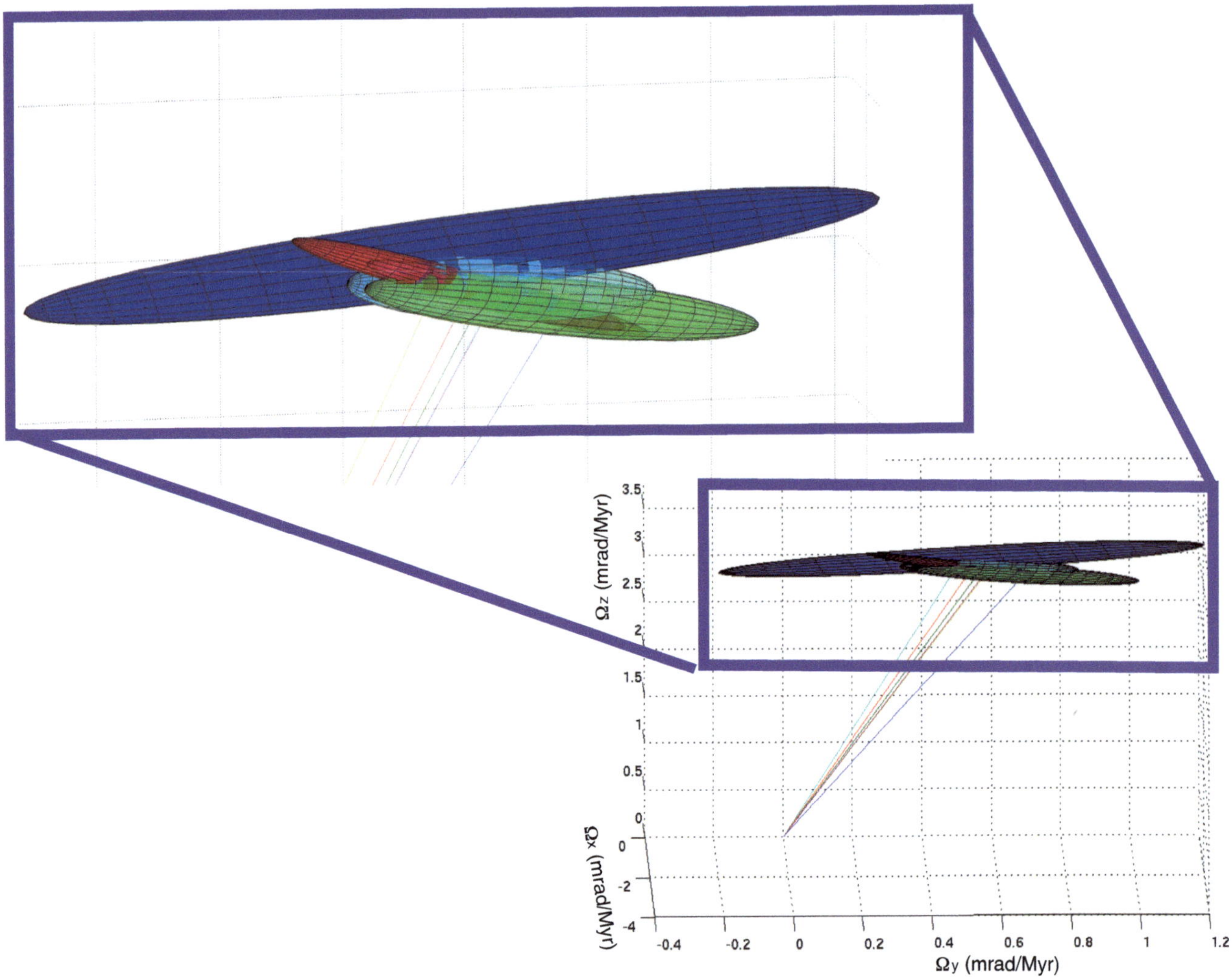

Fig. 3 Error ellipsoids (95% confidence) of six Euler vectors: WEST (*blue*), CENTRAL (*green*), WEST_C (*cyan*), SOUTH_C (*magenta*), SOUTH (*red*), and NUBIA (*black*) in three dimensions, calculated with respect to the IGb08 reference frame. All Euler vector ellipsoids intersect each other. WEST has the largest ellipsoid while NUBIA and SOUTH_C are overlapped completely

6 Annual Signal Amplitude

Many periodic signals affect GPS time series (e.g., satellite orbit configuration, seasonal variation of atmospheric water content, and groundwater storage) and are usually most prominent in the vertical component (Van Dam et al. 2010; Blewitt and Lavallée 2002; Hinderer et al. 2009; Nahmani et al. 2012). Here, we analyze the variation of the annual signal (b in Eq. 1) and how it changes as function of latitude.

The annual variation of the horizontal component ranges between 0.1 and 0.2 mm and is of similar magnitude to the repeatability of the site position. The only stations with large horizontal annual amplitudes are MSKU and KSTD, which do not fit the rigid plate behavior. The amplitude of the annual signal of the vertical component varies from 0.5 to 2.5 mm, and has strong regional variation. In the Southern and driest region, the annual signals have low amplitudes, while sites within the Western region and the Congo and Zambezi basins show large amplitudes (Fig. 4). These annual amplitudes correlate with the climatic variability of the two regions (e.g., Nahmani et al. 2012 and Ramillien et al. 2014). The central part of the West region (5° N to 15° N) has amplitudes of 1.5 to 2.5 mm, and is strongly affected by the West Africa Monsoon (WAM) (Bock et al. 2008). The second largest annual signal is at MAUA, near the Okavango River delta, one the largest inland river deltas with extensive seasonal flooding (McCarthy 1993). The phases of the annual signal also agree with the observations of Nahmani et al. (2012) and Ramillien et al. (2014) and with the water cycle (Crowley et al. 2006; Ramillien et al. 2014): the peak of the annual signal for the West network is completely out of phase with the amplitude at MAUA and the other sites around the Zambezi/Congo basins (Fig. 4). This suggests that

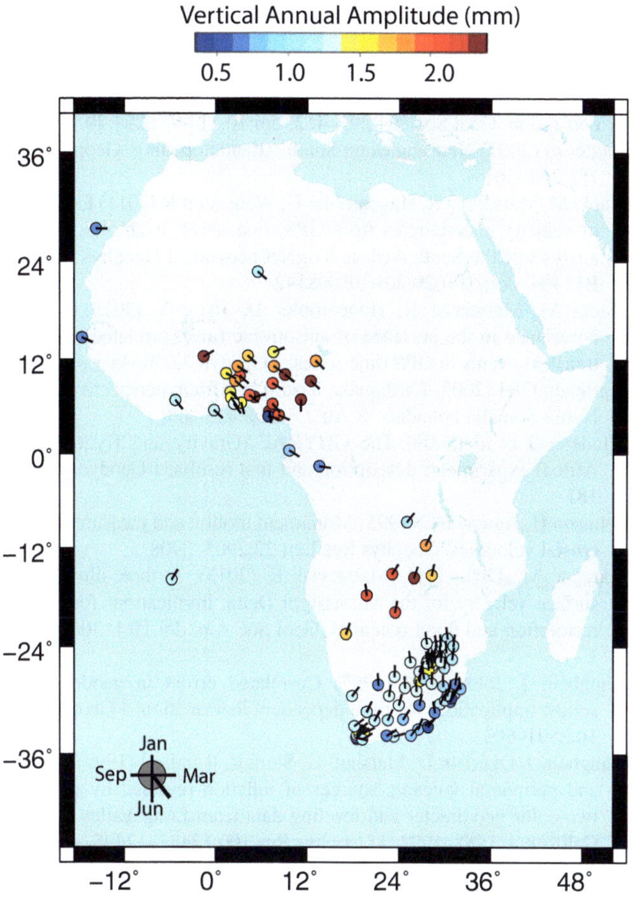

Fig. 4 Signal amplitudes of the vertical GPS component of the Nubia plate. Amplitudes vary with latitude: the South region has the smallest amplitudes while the West and Central regions have the largest amplitudes. *Lines* indicate the phase of the seasonal signal by pointing North if the peak of the signal is in January, and South if it is in June

hydrologic and atmospheric loading are probably the main sources of seasonal variation.

7 Noise Power Spectrum

GPS time series are affected by many sources of noise including GPS monument stability, antenna problems, multipath, and modeling assumptions (e.g., troposphere, ionosphere, oceanic and atmospheric loading, and orbits) (Wyatt 1989; Johnson and Agnew 1995; Langbein et al. 1995; Langbein and Johnson 1997). Some sources of noise are related to the water cycle, and we expect them to be dependent on latitude.

However, our analysis of the power spectrum of the noise component fit by the AVR white and power law model (Hackl et al. 2011), indicates that spectral characteristics do not vary with latitude. The spectral indices for all three GPS components at the sites on the Nubia plate fall between −0.6 and −1.1, with the majority clustering between −0.9 and −1.1 (essentially pure flicker noise). The small geographic variation suggests that the differences are more likely related to local effects (monument type, multipath, or human activities near the site) than to latitudinal variation. As already observed by Hackl et al. (2013), the power spectrum helps identify stations that are problematic or affected by transient behavior by having a spectral index closer to a random walk than to flicker noise.

8 Discussion and Conclusions

Despite geological and geophysical observations suggesting that there is internal deformation within the Nubia plate, our analysis shows that within our current network geometry and uncertainties, the Nubia plate behaves like a rigid block. Thus, the assumption of a rigid Nubia plate would not significantly bias a global reference frame. The Euler vectors calculated in this study indicate that the West region is the only region that may move relative to the rest of the plate. The ellipsoid corresponding to the Euler vector describing the rigid motion of this region is the only one not nested within the others. Nonetheless, it is still compatible with the other Euler vectors within 68% confidence.

Given the geophysical and geological observations of possible deformation along the CVL and swEAR, we suggest that better geometry and denser local networks are needed to identify tectonic signals in those regions.

Even within regions with a denser network like the South African Cape Town region, the GPS network is not sufficient to observe slow tectonic signals. Historically, this area has been affected by moderate to strong earthquakes (Midzi et al. 1999, 2013), but the GPS data do not show significant strain accumulation. Detailed studies accounting for local effects at each station, and a better realization of local reference, are necessary to identify such signals.

Large reduced χ^2 and the magnitude and orientation of residuals suggest that we underestimate uncertainties. Our choice of the error model (a periodic signal combined with white and power law noise) could lead to such underestimation, or we may need longer time series to better estimate the higher correlated noise. Time series are affected by both anthropogenic (e.g. mining, agricultural, water extraction, and damming) and natural (drought, water cycle, and atmospheric) signals that are quasi-periodic. These signals may affect our velocity field and uncertainty estimation because they cannot be fully corrected for using periodic signals. A detailed analysis similar to Karegar et al. (2015) could improve our ability to separate these quasi-periodic signals and to better quantify the velocity field and its uncertainties. Furthermore, a denser and better distributed GPS network is needed for an improved understanding of intraplate deformation and associated hazards.

Acknowledgment We thank the editor and three anonymous reviewers for the interesting and thoughtful comments that have significantly improved the paper. The authors want also to thanks T.H. Dixon, U. Hugentobler, and M. Rodgers for the useful comments and help improving the manuscript.

A.1 Electronic Supplementary Material

Below is the link to the electronic supplementary material.
[INSERT CAPTION HERE] (PDF 310 kb)
[INSERT CAPTION HERE] (PDF 187 kb)
[INSERT CAPTION HERE] (PDF 2584 kb)
[INSERT CAPTION HERE] (PDF 402 kb)
[INSERT CAPTION HERE] (PDF 2562 kb)
[INSERT CAPTION HERE] (PDF 451 kb)
[INSERT CAPTION HERE] (DOCX 24 kb)

References

Altamimi Z, Collilieux X, Legrand J, Garayt B, Boucher C (2007) ITRF2005: a new release of the international terrestrial reference frame based on time series of station positions and Earth orientation parameters. J Geophys Res 112, B09401. doi:10.1029/2007JB004949

Altamimi Z, Collilieux X, Métivier L (2011) ITRF2008: an improved solution of the international terrestrial reference frame. J Geod 85(8):457–473. doi:10.1007/s00190-011-0444-4

Begg GC et al (2009) The lithospheric architecture of Africa: seismic tomography, mantle petrology and tectonic evolution. Geosphere 5:23–50

Bennett RA, Hreinsdóttir S, Velasco MS, Fay NP (2007), GPS constraints on vertical crustal motion in the northern Basin and Range. J Geophys Res Solid Earth 34. doi:10.1029/2007GL031515

Bertiger W, Desai SD, Haines B, Harvey N, Moore AW, Owen S, Weiss JP (2010) Single receiver phase ambiguity resolution with GPS data. J Geod 84:327–337. doi:10.1007/s00190-010-0371-9

Bird P (2003) An updated digital model of plate boundaries. J Earth Sci 1027:1525–2027. doi:10.1029/2001GC000252

Blewitt G, Lavallée D (2002) Effect of annual signals on geodetic velocity. J Geophys Res. doi:10.1029/2001JB000570

Bock O et al (2008) The West African Monsoon observed by ground based GPS receivers during the AMMA project. J Geophys Res 113, D21105. doi:10.1029/2008JD010327

Boehm J, Werl B, Schuh H (2006) Troposphere mapping functions for GPS and very long baseline interferometry from European Centre for Medium-Range Weather Forecasts operational analysis data. J Geophys Res 111, B02406. doi:10.1029/2005JB003629

Calais E, Ebinger CJ, Hartnady C, Nocquet JM (2006) Kinematics of the East African Rift from GPS and earthquake slip vector data. In: Yirgu G, Ebinger CJ, Maguire PKH (eds) The Afar Volcanic province within the East African Rift system, vol 259. Geological Society London Special Publications, London, UK, pp 9–22

Chu D, Gordon R (1999) Evidence for motion between Nubia and Somalia along the Southwest Indian Ridge. Nature 398:64–66

Crowley JW, Mitrovica JX, Bailey RC, Tamisiea ME, Davis JL (2006) Land storage within the Congo Basin inferred from GRACE satellite gravity data. Geophys Res Lett 33, L19402. doi:10.1029/2006GL027070

De Wit M, Stankiewicz J, Reeves C (2008) Restoring Pan-African Brasiliano connections: more Gondwana control, less Trans-Atlantic corruption. Geol Soc 294:399–412. doi:10.1144/SP294.20

Eagles G (2007) New angles on South Atlantic opening. Geophys J Int 168:353–361

Hackl M, Malservisi R, Hugentobler U, Wonnacott R (2011) Estimation of velocity uncertainties from GPS time series: examples from the analysis of the South African TrigNet network. J Geophys Res 116, B11404. doi:10.1029/2010JB008142

Hackl M, Malservisi R, Hugentobler U, Jiang Y (2013) Velocity covariance in the presence of anisotropic time correlated noise and transient events in GPS time series. J Geodyn 72:36–45

Hartnady CJH (2002) Earthquake hazard in Africa: perspectives on the Nubia-Somalia boundary. S Afr J Sci 98:425–428

Hinderer J et al (2009) The GHYRAF (Gravity and Hydrology in Africa) experiment: description and first results. J Geodyn 48:172–181

Johnson H, Agnew DC (1995) Monument motion and measurements of crustal velocities. Geophys Res Lett 22:2905–2908

Karegar M, Dixon TH, Malservisi R (2015) A three-dimensional surface velocity for the Mississippi Delta: implications for Coastal restoration and flood potential. Geol Soc Am. doi:10.1130/G36598.1

Langbein J, Johnson H (1997) Correlated errors in geodetic time series: implications for time-dependent deformation. J Geophys Res 102:591–603

Langbein J, Dzurisin D, Marshall G, Stein R, Rundle J (1995) Shallow and peripheral volcanic sources of inflation revealed by modeling two-color geodimeter and leveling data from Long Valley Caldera, California, 1988–1992. J Geophys Res 100:12487–12495

Lichten S, Border J (1987) Strategies for high-precision global positioning system orbit determination. J Geophys Res Solid Earth 192:12751–12762. doi:10.1029/JB092iB12p12751

Lyard F, Lefevre F, Letellier T (2006) Modelling the global ocean tides: modern insights from FES2004. Ocean Dyn 56(5–6):394–415. doi:10.1007/s10236-006-0086-x

Malservisi R, Ugentobler U, Wonnacott R, Hackl M (2013) How rigid is a rigid plate? Geodetic constraint from the TrigNet CGPS network, South Africa. Geophys J Int 192:918–928. doi:10.1093/gji/ggs081

Malservisi R, Schwartz SY, Voss N, Protti M, Gonzales V, Dixon TH, Jiang Y, Newman AV, Richardson JA, Walter JI, Voytenko D (2015) Multiscale postseismic behavior on a megathrust: the 2012 Nicoya earthquake, Costa Rica. Geochem Geophys Geosyst 16. doi:10.1002/2015GC005794

McCarthy TS (1993) The great inland deltas of Africa. J Afr Earth Sci 17(3):275–291

McKenzie D, Parker R (1967) The North Pacific: an example of tectonics on a sphere. Nature 216:1276–1280

Midzi V, Hlatywayo D, Chapola L, Kebede F, Atakan K, Lombe D, Turyomurugyendo G, Tugume F (1999) Seismic hazard assessment in Eastern and Southern Africa. Ann Geofis 42:1067–1083

Midzi J, Bommer J, Strasser FO, Albini P, Zulu BS, Prasad K, Flint NS (2013) An intensity database for earthquakes in South Africa from 1912 to 2011. J Seismol 17(4):1183–1205

Modisi M (2000) Fault system of the southeastern boundary of the Okavango Rift, Botswana. J Afr Earth Sci 30:569–578

Nahmani S, Bock O, Bouin M, Santamaría-Gómez A, Boy J, Collilieux X, Métivier L, Panet I, Genthon P, de Linage C, Wöppelmann G (2012) Hydrological deformation induced by the West African Monsoon: comparison of GPS, GRACE and loading models. J Geophys Res 117:b05409. doi:10.1029/2011jb009102

Njoroge M (2015) Is Nubia plate rigid? A geodetic study of the relative motion of different cratonic areas within Africa. Master's thesis, University of South Florida

Nocquet JM, Willis P, Garcia S (2006) Plate kinematics of Nubia-Somalia using a combined DORIS and GPS solution. J Geod 80:591–607
Plattner C, Malservisi R, Dixon TH, LaFemina P, Sella GF, Fletcher J, Suarez-Vidal F (2007) New constraints on relative motion between the Pacific Plate and Baja California microplate (Mexico) from GPS measurements. Geophys J Int. doi:10.1111/j.1365-246X.2007.03494.x
Ramillien G, Frappart F, Seoane L (2014) Application of the regional water mass variations from GRACE satellite gravimetry to large scale water management in Africa. Remote Sens 6:7379–7405. doi:10.3390/rs6087379
Rebischung P, Griffiths J, Ray J, Schmid R, Collilieux X, Garayt B (2011) IGS08: the IGS realization of ITRF2008. GPS Solution 16:483–494. doi:10.1007/s10291-011-0248-2
Reeves C, De Wit M (2000) Making ends meet in Gondwana: retracting the transforms of the Indian Ocean and reconnecting continental shear zones. Terra Nova 12:272–280
Roberts EM, Stevens NJ, O'Connor PM, Dirks PHGM, Gottfried MD, Clyde WC, Armstrong RA, Kemp AIS, Hemming S (2012) Initiation of the Western branch of the East African Rift coeval with the Eastern branch. Nat Geosci 5(4): 289–294. doi:10.1038/ngeo1432
Saria E, Calais E, Altamimi Z, Willis P, Farah H (2013) A new velocity field for Africa from combined GPS and DORIS space geodetic solutions: contribution to the definition of the African reference frame (AFREF). J Geophys Res Solid Earth 118:1677–1697. doi:10.1002/jgrb.50137
Selesnick IW, Arnold S, Dantham VR (2012) Polynomial smoothing of time series with additive step discontinuities. IEEE Trans Signal Process 60(12):6305–6318
Shemang E, Molwalefhe L (2011) Geomorphic landforms and tectonism along the eastern margin of the Okavango Rift Zone, Northwestern Botswana, as deduced from geophysical data. In: Sharkov E (ed) New frontiers in tectonics research, general problems, sedimentary basins, and island arcs. Intechopen, Rijeka, pp 169–182
Stamps DS, Calais E, Saria E, Hartnady C, Nocquet JM, Ebinger C, Fernandez R (2008) A kinematic model for the East African Rift. Geophys Res Lett 35:L05304. doi:10.1029/2007GL032781
Tokam KAP (2010) Crustal structure beneath Cameroon (West Africa) deduced from the joint inversion of Rayleigh wave group velocities and receiver functions. PhD thesis, University of Yaoundé
van Dam T, Altamimi Z, Collilieux X, Ray J (2010) Topographically induced height errors in predicted atmospheric loading effects. J Geophys Res 115, B07415. doi:10.1029/2009JB006810
Vanicek P, Krakiwsky E (1987) Geodesy: the concepts, 2nd edn. Elsevier, Amsterdam
Wyatt F (1989) Displacements of surface monuments: vertical motion. J Geophys Res 94:1655–1664
Yu Y, Gao SS, Moidaki M, Reed CA, Liu KH (2015) Seismic anisotropy beneath the incipient Okavango rift: implications for rifting initiation. Earth Planet Sci Lett 430:1–8. doi:10.1016/j.epsl.2015.08.009
Zumberge JF, Heflin M, Jefferson D, Watkins M, Webb F (1997) Precise point positioning for the efficient and robust analysis of GPS data from large networks. J Geophys Res Solid Earth. doi:10.1029/96JB03860

Part IV

Celestial to Terrestrial Frame Transformations

How Consistent are The Current Conventional Celestial and Terrestrial Reference Frames and The Conventional Earth Orientation Parameters?

Robert Heinkelmann, Santiago Belda, José M. Ferrándiz, and Harald Schuh

Abstract

Many applications in geodesy, geodynamics, astronomy and space navigation depend on the availability of accurate Earth orientation parameters (EOP). EOP are the orientational part of the transformation between terrestrial and celestial reference frames. The conventional EOP refer to conventional frames. The current conventional terrestrial reference frame, ITRF2008, and the attached EOP, IERS 08 C04, have been determined combining the Earth rotation parameters derived from the four observing techniques but keeping the celestial pole offsets obtained by VLBI unchanged. This set of EOP refers to ITRF2008, but it does not directly refer to the current conventional celestial reference frame, ICRF2. Therefore, the conventional reference frames and the IERS 08 C04 are not entirely consistent. In the paper we assess this inconsistency by VLBI data analysis. For test purposes, we have to fix coordinates on the frames. This approach causes small systematics of the EOP. These systematics are interpreted as the uncertainty of our assessment method that is about 30 μas and about 3 μas/year. The VLBI-only terrestrial reference frame, VTRF2008, is consistent with ICRF2 at the 10 μas level. We thus interpret the differences between EOP based on this frame and EOP based on ITRF2008 as the inconsistency of IERS 08 C04 w.r.t. ICRF2. The largest difference was found for y_p being −38.8 μas and −18.6 μas/year. Applying our method, we also found differences in $dUT1$ of 11.3 μs when comparing EOP based on ITRF2008 and DTRF2008, an alternative TRF determined at DGFI, Munich. This is astonishing, because both frames are derived from identical input data. The orientation of a terrestrial reference frame depends to a significant part on the choice of stations for the no net rotation constraint. Our conclusion is that a single solution that involves the determination of both fundamental frames, ITRF and ICRF, is the only way to obtain conventional EOP that provide accuracy.

Keywords

Consistency • Earth orientation parameters (EOP) • International Celestial Reference Frame (ICRF) • International Terrestrial Reference Frame (ITRF) • Very Long Baseline Interferometry (VLBI)

R. Heinkelmann (✉) • H. Schuh
Helmholtz Centre Potsdam, GFZ German Research Centre for Geosciences, Potsdam, Germany
e-mail: heinkelmann@gfz-potsdam.de

S. Belda • J.M. Ferrándiz
Universidad de Alicante, Space Geodesy Laboratory, Matemática Aplicada, EPS, Alicante, Spain

1 Definition

During the last years many papers were published about the consistency of reference frames and EOP, e.g. Seitz et al. (2011) or Malkin (2013). In the current IERS conventions 2010 (Petit and Luzum 2010; Chap. 5, p. 43ff) the Earth orientation parameters (EOP) are defined as the arguments of

T. van Dam (ed.), *REFAG 2014*, International Association of Geodesy Symposia 146,
DOI 10.1007/1345_2015_149

the rotation matrices relating the GCRS (Geocentric Celestial Reference System) orientation to the ITRS (International Terrestrial Reference System) orientation:

$$[GCRS] = Q(X, Y, s)\, R(-ERA)\, W\left(-s', x_p, y_p\right) [ITRS] \tag{1}$$

where the EOP are symbolically written as dX, dY – celestial pole offsets (CPO; Seidelmann and Kovalevsky 2002), ERA – Earth rotation angle ($\propto UT1$), and x_p, y_p – pole coordinates. In addition in the text it is stated that

> [... The above equation], as well as the following formulas in this chapter, are theoretical formulations that refer to reference "systems". However, it should be clear that the numerical implementation of those formulas involves the IAU/IUGG adopted realization of those reference systems, i.e. the International Terrestrial Reference Frame (ITRF) and the International Celestial Reference Frame (ICRF), respectively.

The conventional EOP are those that refer to the conventional frames, currently ITRF2008 (Altamimi et al. 2011) and ICRF2 (Fey et al. 2009). The inconsistencies are:

- deviation of ICRS theory and realization
- deviation of ITRS theory and realization
- deviation of EOP theory and realization
- separate estimation of ITRF, ICRF and EOP

The first two have been assessed by us in our previous work (Heinkelmann et al.). The fourth item can be assessed comparing the current EOP to EOP determined together with ITRF and ICRF as it was achieved by Seitz et al. (2014). Here we focus on the third item.

1.1 ICRF Orientation and EOP

The coordinate differences of elements common to two celestial reference frames of type ICRF can result in a misalignment of the orientation of the axes of the frames

$$\begin{pmatrix} \Delta\alpha \\ \Delta\delta \end{pmatrix} = \begin{pmatrix} \tan\delta\cos\alpha * A_1 + \tan\delta\sin\alpha * A_2 - A_3 \\ -\sin\alpha * A_1 + \cos\alpha * A_2 \end{pmatrix} \tag{2}$$

expressed by the three global rotation angles: A_1, A_2 and A_3. A change of the orientation of the celestial frame systematically affects the EOP

$$A_1 \cong -dY; A_2 \cong dX; A_3 \cong dUT1/r' + R_z \tag{3}$$

where $r' \approx 0.997$ denotes the ratio between solar and sidereal time (Aoki et al. 1982) and R_z is the part of the global rotation given with respect to the terrestrial reference frame.

1.2 ITRF Orientation and EOP

The coordinate differences of elements common to two terrestrial reference frames of type ITRF can result in a misalignment of the orientation, of the origin, and of the network scale of the frames

$$\begin{pmatrix} \Delta x \\ \Delta y \\ \Delta z \end{pmatrix} = \begin{pmatrix} -y * R_z + z * R_y \\ x * R_z - z * R_x \\ -x * R_y + y * R_x \end{pmatrix} \tag{4}$$

where R_x, R_y, R_z are small global rotation angles. Translations and network scale changes have been omitted in Eq. (4). A change of the orientation of the terrestrial frame systematically affects the EOP

$$R_x \cong -dy_p;\ R_y \cong dx_p;\ R_z \cong A_3 - dUT1/r' \tag{5}$$

2 Consistency Among Reference Frames and EOP

2.1 VLBI Data and Data Analysis

In this paper we assess the level of inconsistency of the conventional frames, ITRF2008 and ICRF2, and hence the level of inconsistency of the IERS 08 C04 w.r.t. the conventional EOP defined by Eq. (1). To assess the inconsistency among terrestrial and celestial frames, and the involved EOP, the only technique that can be applied is VLBI because it provides the direct link between Earth crust fixed observatories (ITRF) and extragalactic radio sources (ICRF). The ICRF is directly realized by VLBI, whereas the ITRF contains VLBI but its orientation rests predominantly upon GPS (Navstar Global Positioning System). Applying VLBI, we want to assess the level of inconsistency this choice produces. Consequently, the uncertainty of our assessments is limited to the quality of the VLBI data analysis. Hence we first quantify this quality. The VLBI solutions used here are based on the GFZ contribution to ITRF2013 (now ITRF2014) (Heinkelmann et al. 2014) that fully adheres to IERS Conventions (2010). Apart from the solution provided for ITRF computation, for the solutions compared here non-tidal atmospheric loading was in addition accounted for using the GSFC (NASA Goddard Space Flight Center) atmospheric pressure loading model[1] (Petrov and Boy 2004). In addition, the number of sessions was limited to the geometrically stable sessions with more than three network stations (about 3050 sessions). The VLBI solutions cover a period from 1984-07-09 until 2013-12-31. For each of the comparisons the number of sessions was additionally limited to those sessions

[1] http://gemini.gsfc.nasa.gov/aplo.

with a chi-squared per degree of freedom ≤ 3. The EOP were estimated as unconstrained offsets individually for each VLBI session.

2.2 EOP Theory and Realization Through IERS 08 C04

The conventional EOP are the EOP determined together with the conventional terrestrial and celestial reference frames. As such the time series of session-wise EOP parameters, the conventional EOP, are consistent with parameters valid for all sessions, such as constant angles or positions and linear velocities or piece-wise linear models. IERS 08 C04 (Gambis 2004; Bizouard and Gambis 2009) is an EOP series that is based on a weighted combination of weekly SINEX files provided by the IAG Services. This operational solution is then made consistent with the EOP that were determined together with the multi-session positions and velocities in the ITRF2008. Consequently, if we want to assess the long-term consistency, we have to assess the consistency of IERS 08 C04 w.r.t. the frames using the multi-session coordinates given in the catalogues, i.e. fixing the coordinates on a priori values. However, if station positions are fixed on the catalogue values the neglected non-linear station displacements affect the EOP (Bloßfeld et al. 2014). This approach is usually not applied for precise analyses but it is necessary here for the consistency assessment. To assess this effect we determine two identical solutions: one solution with fixed coordinates and one with adjusted coordinates; in both cases the a priori values are taken from ITRF2008. For the solution with adjusted coordinates we applied no net rotation (NNR) and no net translation constraints on the session level including those stations reported in ITRF2008. Comparing the estimated EOP offsets w.r.t. IERS 08 C04 (Table 1) we find a maximal shift difference of $-33.5\,\mu$as and a maximal drift difference of $2.9\,\mu$as/year both for y_p. We take these values for the uncertainty of our VLBI based consistency assessment.

2.3 Consistency Assessment

Our approach for the assessment of the consistency of ICRF2 and ITRF2008 is comparing ITRF2008 with VTRF2008 (Böckmann et al. 2010), the VLBI input to ITRF2008. Both catalogues are based on the same VLBI data but their orientations are realized in different ways. ITRF2008 contains in addition to VLBI the results of the three satellite geodetic techniques: DORIS (Doppler Orbitography and Radiopositioning Integrated by Satellite), GNSS (Global Navigation Satellite Systems), and SLR (Satellite Laser Ranging). VTRF2008 is consistent with ICRF-Ext.2 (Fey et al. 2004) and thus we have to quantify the effects when fixing the defining sources on ICRF-Ext.2 instead of on ICRF2. Therefore, we compare two identical VLBI solutions where station coordinates were fixed to ITRF2008 and where the radio source coordinates of the defining sources are fixed on the values given in the two different catalogues (Table 2). The maximal EOP differences obtained by this comparison appear at celestial pole offsets and are a shift of $9.9\,\mu$as in X and a drift of $1\,\mu$as/year in Y. Those are at the accuracy level of the ICRF2 axes of about $10\,\mu$as and can be considered negligible. Therefore we can assume that the ICRF-Ext.2 and ICRF2 orientations are identical within this uncertainty.

Next, we determine EOP with identical solutions where the defining source coordinates are fixed on ICRF2 catalogue values and where station positions are fixed on various terrestrial reference frames, namely ITRF2005 (Altamimi et al. 2007), VTRF2008, or DTRF2008 [an alternative TRF determined at the IERS Combination Center at DGFI, Munich Seitz et al. (2012)] and compare them to those determined by fixing station coordinates on ITRF2008 (Table 3). ITRF2005 is included in the comparison because the orientation of ITRF2008 is realized by NNR constraint with respect to the orientation of ITRF2005 and thus one would expect negligible or very small rotational differences. It is noticeable that although the ITRF2008 and the ITRF2005 are constrained to be kinematically non-rotating we find significant rotations for the VLBI subset of stations, in particular in the pole

Table 1 VLBI EOP estimates when fixing the defining sources to ICRF2 and fixing or not fixing the station positions to ITRF2008 catalogue values

EOP	ITRF2008 fixed		ITRF2008 free	
	Shift	Drift	Shift	Drift
x_p	−17.4(10.8)	−2.5(0.5)	−28.2(14.1)	−2.5(0.5)
y_p	−0.6(9.5)	1.2(0.4)	−34.1(14.9)	4.1(0.5)
$UT1$	5.7(0.6)	−0.2(0.03)	6.0(0.8)	−0.2(0.03)
dX	−1.2(5.6)	0.6(0.2)	0.7(5.8)	0.4(0.2)
dY	14.5(5.8)	−3.2(0.2)	15.2(5.9)	−3.3(0.2)

Shift at epoch 2000.0 (μas, μs for UT1), drift (μas/year, μs/year for UT1) with respect to a priori EOP (IERS 08 C04). Formal errors are given in brackets

Table 2 VLBI EOP estimates when fixing station coordinates on ITRF2008 and fixing defining sources on ICRF2 or ICRF-Ext.2

EOP	ICRF2 fixed		ICRF-Ext.2 fixed	
	Shift	Drift	Shift	Drift
x_p	−17.0(10.8)	−2.6(0.5)	−22.3(11.1)	−2.9(0.5)
y_p	−0.4(9.5)	1.2(0.4)	8.0(9.8)	0.6(0.5)
$UT1$	5.8(0.7)	−0.2(0.03)	4.7(0.7)	−0.2(0.03)
dX	−1.2(5.6)	0.7(0.3)	8.7(6.2)	1.3(0.3)
dY	14.5(5.8)	−3.1(0.3)	10.0(6.7)	−2.1(0.3)

Shift at epoch 2000.0 (μas, μs for UT1), drift (μas/year, μs/year for UT1) with respect to a priori EOP (IERS 08 C04). Formal errors are given in brackets

Table 3 VLBI EOP estimate differences when fixing defining source coordinates on ICRF2 and station coordinates on different terrestrial reference frames in comparison to fixing station coordinates on ITRF2008; top: shift at epoch 2000.0 (μas, μs for UT1), below: drifts (μas/year, μs/year for UT1). Formal errors in brackets

EOP shift	ITRF2005	VTRF2008	DTRF2008
x_p	−11.3(4.5)	−48.9(1.4)	30.8(1.4)
y_p	−31.4(5.7)	−27.9(1.5)	−9.0(1.7)
$UT1$	7.1(0.3)	6.2(0.06)	11.5(0.07)
dX	−2.2(1.0)	−1.3(0.7)	−1.0(0.6)
dY	0.7(1.0)	1.6(0.7)	0.5(0.7)
EOP drift	ITRF2005	VTRF2008	DTRF2008
x_p	7.6(0.4)	−1.2(0.1)	−1.8(0.1)
y_p	−8.8(0.6)	−19.9(0.1)	0.5(0.1)
$UT1$	−0.5(0.02)	−0.1(0.01)	−0.2(0.01)
dX	0.0(0.1)	0.2(0.1)	0.1(0.03)
dY	−0,1 (0.1)	−0.3(0.1)	−0.1(0.03)

coordinates, e.g. −31.4 μas and −8.8 μas/year for y_p. The kinematic non-rotation holds only for the set of stations that are included in the NNR constraint. Obviously the set of ITRF2008 NNR stations and the VLBI subset of stations differ, so that the non-rotation is not given for the VLBI part of ITRF2008. Fixing coordinates on VTRF2008, the VLBI-only frame that is consistent with ICRF2 at the 10 μas uncertainty level, results in large shifts and drifts of the pole coordinates and $UT1$ w.r.t. IERS 08 C04, in particular for x_p (−48.9 μas) and for y_p (drift of −19.9 μas/year). The EOP differences when fixing on DTRF2008 in comparison to ITRF2008 are also significant, in particular the shifts; the EOP determined with station coordinates fixed on the two catalogues differ most significantly in terms of $UT1$ (11.5 μs), but the effect on x_p (30.8 μas) is also noticeable.

3 Summary, Discussion and Conclusions

We assessed the consistency of ITRF2008, IERS 08 C04, and ICRF2 by VLBI data analysis. We found significant effects for y_p between ITRF2008 and ITRF2005. The orientation of ITRF2008 is realized with NNR constraint w.r.t. ITRF2005 but the set of stations involved in the constraint differs from the VLBI subset of stations. This circumstance downgrades the application of ITRF2008 for EOP determination using the VLBI technique. The inconsistency is at the same level as the uncertainty of our approach (fixing or not fixing coordinates on ITRF2008). However, the results obtained using VLBI data analysis are independently confirmed through the EOP[2] that were consistently determined together with ITRF2005 (IERS 05 C04) and ITRF2008 (IERS 08 C04), respectively (Fig. 1). The shifts and drifts between IERS 05 C04 and IERS 08 C04 (for y_p 55.9 μas and 6.5 μas/year) show how much the EOP are affected even though the one frame is NNR constrained to the other frame.

Considering the uncertainty of our approach, one significant inconsistency is the drift difference of about −19.9 μas/year between ITRF2008 and VTRF2008 for y_p. The effects causing this drift need to be identified. Most likely is the explanation that the drift comes through the multi-technique combination. The orientation of the VLBI ground network that refers to ICRF2 via the VLBI EOP is allowed to rotate during the ITRF combination. Because of this fact the reference to ICRF2 is lost. Another explanation could be that the unconsidered geophysical models (e.g. hydrological loading) cause effects that differ between the space geodetic techniques, e.g. due to the different geometries of the station networks. The fact that the kinematically NNR constraint depends on the subset of points must also be considered in this context. When choosing the stations for the NNR constraint of ITRF, care should be taken that the non-rotation appropriately holds for the subsets of stations of the individual techniques as well. The difference could be caused by the fact that the ITRF2008 NNR constraint in contrast to the VTRF2008 NNR constraint does not provide kinematical non-rotation for the VLBI subset of points.

The other significant inconsistency is in $UT1$ between all the tested TRFs but in particular between ITRF2008 and DTRF2008, where it reaches 11.5 μs equivalent to more than 5 mm on the Earth equator. Seitz et al. (2012) also find large differences between DTRF2008 and ITRF2008 at the level of $\Delta R_z = 5.3$ mm. Both frames, ITRF2008 and DTRF2008, are based on identical input data and differ only in terms of the combination procedure. The differences in the combination procedures that lead to the different orientations should be exposed. The most likely explanation is the fact that the subset of datum stations of ITRF2008 involves geographically well distributed sites of all techniques, whereas the DTRF2008 subset is based on IGS08 only, the GNSS contribution to ITRF2008. Furthermore, the combination algorithm of DTRF2008 includes celestial pole offsets, while the combination algorithm of ITRF2008 does not. In other words, the Earth rotation parameters during creation of ITRF2008 have a loose end in UT1 (unspecified angle A_3 in Eq. (5) and truncated CPO). Consequently, the Earth rotation parameters determined with ITRF2008 – and so the IERS 08 C04 on the long-term – do not refer to ICRF2 but to the dynamic satellite orbit frames of the involved satellite

[2]http://www.iers.org/IERS/EN/DataProducts/EarthOrientationData/eop.html.

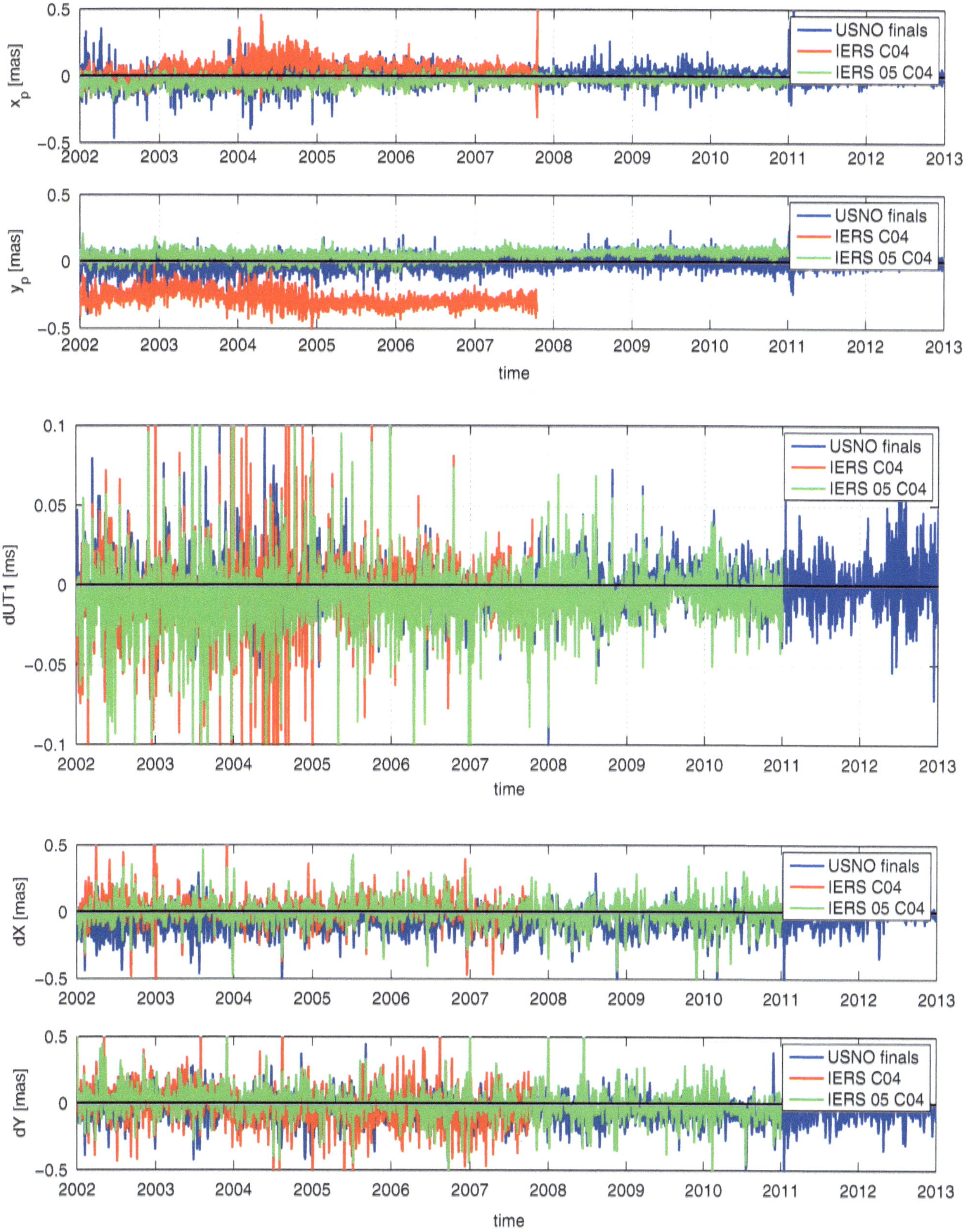

Fig. 1 Difference of EOP prior to ITRF2005 ("IERS C04") and consistent to ITRF2005 ("IERS 05 C04") with those consistent to ITRF2008. The difference to the United States Naval Observatory (USNO) finals is included as well for an estimate of the effect of a different combination procedure on EOP

techniques, predominantly GPS, to which the CPO are (later) added without considering the correlations between the EOP. This might partly cause the large difference in UT1.

The celestial pole offsets associated with IERS 08 C04 disagree with the United States Naval Observatory (USNO) final series (Fig. 1), a VLBI solution independent of our approach. The mean differences between the celestial pole offsets of IERS 08 C04 and USNO finals are $-61.1\,\mu$as for X and $50.6\,\mu$as for Y. With our consistency assessment we found no effects on celestial pole coordinates at this level. Consequently, the differences must come from the different combination procedures. The celestial pole offsets published with IERS 08 C04 were later added based on weekly SINEX. This approach introduces inconsistencies because the correlations between the EOP are suppressed. We expect the consistency between the EOP and the accuracy of the celestial pole offsets associated with IERS 08 C04 to improve, if celestial pole offsets are among the EOP that are determined together with ITRF. Concluding, the IERS 08 C04 cannot be considered conventional according to the definition (Eq. (1)). Furthermore, we do not recommend using IERS 08 C04 for accuracy assessments, in particular for VLBI EOP analyses. The IERS 08 C04 series are continuously updated based on the incoming results. This process causes additional inconsistencies that remain to be assessed. Since the analysis of space geodetic techniques requires these updates, this inconsistency between reference frames and EOP cannot be omitted and remains also in the case of conventional EOP.

The conventional EOP are consistent with the multi-session regularized station coordinates and hence contain neglected non-linear station displacements. To avoid the degradation of EOP due to uncorrected non-linear station displacements, all significant effects have to be considered on the observation level. Alternatively, the concept of multi-session coordinates could be replaced by time series or epoch reference frames (Bloßfeld et al. 2014).

IUGG Res. 3 (2011) addresses that

> [...] highest consistency between the ICRF, the ITRF, and the EOP as observed and realized by the IAG and its components such as the IERS should be a primary goal in all future realizations of the ICRS..

Beyond this resolution, to ensure consistency of ITRF and ICRF, all observations of all involved techniques have to be included in a single solution where the EOP of the satellite techniques are determined referring to the involved celestial pole offsets and UT1 provided by the VLBI technique. The EOP determined with this approach would follow the definition (Eq. (1)) and would be conventional EOP.

As we have shown in a previous paper (Heinkelmann et al. 2015), the orientation of ICRF2 is the most accurately known orientation currently realized by any global frame. The dynamical reference frames realized by the satellite configurations are not comparably accurate and thus, we expect a significant improvement of accuracy of the EOP, if the full set of EOP is considered during ITRF and ICRF computation.

Acknowledgements José M. Ferrándiz and Santiago Belda acknowledge partial support of the Spanish MINECO under grants AYA2010-22039-C02-01 and CGL2010-12153-E.

References

Altamimi Z, Collilieux X, Legrand J, Garayt B, Boucher C (2007) ITRF2005: a new release of the International Terrestrial Reference Frame based on time series of station positions and Earth Orientation Parameters. J Geophys Res 112:B09401

Altamimi Z, Collilieux X, Métivier L (2011) ITRF2008: an improved solution of the international terrestrial reference frame. J Geod 85:457–473

Aoki S, Guinot B, Kaplan GH, Kinoshita H, McCarthy DD, Seidelmann PK (1982) The new definition of universal time. Astron Astrophys 105:359–361

Bizouard C, Gambis D (2009) The combined solution C04 for Earth Orientation Parameters, recent improvements. In: Drewes H (ed) Proceedings of the IAG symposium "Geodetic reference frames - REFAG2006", Munich, Germany, 9–14 October 2006. International Association of Geodesy Symposia, vol 134, pp 265–270

Bloßfeld M, Seitz M, Angermann D (2014) Non-linear station motions in epoch and multi-year reference frames. J Geod 88:45–63

Böckmann S, Artz T, Nothnagel A (2010) VLBI terrestrial reference frame contributions to ITRF2008. J Geod 84:201–219

Fey AL, Ma C, Arias EF, Charlot P, Feissel-Vernier M, Gontier A-M, Jacobs CS, Li J, MacMillan DS (2004) The second extension of the international celestial reference frame: ICRF-Ext.2. Astron J 127:3587–3608

Fey AL, Gordon D, Jacobs CS (eds) (2009) The second realization of the international celestial reference frame by very long baseline interferometry, IERS Technical Note 35, 204. Verlag des Bundesamts für Kartographie und Geodäsie, Frankfurt am Main (2009)

Gambis D (2004) Monitoring earth orientation using space-geodetic techniques: state-of-the-art and prospective. J Geod 78(4–5):295–303

Heinkelmann R, Karbon M, Nilsson T, Raposo-Pulido V, Soja B, Schuh H. Reference frame-induced errors in VLBI Earth orientation determinations. In: Proceedings of the IAG Scientific Assembly, Potsdam, Germany, 1–6 September 2013. International Association of Geodesy Symposia (accepted)

Heinkelmann R, Nilsson T, Karbon M, Liu L, Lu C, Mora-Diaz JA, Parselia E, Raposo-Pulido V, Soja B, Xu M, Schuh H (2014) The GFZ VLBI solution: characteristics and first results. In: Behrend D, Baver KD, Armstrong KL (eds) IVS 2014 General Meeting Proceedings "VGOS: The New VLBI Network". Science Press, Beijing, pp 330–334

Malkin Z (2013) Impact of seasonal station motions on VLBI UT1 intensives results. J Geod 87:505–514

Petit G, Luzum B (eds) (2010) IERS Conventions (2010), IERS Technical Note 36, 179. Verlag des Bundesamts für Kartographie und Geodäsie, Frankfurt am Main. http://www.iers.org/IERS/EN/Publications/TechnicalNotes/tn36.html

Petrov L, Boy J-P (2004) Study of the atmospheric pressure loading signal in very long baseline interferometry observations. J Geophys Res 109:B0340

Seidelmann PK, Kovalevsky J (2002) Application of the new concepts and definitions (ICRS, CIP and CEO) in fundamental astronomy. Astron Astrophys 392:341–351

Seitz M, Heinkelmann R, Steigenberger P, Artz T (2011) Common realization of terrestrial and celestial reference frame. In: Alef W, Bernhart S, Nothnagel A (eds) Proceedings of the 20th meeting of the European VLBI group for geodesy and astrometry, Universität Bonn, Insitut für Geodäsie und Geoinformation, Schriftenreihe 22, ISSN 1864-1113, pp 123–127

Seitz M, Angermann D, Bloßfeld M, Drewes H, Gerstl M (2012) The 2008 DGFI realization of the ITRS: DTRF2008. J Geod 86:1097–1123

Seitz M, Steigenberger P, Artz T (2014) Consistent adjustment of combined terrestrial and celestial reference frames. In: Rizos C, Willis P (eds) Proceedings of the IAG General Assembly, Melbourne, Australia, 28 June - 2 July 2011. International Association of Geodesy Symposia, vol 139 "Earth on the Edge: Science for a Sustainable Planet", pp 215–221

The Effects of Simulated and Observed Quasar Structure on the VLBI Reference Frame

Stanislav S. Shabala, Lucia Plank, Robert G. Schaap, Jamie N. McCallum, Johannes Böhm, Hana Krásná, and Jing Sun

Abstract

Radio-loud quasars making up the Celestial Reference Frame are dynamic objects with significant structure that changes on timescales of months and years. This is a problem for geodetic VLBI, which has so far largely treated quasars as point sources in analysis. We quantify the effects of various levels of source structure on the terrestrial (TRF) and celestial (CRF) reference frames using the source structure simulator recently implemented in the Vienna VLBI Software (VieVS) package. We find that source structure affects station positions at the level of 0.2–1 mm. While quasar structure contributes only $\sim$10% to the total TRF error budget, which is dominated by tropospheric turbulence; the effect of quasar structure on the CRF is discernible even in present-day observations.

Astrophysical properties of quasars are related to their structure and geodetic stability, and we discuss several quasar structure mitigation strategies. These include: (1) astrophysically-based quasar selection techniques; (2) scheduling sources by taking into account source structure; and (3) analyzing geodetic observations using knowledge of source structure. We find that for observed highly variable quasars, flux density is strongly anti-correlated with structure and position stability, suggesting that such quasars should preferentially be observed in their bright phase. We use simulations to investigate new scheduling strategies which avoid unfavourable jet—baseline orientations. Improvement is seen at the millimetre level on the longest baselines when our new scheduling strategy is used in simulations that only include quasar structure. This improvement disappears in the full simulations including the troposphere, because we are compromising sky coverage in order to mitigate source structure effects. This again confirms that, at present, tropospheric turbulence dominates the accuracy of TRF determination. However, the contribution of quasar structure will become more important as tropospheric effects decrease in future broadband observations.

Keywords

Astrometry • Celestial reference frame (CRF) • Geodesy • Quasar structure • Terrestrial reference frame (TRF) • Very long baseline interferometry (VLBI)

S.S. Shabala (✉) • L. Plank • R.G. Schaap • J.N. McCallum
School of Physical Sciences, University of Tasmania, Private Bag 37, Hobart, TAS 7001, Australia
e-mail: Stanislav.Shabala@utas.edu.au

J. Böhm • H. Krásná
Department of Geodesy and Geoinformation, Vienna University of Technology, Gusshausstrasse 27-29, 1040 Vienna, Austria

J. Sun
Shanghai Astronomical Observatory, 80 Nandan Road, Xuhui, Shanghai, China

T. van Dam (ed.), *REFAG 2014*, International Association of Geodesy Symposia 146,
DOI 10.1007/1345_2015_219

1 Introduction

Very Long Baseline Interferometry (VLBI) is the only space geodetic technique that ties the Terrestrial Reference Frame (TRF) to a quasi-inertial Celestial Reference Frame (CRF), realised by distant quasars. In this way, VLBI can uniquely simultaneously measure station positions on Earth, quasar positions on the sky, and the Earth Orientation Parameters (EOPs) connecting the two reference frames. The fundamental VLBI observable is the *group delay*, a difference in signal arrival times measured by cross-correlating signals at two radio telescopes observing the same quasar. Because of the finite nature of the speed of light, group delay measurements can be directly translated to a *baseline length*, the distance between the two telescopes. By repeating these measurements for a large set of baselines and quasars, VLBI observations allow the TRF, CRF and EOPs to be determined.

An ideal VLBI quasar is a perfect point source with a fixed position. In reality, quasars are regions of intense electromagnetic radiation associated with black holes at the centres of distant galaxies (Fig. 1). Radio emission in quasars comes from synchrotron-emitting plasma located at some distance from the black hole, and in many cases extended jets are seen in addition to compact cores (e.g. Lister et al. 2009). This departure from being point-like introduces an extra contribution to the measured group delay. Charlot (1990) and Fey and Charlot (2000) introduced the concept of *structure index*, $SI = 1 + 2 \log \tau_{\text{median}}/\text{ps}$, a logarithmic quantity defined in terms of the median contribution by quasar structure to group delay over all Earth-bound baselines. A structure index of 3 corresponds to a median group delay of 10 ps, or equivalently 3 mm in distance. Quasars with structure indices in excess of 3 are therefore not recommended for use in geodesy (Ma et al. 2009).

Importantly, the structure of quasars can vary significantly on timescales of months and years (Fig. 2) due to the complex interaction of infalling matter and magnetic fields close to the central black hole. This variability is manifested by appearance, disappearance and changes in morphology of the radio-emitting quasar jets. This is potentially problematic for two reasons. First, as we discuss in Sect. 4.1, quasar images can in principle be used to correct the group delay measurements. However, meaningful corrections require these images to be sufficiently contemporaneous (typically no more than 6 months apart) with the geodetic observations that are being corrected. Second, this variability poses problems with selecting geodetic-quality quasars based on structure index. For example, the median structure index of ICRF2 quasars is 2.75 (i.e. below the nominal SI = 3 cutoff), however the median standard deviation in this value for quasars imaged in 10 or more sessions is 0.53 due to their temporal variability (A. Collioud, private communication). In other words, "good" quasars can sometimes turn "bad" for at least some of the time. On the other hand, as we discuss in Sect. 4.2, quasar variability can be turned into an advantage with a suitable selection strategy.

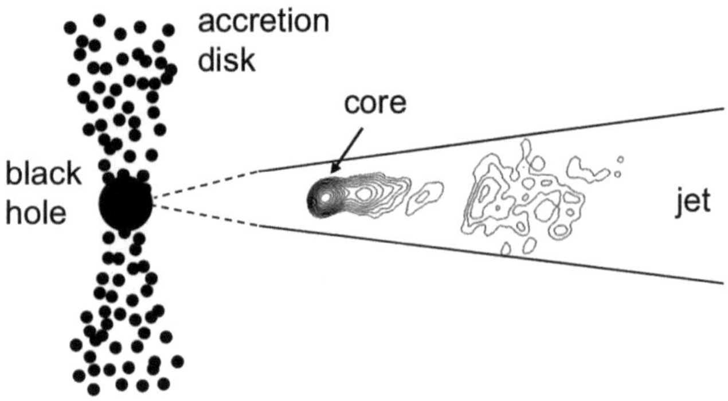

Fig. 1 Model of a quasar. Optical emission usually comes from regions close to the black hole accretion disk (although some optical synchrotron emission can also come from the jet). Radio emission is associated with the jet of synchrotron-emitting plasma. The location of the peak in radio emission (the "core") is frequency-dependent due to synchrotron self-absorption, and always some distance from the black hole. While jet direction typically remains fixed, the amount of structure seen in the radio images can vary significantly on timescales of months and years

In this contribution, we first quantify the effects of quasar structure on geodetic parameters derived from VLBI sessions; this is done through simulations. We then outline several options for mitigating the effects of quasar structure, and discuss the advantages and difficulties of each approach.

2 Source Structure Simulator

The wet component of the troposphere has long been known to be a major source of error in present-day VLBI measurements. Recent extensive simulations of Petrachenko et al. (2009) confirmed this, and provided a major motivation for the move to so-called VLBI Global Observing System (VGOS) observing using broadband (2–14 GHz) receivers on fast-slewing, relatively small 12 m—class antennas that is currently underway around the world. These simulations considered tropospheric turbulence, clock errors, and instrumental errors (e.g. Pany et al. 2011), but not the structure of quasars. Recently, we have extended the capability of the Vienna VLBI Software (VieVS; Böhm et al. 2012) to include the effects of source structure (Shabala et al. 2015). The latest version of VieVS has the capacity to both simulate source structure effects, and correct VLBI observations using quasar images. Because quasars are variable, for initial investigations we have focused on simulated (rather than real) quasar structure. The advantage of using the simulator

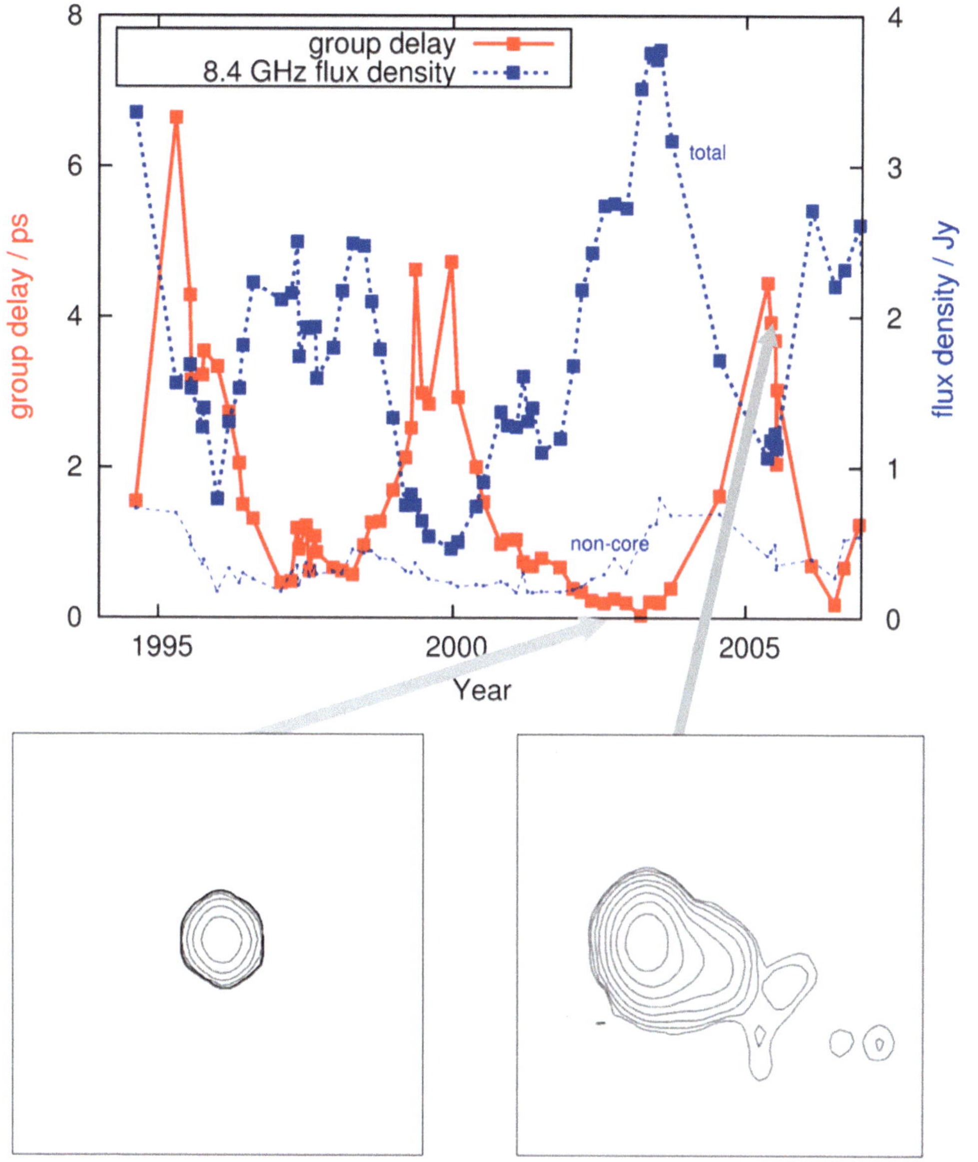

Fig. 2 Quasar structure varies on timescales of months to years. *Top* total source brightness variability (*blue*) is anti-correlated with the amount of structure (*red*) for the ICRF2 quasar 0059 + 581 (from Schaap 2014). The group delay shown is the median group delay over all Earth-bound baselines, calculated from the structure index which is derived using VLBI images of this source at each epoch. *Bottom* morphology of radio emission, with a faint jet component sometimes present in addition to the bright core (images courtesy of A. Pushkarev (*left*) and Y. Kovalev (*right*)). Each box is 10 milliarcseconds in size; the lowest contour is 0.5% of peak flux density. The astrometric position of this source is significantly more stable in the high flux density state (position rms of 288 μas) than the low flux density state (433 μas)

module of VieVS is in its ability to quantify the effects of source structure on source and station positions and the Earth Orientation Parameters in "clean" simulations, free from other important effects such as the clocks, wet troposphere, measurement error and technical problems at observatories. Once the quasar contribution is understood, we perform full simulations that include clocks, troposphere and measurement error in addition to quasar structure. These simulations are briefly described in Sect. 3. We refer the interested reader to Shabala et al. (2015) for full details of our software, simulation strategy and results.

3 Effects of Quasar Structure

3.1 Terrestrial Reference Frame

We used our simulator to quantify the effects of source structure on station positions. For details of our simulation strategy and results we refer the reader to Shabala et al. (2015). In brief, we constructed six catalogues with different levels of quasar structure, ranging from SI = 1, 2, 3, 4 to a distribution of structure indices mimicking either the full

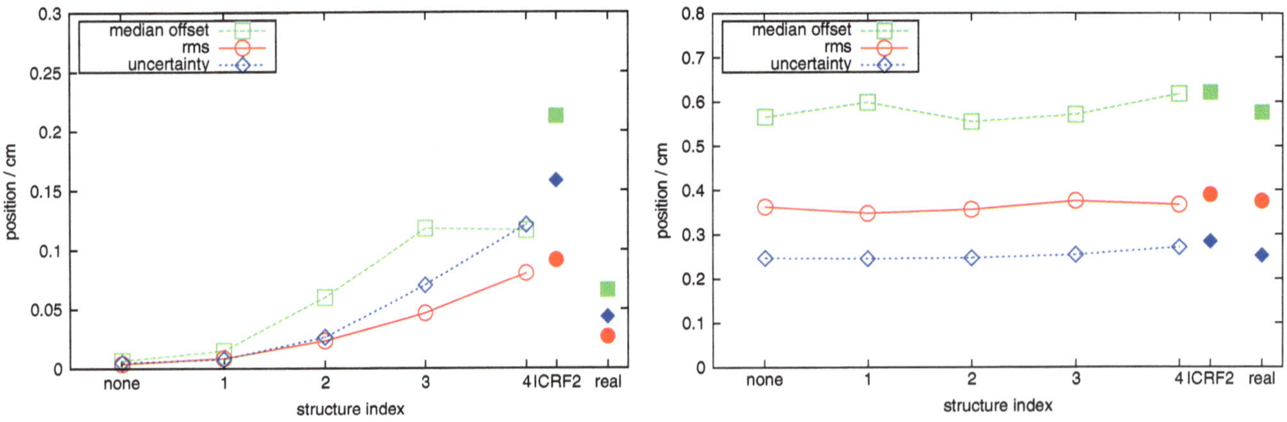

Fig. 3 Effects of source structure on station coordinate offsets (measured—true value, shown in *green*), debiased rms (*red*) and median formal uncertainty (*blue*) over 15 days of CONT11. *Left panel* structure-only simulations; *right panel* full simulations including clocks and troposphere. Median values over all stations are shown. *Open symbols* represent simulations in which all sources have the same structure index (none, 1, 2, 3 or 4), while for *filled symbols* we simulated a distribution of structure indices. Contributions from the wet troposphere and clocks mask the effects of quasar structure on station positions. From Shabala et al. (2015)

ICRF2 catalogue (median SI = 2.75, with some sources having structure indices as large as 5), or the sources that were observed in the CONT11 campaign (a lower median SI of 2.2), our chosen testbed. CONT11 was a global 15-day observing campaign run in September 2011 with 13–14 antennas widely distributed around the world. We chose to simulate these schedules because of the excellent station position repeatability derived from the real data collected during this campaign (Lovell et al. 2013), most likely due to the high number of observations and a fixed network compared to regular International VLBI Service (IVS) sessions.

We ran two types of simulations: structure-only, and structure plus noise contributions from the wet troposphere, station clocks and measurement errors. Unlike quasar structure which is a systematic effect, these other error sources are stochastic. Each 24-h session was simulated 30 times. We find that, on average, station positions are affected at or below the millimetre level, depending on the amount of simulated quasar structure. In particular, using the "real" distribution of structure indices results in station position errors of between 0.2 and 1.3 mm.

Figure 3 shows the median effects of different levels of quasar structure over all stations. The left panel shows results for structure-only simulations. It is clear that quasar structure affects station positions in three ways: (1) the median offset from the true station position; (2) the formal uncertainty in the estimated station position; and (3) the rms between position measurements made on different days; all increase with the amount of structure. The structure effects are significant at the millimetre level, as predicted by analytical work (e.g. Charlot 1990).

The right panel of Fig. 3 presents results of the same simulations but with the stochastic wet troposphere and clocks included. It is clear that now quasar structure effects are barely discernible, with the troposphere dominating. We find that source structure contributes ~10% of the total error budget. This is consistent with the measured present-day precision of station positions at the level of ~1 cm, compared to the millimetre-level effects of source structure seen in our structure-only simulations.

3.2 Celestial Reference Frame

One may expect the effects of quasar structure to be most apparent in the estimated quasar (rather than station) positions. To test this, we estimated session-wise source positions for each realisation of each day of CONT11. As with station positions, different levels of source structure were simulated.

As a first metric, we considered the median offset in estimated source positions. As previously discussed by a number of authors (e.g. Charlot 1990; Fey and Charlot 2000), the choice of the reference point is important for quantifying structure effects. However, defining a source position is not trivial for an extended, asymmetric source. Here, we used the quasar core for a reference position, and compared the astrometry of point-like and extended sources. An example of where such a comparison makes sense is for highly variable sources such as shown in Fig. 2, where the source is sometimes very point-like, with the core completely dominating any jet emission; and at other times the core and jet components have more comparable flux densities. Figure 4 shows the estimated positions of one quasar observed during the CONT11 campaign in the absence (left panel) and presence (right panel) of source structure. In a detailed study (Plank et al. 2015) we found that the magnitude of the median offset is strongly dependent both on the exact quasar structure (i.e. not just the structure index)

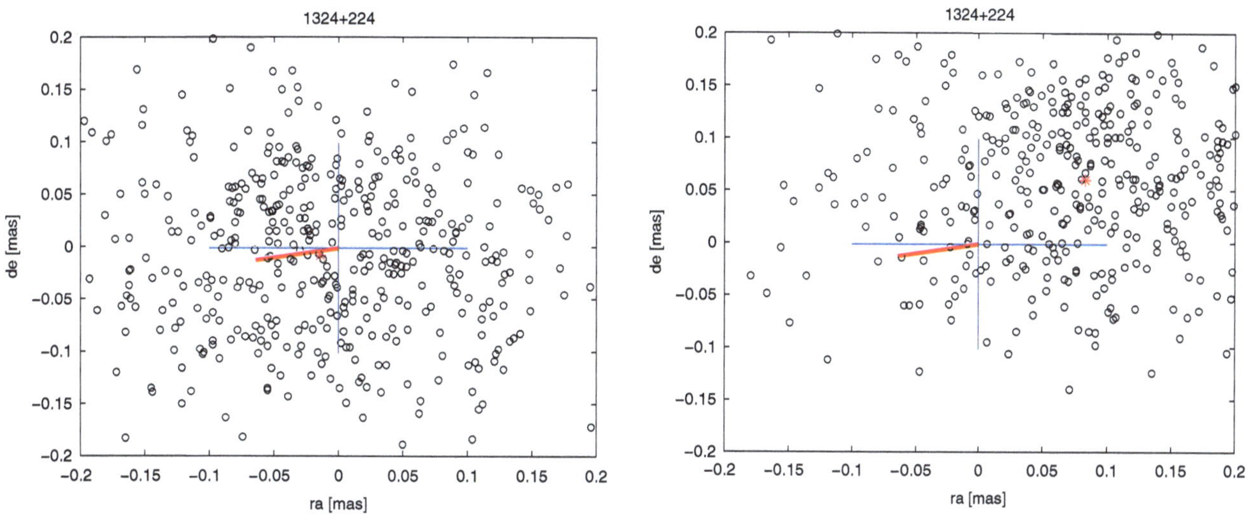

Fig. 4 Errors in estimated global source positions due to simulated quasars structure. *Left* no structure; *right* SI = 3 simulation. Thirty realisations are plotted for each of the 15 days of CONT11 (*black points*). *Red asterisk* shows the median offset of estimated position from the position of the quasar core, and *red line* the direction of the quasar jet. A systematic offset in median estimated position due to quasar structure is seen

as well as details of the observing schedules. The offsets were found to be at the at the 20–60 μas level for commonly observed quasars with structure indices between 2 and 3, comparable to the present ICRF2 noise floor of ~60 μas; and in some cases as large as 300 μas. Interestingly, in that study we found that sources with relatively low levels of structure (SI ~ 2) can result in significant, systematic offsets in quasar positions; while sources with larger structure provide more "noise"-like contributions to source positions. We refer the interested reader to Plank et al. (2015) for more details.

It is important to note that the median offset metric is likely not a good measure of structure effects for quasars that are not highly variable. Alternative metrics such as de-biased rms (as shown for station positions in Fig. 3) may be more appropriate; future work should investigate this and other metrics in detail. We note that Schaap et al. (2013) found that observed sources with structure indices in excess of 3 have significantly worse position repeatability (measured by a metric closely related to a de-biased rms) than quasars with lower levels of structure. Similar findings have also been reported by Ma et al. (2009) in their comparison of structure index and stability index (see their Figure 37).

4 Mitigation Strategies

Broadly speaking, there are three possible ways of mitigating the effects of quasar structure. First, quasars can be selected to have little structure; this approach has traditionally been used in IVS observations, and is the reason for the smaller effect due to the structure of "real" frequently observed sources than those drawn from the full ICRF2 catalogue (which includes a much higher fraction of non-compact quasars) in Fig. 3. The second approach involves re-analysing existing observations with structure corrections included; however this approach is difficult because quasars are highly variable and in many cases the required imaging data is simply not available. Finally, clever scheduling and/or analysis based on the structure of quasar jets is possible. We discuss each of these options briefly below.

4.1 Structure Corrections

Simulation results above suggest that the effects of source structure should contribute around 10% to station position error. We considered two solutions for the CONT11 campaign, with and without corrections for source structure using available quasar images. We found tentative signs of improvement on the longest (>10,000 km) baselines, however the difference between the two solutions was marginal at the level <0.5 mm. As also seen in simulations, these findings confirm that source structure effects are not the dominant source of error in present-day VLBI station position measurements.

4.2 Quasar Variability

An alternative approach is to select quasars which show low levels of structure. Although quasars are often included in IVS programs based on their structure index, this quantity can (and does) vary significantly, with jet components appearing and disappearing on timescales of months.

Shabala et al. (2014) used multi-frequency geodetic VLBI data at S and X-bands to investigate the effects of quasar variability on source positions. Schaap (2014) took this approach a step further, and investigated the relationship between quasar brightness and structure by considering nine well observed, variable quasars. These sources were selected to have been observed in at least 20 IVS sessions between 1995 and 2010, and to exhibit clear variability in source structure. Schaap (2014) found a strong anti-correlation between the structure index and X-band flux density for all nine quasars. An example is shown in Fig. 2. At their brightest, quasars therefore have the *least* structure. The reason for this is that the initial quasar brightening happens in the compact core, at which point the ratio of core flux density to that of the extended jet increases, resulting in a lower structure index. Eventually the newly-born component in the core moves into the jet, and gets dimmer; at this point the core flux density becomes less dominant than previously, yielding a larger structure index. Schaap (2014) compared astrometric stability (i.e. the stability of estimated source coordinates) for quasars in different flux density states, and found the consistent result that at their brightest these variable quasars have the most stable source positions. In most cases the improvement in position stability exceeded a factor of two. While in a large part this is due to the low source structure in the high flux density state, it must also be kept in mind that bright quasars have higher signal-to-noise ratios for a fixed integration time, and therefore smaller formal uncertainties associated with the measured group delays. Hence, there are at least two reasons for observing quasars near the peak of their flux density. We note that these findings come from only nine frequently observed, variable IVS quasars, and a larger sample would be desirable to thoroughly investigate these relations. In the future, development of additional quasar quality flags based on their current flux density state may be useful in scheduling geodetic sessions.

4.3 Scheduling with Respect to Source Structure

While the structure of quasars can vary significantly, the overall *direction* in which this happens generally remains fixed, set by the astrophysics of the black hole accretion disk. This offers a number of possible new strategies with respect to both observations and analysis. One way of reducing the effects of quasar structure is to preferentially observe with baselines that are close to orthogonal to the jet direction. This approach has been previously suggested by Porcas (2010).

The two relevant quantities are the relative angle between the observing baseline and quasar jet, and the projected length of the observing baseline. The top two panels of Fig. 5 show the distributions of simulated group delays as a function of these two quantities over all observations for a representative sample of nine R1/R4 IVS sessions observed in October 2013. Here, we have used the VieVS source-based scheduling module (Sun et al. 2014) to re-schedule these sessions, and only simulated quasar structure. All quasars were simulated using two-component models with a structure index of 3, and we have allowed the jet direction on the sky to be arbitrary. The top left panel of Fig. 5 shows that observations where the relative angle between the jet and observing baseline is close to 90° have the lowest structure delays. It also shows that, as expected, there is a uniform distribution of such relative angles, with every angle between 0 and 90° equally likely. The top right panel considers a metric which combines baseline length with the baseline—jet angle: the "uv range" is a scalar product of the projected baseline and jet direction. Small values of *uv range* correspond to short projected baselines and/or a close-to-orthogonal jet; such configurations correspond to lower structure delays than larger *uv range* values.

It is immediately clear that the effects of source structure can be mitigated if observations with small baseline—jet angles and large *uv range* values are avoided. However, there are two immediate issues with such an approach. First, antennas very rarely observe in pairs. A more typical scenario is three or more telescopes observing the same quasar in a given scan. Thus, it is not possible to achieve the ideal relative baseline—jet orientation for all baselines simultaneously. On the other hand, because baseline length also plays an important role in the magnitude of the structure effect, it should be possible to optimise the observing strategy for such a sub-network of antennas. For example, long baselines may preferentially be scheduled close to orthogonal to the jet direction, while shorter baselines may still do ok with a less favourable jet orientation. The bottom two panels of Fig. 5 show the distribution of structure corrections as a function of the median angle and *uv range* for all baselines in a given scan.

We used this information to re-schedule the simulated R1/R4 observations using the VieVS scheduling software, this time introducing source structure as an extra constraint in addition to the usual requirements of optimal sky coverage. While *uv range* is a useful metric for quantifying the effects of source structure, large values of this parameter usually correspond to long baselines. Therefore, downweighting or eliminating from the schedule observations with large *uv range* values will preferentially bias against long baselines. Generation of some test schedules confirmed this suspicion. We therefore adopted the median jet angle over a scan (bottom left panel of Fig. 5) as the additional scheduling constraint. In particular, we rescheduled the nine R1/R4 sessions with the additional constraint that scans with the

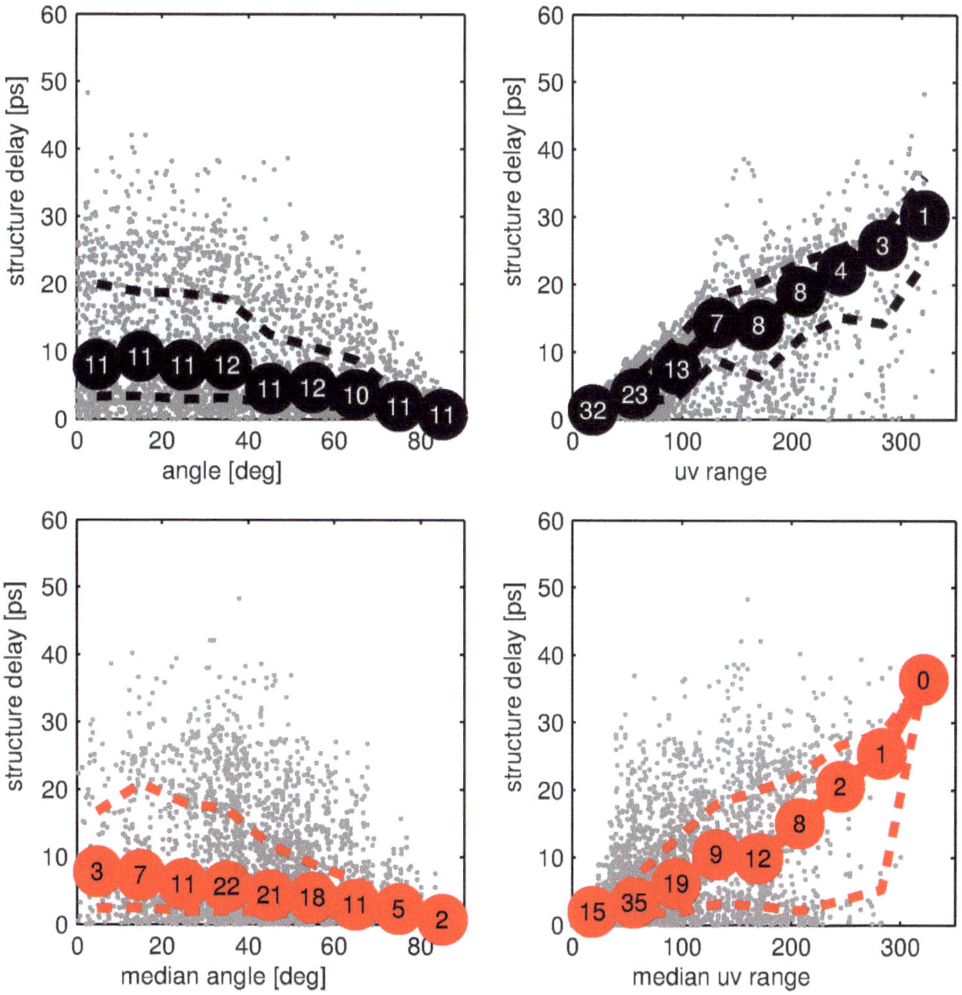

Fig. 5 Simulated delays due to source structure in nine R1/R4 sessions in October 2013, scheduled with VieVS. All simulated quasars have SI = 3. *Top panels* show the distribution of structure delays as a function of relative angle between the observing baseline and quasar jet (*left*) and *uv range* (*right*) per baseline. *Bottom panels* plot similar quantities, but the abscissa are median values over a scan (i.e. multiple baselines observing the same quasar). *Black/red circles* represent the median values in each of the nine bins, with the number indicating the percentage of the total number of observations. *Dashed lines* show the 25th and 75th percentiles

median jet angle of <40° be avoided if there are suitable alternatives.

Figure 6 shows the simulation results for these rescheduled sessions. There is a clear reduction in the number of observations (top left panel) and scans (bottom left) with unfavourable (i.e. small) jet angles. In the original schedules, 43% of scans and 45% of observations had median jet angles less than 40°. In the new schedules, these values are reduced to 19 and 31%, respectively. We note that the distribution of *uv range* values (bottom right panel) is also improved, without the loss of long baselines; in other words, this improvement is purely due to the exclusion of scans with unfavourable relative angles between the projected baseline and the quasar jet.

We quantify the improvement in baseline length repeatability due to these new schedules in the left panel of Fig. 7. We find that repeatability of 18 baselines is improved by our new scheduling strategy, with a median improvement of 0.7 mm. Conversely, 11 baselines show worse repeatability albeit at a lower level of 0.4 mm in the median. Importantly, a clear improvement is seen for the longest (>10,000 km) baselines.

It is important to stress that here we have presented simulations that only include the effects of quasar structure. An immediate concern is that, at present, troposphere is the single most important factor limiting the accuracy of IVS observations. Because of this, scheduling software are written to optimise sky coverage above each antenna, and any additional constraints (such as source structure information) is likely to degrade the troposphere solutions. This is indeed what we find when we run full simulations that include the effects of the turbulent troposphere as well as clocks and instrumental errors, with the overall baseline length repeatability decreasing by 1.5 mm (right panel of

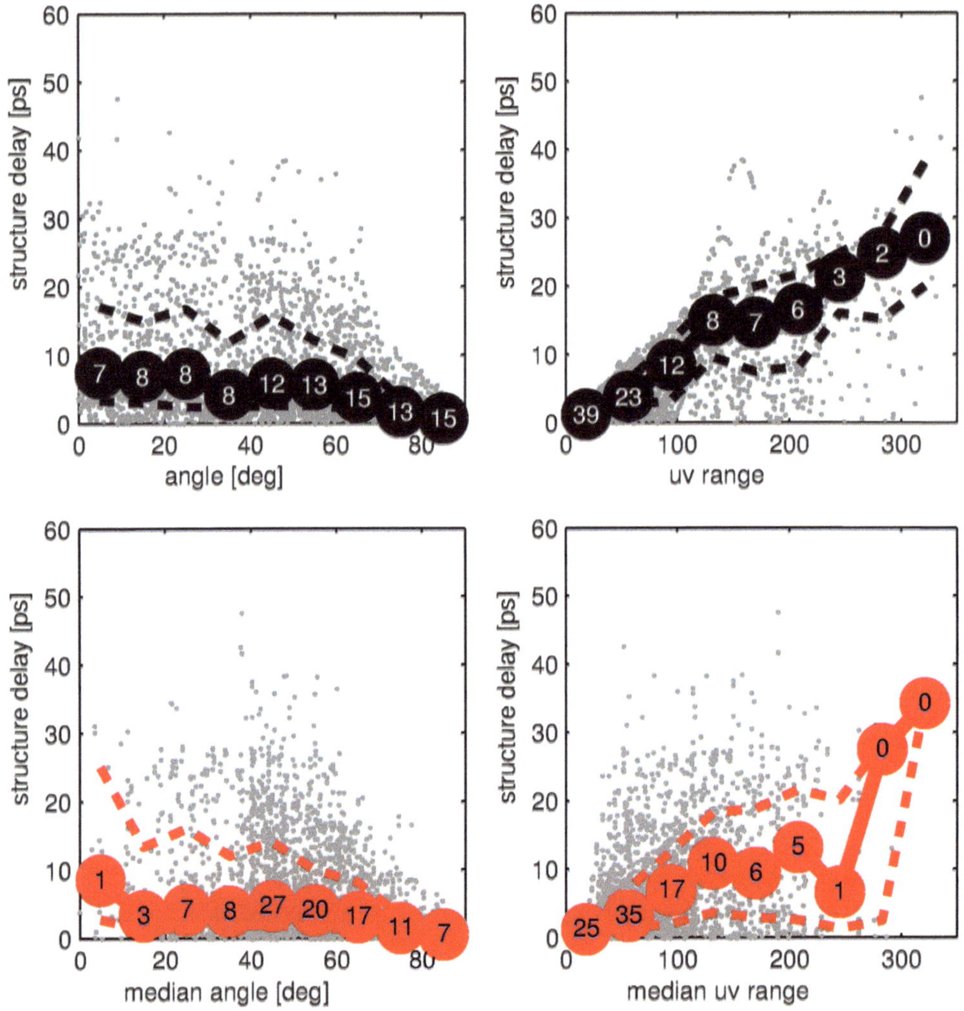

Fig. 6 As in Fig. 5 but for schedules that take source structure into account

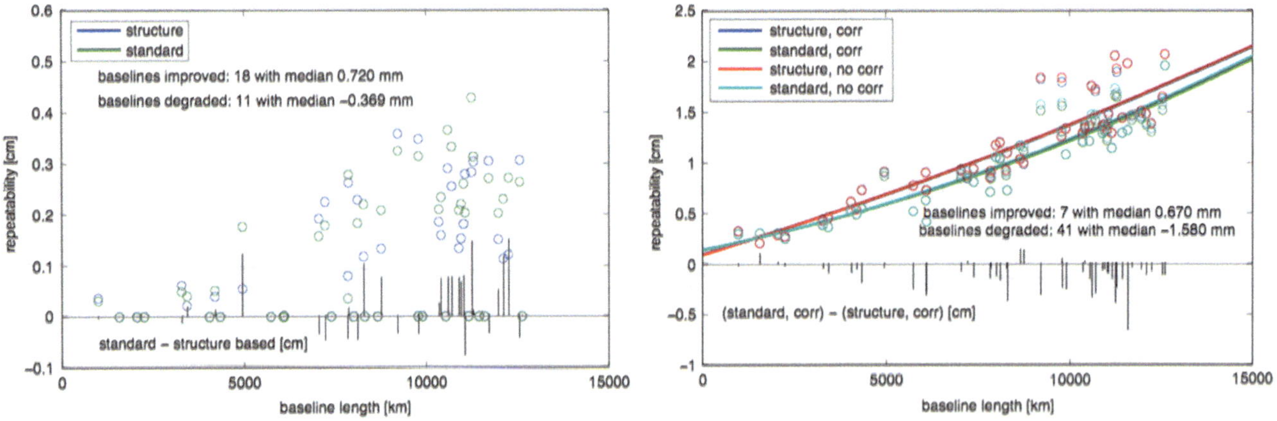

Fig. 7 Comparison of baseline length repeatabilities using conventional schedules and new scheduling that accounts for source structure. *Left panel* structure-only simulations. *Green points* are for conventional scheduling, and *blue points* for schedules with the source structure constraint. *Right panel* full simulations. Turquoise points are for conventional scheduling; *green points* are conventional scheduling with source structure corrections. *Red points* are for new schedules taking into account source structure but without structure corrections; *blue points* are for structure-based schedules with corrections. *Best-fit lines* are shown for guidance. Scheduling with respect to source structure shows a marginal improvement over classical scheduling for structure-only simulations, especially on long baselines. This improvement disappears for the full simulation, where the extra scheduling constraints make solutions worse because the difference in the sampling of the troposphere above each station is much larger than the source structure effect

Fig. 7). This degradation is due to the decreased number of total observations (25,434 instead of 28,476 in the original schedules) induced by the additional scheduling constraint related to source structure.

Nevertheless, scheduling with respect to source structure may become a feasible proposition in the VGOS era, with stochastic sources of error such as the troposphere, clocks and instrumentation reduced significantly compared to the present day.

4.4 Structure-Dependent Weighting in Analysis

An alternative to rescheduling is incorporating structure information in analysis. In particular, we investigated the possibility of downweighting observations with high values of *uv range* (top right panel of Fig. 5). This amounted to reducing the weights of the 24% of observations with *uv range* >150 by a factor of 4. In structure-only simulations, we found that this effect improved station position repeatability by 22%, and formal uncertainties by 30%. However, this effect disappeared in full simulations—preferentially downweighting a large number of long baseline observations in fact degraded our solutions, again consistent with the idea that tropospheric turbulence dominates the accuracy of present day station position determination. The quasi-linear relationship between *uv range* and median structure delay suggests that a more sophisticated weighting algorithm than the simple step function adopted here (e.g. weighting that is linear with *uv range*) may be worth investigating. Regardless of this, even our simple downweighting with respect to source structure may offer a promising way forward in source (rather than station) position estimation.

4.5 Re-parametrization of Source Positions

A final possibility is to analyse all existing data, without downweighting, in a way that takes account of quasar structure. Because the direction of each quasar's jet is well defined, it reasonable to expect the location of the estimated quasar position to vary along the jet, but not perpendicular to it. We have recently implemented this approach in the VieVS software. In the new solution, instead of estimating one global position in each of the Right Ascension and Declination coordinates, the position perpendicular to the jet direction is estimated as a global parameter, while the position along the jet (which will change with the projected baseline) is left as a free, or "arc" parameter. Our initial simulations (Plank et al. 2015) suggest that source position errors may be improved by as much as a factor of two using this new parametrisation. We defer a detailed discussion of this approach to a future contribution.

Acknowledgements We thank Chris Jacobs and the two anonymous referees for considerate comments which have significantly improved the manuscript. It is a pleasure to receive such constructive feedback from expert colleagues. We thank Arnaud Collioud for sharing data on quasar structure variability from the Bordeaux VLBI Image Database. SSS thanks the Australian Research Council for an Early Career Fellowship DE130101399. JNM and LP thank the Australian Research Council for Super Science Fellowships FS100100037 and FS110200045. RGS thanks the University of Tasmania for an Honours scholarship. HK works within the FWF Hertha Firnberg position T 697-N29. We are grateful to Jim Lovell and Bill Petrachenko for useful discussions.

References

Böhm J et al (2012) IAG Symp Proc 136:1007
Charlot P (1990) Astron J 99:1309
Fey AL, Charlot P (2000) Astrophys J Suppl Ser 128:17
Lister ML et al (2009) Astron J 138:1874
Lovell JEJ et al (2013) J Geod 87:527
Ma C et al (2009) IERS Technical Note 35, 1. http://adsabs.harvard.edu/abs/2009ITN....35....1M
Pany A et al (2011) J Geod 85:39
Petrachenko B et al (2009) IVS 2008 Ann Rep 13. http://adsabs.harvard.edu/abs/2009vlbi.rept....1P
Plank L et al (2015) Mon Not R Astron Soc 455:343. http://adsabs.harvard.edu/abs/2016MNRAS.455..343P
Porcas R (2010) IVS GM Proc, p 8. http://adsabs.harvard.edu/abs/2010ivs..conf....8P
Schaap RG (2014) U. Tasmania Honours thesis. University of Tasmania
Schaap RG et al (2013) Mon Not R Astron Soc 434:585
Shabala SS et al (2014) J Geod 88:575
Shabala SS et al (2015) J Geod 89:873
Sun J et al (2014) J Geod 88:449

Towards Improved Lunar Reference Frames: LRO Orbit Determination

Anno Löcher, Franz Hofmann, Philipp Gläser, Isabel Haase, Jürgen Müller, Jürgen Kusche, and Jürgen Oberst

Abstract

Lunar reference systems are currently realized by sets of coordinates of the few laser reflectors deployed by Apollo astronauts and unmanned Soviet spacecrafts. Expanding this coordinate knowledge to other features identifiable in images of the lunar surface requires highly accurate orbits of the acquiring spacecraft. To support such activities using images and altimetry data from the Lunar Reconnaissance Orbiter (LRO), an independent processing facility for tracking observations to LRO has been established. We present orbits from 1 year radio Doppler, radio ranging and laser ranging data obtained by different combinations of data types. To obtain an external confirmation for the achieved orbit accuracy, coordinates of the Apollo 15 reflector were measured in LRO images by photogrammetric techniques and compared to reference values from Lunar Laser Ranging (LLR). Coordinate differences were found to be at the 10 m level.

Keywords

Lunar Laser Ranging • Lunar Reconnaissance Orbiter • Precise orbit determination

1 Introduction

The Lunar Reconnaissance Orbiter (LRO, Chin et al. 2007) launched in 2009 by the National Aeronautics and Space Administration (NASA) still orbits the Moon in a low polar orbit, currently in a 200 × 30 km ellipse with the periselene over the lunar south pole. The main objective of the mission is the detailed exploration of the lunar surface by means of the Lunar Orbiter Laser Altimeter (LOLA) and three cameras bundled in the Lunar Reconnaissance Orbiter Camera (LROC) unit. Referring these observations to a Moon-fixed reference frame requires the computation of highly accurate and consistent orbits. In the mission requirements, the targeted accuracy was defined to be 50–100 m in total position and 1 m in the selenocentric radial direction (Vondrak et al. 2010). The primary source for the determination of such accurate orbits are two-way ranging and Doppler data from the NASA station White Sands, New Mexico, and four radio stations in Hawaii, Australia and Europe associated with the commercial Universal Space Network (USN). To compensate for the presumed lower performance of the USN stations, LRO was further equipped with an Earth-directed optics connected with the LOLA unit which enables one-way optical laser measurements from specially adapted sites. Currently ten stations of the International Laser Ranging

A. Löcher (✉) • J. Kusche
Institut für Geodäsie und Geoinformation, Universität Bonn, Nussallee 17, 53115 Bonn, Germany
e-mail: loecher@geod.uni-bonn.de

F. Hofmann • J. Müller
Institut für Erdmessung, Leibniz Universität Hannover, Schneiderberg 50, 30167 Hannover, Germany

P. Gläser • I. Haase
Institut für Geodäsie und Geoinformationstechnik, Technische Universität Berlin, Straße des 17. Juni 135, 10623 Berlin, Germany

J. Oberst
Institut für Geodäsie und Geoinformationstechnik, Technische Universität Berlin, Straße des 17. Juni 135, 10623 Berlin, Germany

Moscow State University for Geodesy and Cartography (MIIGAiK), Moscow, Russia

T. van Dam (ed.), *REFAG 2014*, International Association of Geodesy Symposia 146,
DOI 10.1007/1345_2015_146

Service (ILRS) participate more or less regularly in this campaign.

While instrument data from LRO are analyzed at various institutions, orbit determination is almost exclusively performed by NASA so far. Two types of orbits are routinely generated and released by NASA branches: For navigation purposes, a preliminary orbit is computed at the Flight Dynamics Facility (FDF) at Goddard Space Flight Center (GSFC) satisfying a more relaxed accuracy requirement of 500 m (Nicholson et al. 2010). Based on this coarse position knowledge, definitive orbits are computed by the LRO science team at GSFC using the software package GEODYN (Mazarico et al. 2012). The LRO science orbits have been reprocessed several times, mainly due to the progress in modelling the lunar gravity field as a result of the Gravity Field and Interior Laboratory (GRAIL, Zuber et al. 2013) mission. The accuracy of the more recent releases is reported to be in the order of 10 m in total position and 0.5 m radially (Mazarico et al. 2013). Errors roughly twice this size were obtained by Maier et al. (2014) who computed LRO orbits independently, but using the same software GEODYN.

The official LRO orbits thus seem to be safely within the mission requirements, but have not been validated by an independent solution so far. At least theoretically, there seems even to be room for improvements given the fact that the NASA orbit releases are based exclusively on radiometric data, supplemented by altimetric crossovers, while the laser observations are analyzed only internally. To close these gaps, an analysis scheme for all types of LRO tracking data has been developed at the University of Bonn resulting in a completely independent software application which is now gaining operational status. This work is part of a project on lunar reference frames supported by the German Research Foundation (DFG) with partners engaged in the analysis of Lunar Laser Ranging (LLR) and the processing of LRO imaging and altimetry. The aim of this project is to create a consistent dataset of lunar ephemeris, lunar rotation and coordinates of features which can be identified in LRO images of the lunar surface. The LRO orbits are the essential link within this project. They are the prerequisite to reference the LRO images and refer themselves to the models for lunar motion which have to be applied when deriving the orbits from Earth-based observations.

The present paper reports results of the project, with an emphasis on LRO orbit determination. Section 2 outlines the processing strategy applied at Bonn, in Sect. 3 series of solutions from 1 year tracking data are presented and compared with the NASA orbits. Section 4 documents a joint effort of the project partners to locate the laser reflector left by Apollo 15 astronauts by combining LLR and LRO data.

2 Orbit Determination Strategy

LRO orbit computations at Bonn are performed using an extension of the software GROOPS originally designed for gravity field determination from short orbit arcs (Mayer-Gürr 2008). The LRO extension adopts many functionalities of the heritage software but not the core algorithm since tracking data to a lunar orbiter cannot be efficiently processed by a short arc approach. The developments for LRO thus started with the implementation of the classical orbit determination procedure based on an iterating numerical integration. The partial derivatives for initial values and force parameters are obtained by simultanously solving the variational equations.

In processing an orbit arc, first the tracking data are converted to the geometrical information needed in the orbit determination step. The primary output of this step are the observed range and the station coordinates at observation time expressed in an inertial reference frame. For LRO, the station coordinates are shifted to a Moon-centered frame using the lunar ephemeris from DE421 (Williams et al. 2008) or an alternative ephemeris from the project.

As indicated above, all tracking data are converted to ranges. This is the common way for the runtime measurements provided by the radio stations and for the laser observations, but not for the Doppler counts, usually considered as the backbone of radiometric tracking. This data type is most often converted to range-rates, in case of S-band tracking using the relation (GSFC 2010)

$$\bar{\dot{r}} = \frac{c}{B}\left(\frac{N(t_i) - N(t_{i+1})}{t_{i+1} - t_i} + 2.4 \cdot 10^8\,\mathrm{s}^{-1}\right), \qquad (1)$$

where c denotes the speed of light, B the transmit frequency multiplied by a constant and $N(t_i)$ and $N(t_{i+1})$ the readings of the Doppler counter at two subsequent epochs. Since $\bar{\dot{r}}$ is the mean range-rate in the interval $[t_i, t_{i+1}]$, it is obvious that only a slight rearrangement of Eq. (1) is needed to obtain the range increment

$$\int_{t_i}^{t_{i+1}} \dot{r}(t)\,dt = \bar{\dot{r}}(t_{i+1} - t_i)$$
$$= \frac{c}{B}\left(N(t_i) - N(t_{i+1}) + 2.4 \cdot 10^8\,(t_{i+1} - t_i)\right). \qquad (2)$$

Summing up these increments then provides us with a biased range. In doing so, no errors are accumulated because in summing up the increments from, e.g., t_0 to t_n all Doppler

Table 1 Models used for LRO orbit determination

Modelled effect	Model
Applied to station coordinates	
Earth tides	IERS Conventions 2010 (Petit and Luzum 2010)
Pole tides	IERS Conventions 2010
Ocean tides	IERS Conventions 2010, tide model FES2004
Applied to ranges	
Geocentric light-time correction	IERS Conventions 2010 (for Earth rotation)
Relativistic light-time correction	IERS Conventions 2010 (Sun, Earth, Moon)
Ionospheric delay in radio data	IONEX TEC maps from International GNSS Service (IGS)
Tropospheric delay in radio data	Vienna mapping function (Böhm et al. 2006)
Tropospheric delay in laser data	Mendes-Pavlis model (Mendes and Pavlis 2004)
Mass center offset	$-1/0/-3$ m in S/C system, fixed (estimated from Tooley 2009, p. 28)
Force models	
Gravity	GL0660B (Konopliv et al. 2013) up to d/o 300
Direct tides	Sun, Earth, Planets
Solid body tides	k_2 from gravity field model
Solar radiation pressure	Direct and indirect effect, Moon albedo from Floberghagen et al. (1999), LRO macro model from Smith et al. (2008)
Relativity	Schwarzschild acceleration

counts at the inner interval limits cancel out and the resulting range only depends on $N(t_0)$ and $N(t_n)$. For a series of subsequent ranges, the situation is still more favourable because the error in $N(t_0)$ affects all ranges in the same way and is therefore fully absorbed by the bias to be estimated.

Using ranges from all data types greatly facilitates further processing. To all observations, the same corrections can be applied for mass center offset, Earth rotation during signal travelling and relativistic effects. In accounting for the tropospheric delay, only one model is needed for both types of radio observations (in the following referred to as "radio ranges" and "Doppler ranges"). The same applies for the effect of ionospheric refraction which has the same absolute value in radio and Doppler ranges, but opposite signs. For both media effects, we use state-of-the-art models from GNSS, for the troposphere the Vienna Mapping Function, for the ionosphere the IONEX TEC maps. A comprehensive summary of the models used in our processing chain is given in Table 1.

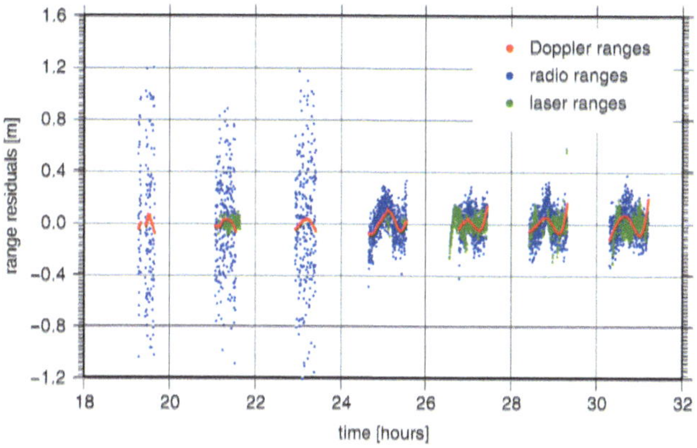

Fig. 1 Range residuals from LRO orbit determination (8.12.2011, extracted from a longer arc). The radio data of the first three passes are from the USN stations Dongara and Kiruna, the following ones from the NASA station White Sands

All observation types are ingested into the orbit determination process with similar weights. For the laser ranges, the individual errors of the normal points are adopted which are typically in the range of 5–20 cm. For the radio data, being raw data with no stochastic information available, the accuracy is set to 10 cm. This value fits rather well the residuals of the Doppler ranges and the radio ranges from White Sands (Fig. 1). The radio ranges from the USN stations have typically much larger residuals, but for practical reasons, the same accuracy was applied to all stations. A more refined weighting will be subject of further study.

In setting up the parameter scheme, a variety of biases must be accounted for. According to, e.g., Slojkowski (2014), the Doppler range-rates from all USN stations are biased. For the integrated Doppler counts, it follows that we have to estimate bias and drift parameters for each station pass. The radio ranges are biased as well due to the transponder delay occuring in two-way measurements with radio waves. In addition, the radio ranges suffer from biases in the time stamps which must also be absorbed by appropriate parameters. The laser ranges finally are the result of one-way measurements and are therefore fully affected by the error of the LRO clock. The most obvious variations can be removed by estimating a polynomial for which degree 2 seems to be sufficient.

For the dynamic modelling of the LRO orbit, utilizing a gravity field model from GRAIL is now mandatory. Here we use the model GL0660B (Konopliv et al. 2013) up to degree 300. Tidal forces are modelled using Sun, Earth and Planets, the solid body tides are computed with the Love number k_2 associated with the gravity field model (0.02405). From the solar radiation pressure the direct and the indirect effect are taken into account, the latter one according to the lunar albedo model from Floberghagen et al. (1999). Both the direct and the indirect effect are computed using the LRO macro model published in Smith et al. (2008).

3 Orbit Determination Results

Applying the strategy described above we processed the LRO tracking data from 16 September 2011 to 16 September 2012, the second year of LRO's science mission phase. In order to facilitate the comparison with the NASA orbits we adopted some of the processing rules of the LRO science team (Mazarico et al. 2012). The data set was thus divided in arcs of a length of approximately 2.5 days, each beginning und ending with an observation period of the station White Sands. The White Sands observations with a total duration of 8–10 h also defined the overlaps between subsequent arcs which are needed for the assessment of the inner accuracy of the orbits. In total, 170 arcs were processed with 141 overlaps, the difference caused by orbit maneuvers which force to stop the integration. For each arc, the parameter set included the initial position and velocity and the measurement biases as given above. Additionaly a scale factor for the solar radiation pressure was estimated and an empirical acceleration in along track direction. The start values for the initial state vector and the coarse positions needed in the data preprocessing were taken from the FDF orbits.

To get information about the contribution of each observation type, the computations were carried out with all combinations of datasets which include the Doppler observations. The results are summarized in Tables 2 and 3, Table 2 showing the mean differences in the overlapping sections and Table 3 the differences to the NASA orbits. Both statistics show similar systematics (with generally higher values in Table 2) and lead to the same conclusions. It becomes clear that the Doppler ranges alone are not sufficient to guarantee highly accurate orbits. Best results are obtained when combining them with the radio ranges, which reduces the total overlap error only slightly from 12.29 m to 11.80 m, but leads to a significant improvement in the radial direction, from 1.49 m to 0.74 m. This means that both types of radio data do not contain identical information though collected at the same epochs at the same stations. An explanation might come from the different bias schemes: While the Doppler ranges are affected by a drift but not biased in the epochs, the situation is inverse for the radio ranges. Despite of their different noise levels, both data types seem to be complementary in this regard.

With the same clarity, the statistics show that the laser ranges contribute little to our solutions. Combining them with the Doppler ranges or both types of radio observations even leads to a slight increase of the error figures. This result cannot be explained by the reduced number of laser observations. It is not unlikely that a more refined modelling of the LRO clock error solves the problem, but this issue needs additional research. Until further insight, we must fully confirm the strategy of NASA to rely only on the radiometric data.

Focusing on the radio-only combination we can conclude from the overlap analysis that our orbits do not yet attain the accuracy of the NASA orbits but could reach competitiveness in future. Further progress is expected from refinements in the radiation pressure modelling which does not yet account for the shadow thrown from one spacecraft component onto another. The estimates for the respective scale factor give indeed a hint that the present modelling is not optimal since they scatter largely around the expected value (one), with outliers and jumps (Fig. 2). Another advantage in the NASA processing is the use of altimetric crossovers for orbit improvement which is also an open issue in our software.

Table 2 Mean RMS of differences in orbit overlaps, in meters

Data types used	Total	Along	Cross	Radial
Doppler ranges	12.29	9.86	5.26	1.49
Doppler and radio ranges	11.80	9.07	5.79	0.74
Doppler and laser ranges	13.40	10.61	6.02	1.61
Doppler, radio and laser ranges	11.97	9.13	5.90	0.79

Table 3 Mean RMS of differences to NASA science orbits, in meters

Data types used	Total	Along	Cross	Radial
Doppler ranges	10.53	8.89	4.22	1.05
Doppler and radio ranges	10.05	8.34	4.49	0.64
Doppler and laser ranges	11.24	9.45	4.56	1.21
Doppler, radio and laser ranges	10.21	8.52	4.48	0.68

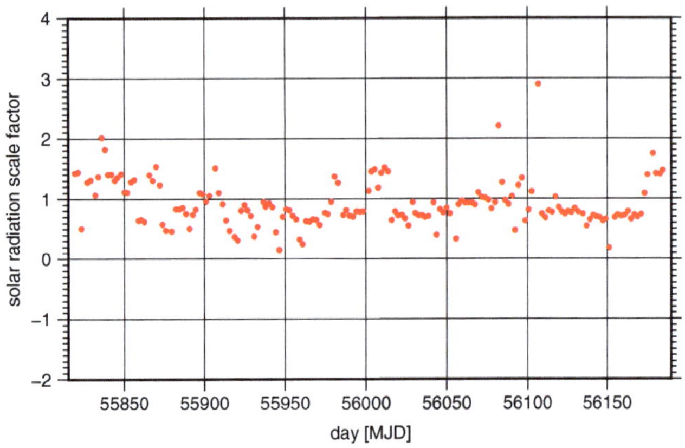

Fig. 2 Estimated scale factors for solar radiation pressure

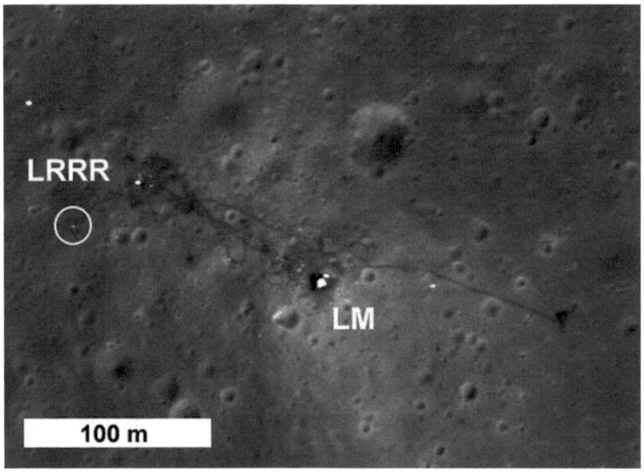

Fig. 3 Apollo 15 landing site with lunar module (LM) and lunar ranging retro reflector (LRRR). Image courtesy of NASA

4 Locating the Apollo 15 Laser Reflector

The outstanding quality of LRO imaging has allowed for discovering details hidden to any previous lunar mission. In images of the LROC Narrow Angle Camera (NAC), lander equipment of US and Soviet missions were identified and even tracks of the Apollo astronauts could be resolved. A feature of particular interest for LRO orbit determination are the reflectors for Lunar Laser Ranging (LLR) left by the Apollo missions 11, 14 und 15. Due to LLR analysis, they are amongst the few points on the lunar surface having coordinates at the submeter level. By locating them in LROC NAC images the opportunity arises to tie the LRO orbits into the reference frame defined by the reflector coordinates.

For the required image analysis, we apply a procedure developed at the Technical University of Berlin (Haase et al. 2012). The LROC NAC images are ortho-rectified and then map-projected onto the lunar surface using the most recent pointing information of the camera, an LRO orbit, a lunar ephemeris and lunar rotational parameters. To precisely intersect the camera axis with the lunar surface, an LROC NAC based digital terrain model is used.

For the present study, the analysis was confined to the Apollo 15 reflector which could be identified in two LROC NAC images taken on 30 October 2009 (Image 1, see Fig. 3) and 6 November 2011 (Image 2). The computations were carried out with different input, including a refined LLR solution currently under progress at the Institut für Erdmessung (IfE) in Hannover. Based on a revised set of observations, this solution provides precise reflector coordinates and an accurate lunar ephemeris, the latter computed by simultaneous numerical integration of the major solar system bodies and the lunar rotation (Müller et al. 2014).

Table 4 Mean equator (ME) coordinates of the Apollo 15 laser reflector derived from LROC NAC images

	Longitude [°E]	Latitude [°N]	Difference to reference value [m]
Solutions using DE421			
DE421 reference value (Williams et al. 2008)	3.628507	26.133396	
Image 1, NASA orbit	3.628176	26.133444	9.12
Image 1, IGG orbit	3.628121	26.133659	13.20
Image 2, NASA orbit	3.628424	26.133271	4.37
Image 2, IGG orbit	3.628626	26.133502	4.60
Solutions using IfE ephemerides			
IfE reference value	3.628177	26.133063	
Image 1, IGG orbit	3.627791	26.133346	13.56
Image 2, IGG orbit	3.628296	26.133197	5.20

If using the IfE ephemerides, the analysis required the following steps by the three teams:

1. Centered around the image acquisition times, ephemeris arcs and rotational parameters were computed from LLR data (IfE Hannover).
2. Using these lunar ephemerides, LRO orbit arcs were estimated from radio and Doppler ranges (IGG Bonn).
3. Based on the ephemerides from Step 1 and the orbits from Step 2, the reflectors were located in a lunar reference system (TU Berlin).

To assess the benefit of the IfE ephemerides, Steps 2 and 3 were repeated using the lunar ephemerides from DE421. In addition, Step 3 was carried out using the NASA science orbits in combination with DE421.

The resulting coordinates are given in Table 4 together with the differences to the previously known coordinates. In order to account for the correlation between the models for lunar motion and the reflector coordinates, the comparisons refer to the coordinates associated with the ephemeris that was used. When using DE421 and the NASA orbits, coordinate differences of 9.12 m and 4.37 m appear. With the IGG orbits based on DE421, the differences are degraded by about 4 m for Image 1 and 25 cm for Image 2. These results fully comply with the assumed accuracy of the NASA orbits and the slightly higher errors we assessed for our orbits. Using the IGG orbits based on the IfE Ephemerides, the differences are nearly unchanged compared to the case IGG/DE421, increased only by a few decimeters. This shows that both types of ephemerides are at the same level of accuracy. Further conclusions are difficult to draw since it is obvious that the differences are mainly caused by the orbit errors. At present stage, the accuracy of the LLR coordinates (<1 m) is clearly missed with any orbit. This accuracy is probably unattainable by the technique applied, but further improvements should be possible.

5 Conclusions

We presented first results of LRO orbit determination using software completely independent from the NASA processing. The accuracies assessed by an overlap analysis confirm the NASA strategy to rely only on radiometric tracking. The accuracy of our radio-only solution is currently 12 m in total position and 80 cm radially which is still slightly above the errors of NASA orbits (10 m and 50 cm, respectively). Based on our orbits, the coordinates of the Apollo 15 reflector could be recovered from LROC NAC images resulting in differences to the reference values from LLR of 5–13 m compared to 4–9 m using NASA orbits. These results fully comply with the overlap analysis. Further progress is expected from an enhanced modelling of solar radiation pressure and the use of laser track crossovers to further constrain the orbit solutions.

Acknowledgements This research was funded by the German Research Foundation (DFG) within the research unit FOR 1503 "Space-Time Reference Systems for Monitoring Global Change and for Precise Navigation in Space". In addition, J. Oberst was hosted by MIIGAiK and supported by Russian Science Foundation, project #14-22-00197.

References

Böhm J, Werl B, Schuh H (2006) Troposphere mapping functions for GPS and very long baseline interferometry from European Centre for Medium-Range Weather Forecasts operational analysis data. J Geophys Res 111:B02406. doi:101029/2005JB003629

Chin G, Brylow S, Foote M, Garvin J, Kasper J, Keller J, Litvak M, Mitrofanov M, Paige D, Raney K, Robinson M, Sanin A, Smith D, Spence H, Spudis P, Stern S, Zuber M (2007) Lunar Reconnaissance Orbiter overview: the instrument suite and mission. Space Sci Rev 129:391–419. doi:101007/s11214-007-9153-y

Floberghagen R, Visser P, Weischede F (1999) Lunar albedo force modeling and its effect on low lunar orbit and gravity field determination. Adv Space Res 23(4):733–738. doi:101016/S0273-1177(99)00155-6

GSFC (2010) Station raw tracking data. http://imbrium.mit.edu/LRORS/DOCUMENT/LRO_DESC_TRK.TXT

Haase I, Oberst J, Scholten F, Wählisch M, Gläser P, Karachevtseva I, Robinson M (2012) Mapping the Apollo 17 landing site area based on Lunar Reconnaissance Orbiter camera images and Apollo surface photography. J Geophys Res 117. doi:101029/2011JE003908

Konopliv A, Park R, Yuan D, Asmar S, Watkins M, Williams J, Fahnestock E, Kruizinga G, Paik M, Strekalov D, Harvey N, Smith D, Zuber M (2013) The JPL lunar gravity field to spherical harmonic degree 660 from the GRAIL primary mission. J Geophys Res 118(7). doi:101002/jgre20097

Maier A, Baur O, Krauss S (2014) POD of LRO and inferred gravity field information. Poster presented at European Planetary Science Congress, Cascais

Mayer-Gürr T (2008) Gravitationsfeldbestimmung aus der Analyse kurzer Bahnbögen am Beispiel der Satellitenmissionen CHAMP und GRACE. Dissertation, Bonn

Mazarico E, Rowlands D, Neumann G, Smith D, Torrence M, Lemoine F, Zuber M (2012) Orbit determination of the Lunar Reconnaissance Orbiter. J Geodesy 86(3):193–207

Mazarico E, Goossens S, Lemoine F, Neumann G, Torrence M, Rowlands D, Smith D, Zuber M (2013) Improved orbit determination of lunar orbiters with lunar gravity models obtained by the GRAIL mission. In: 44th Lunar and Planetary Science Conference, The Woodlands

Mendes V, Pavlis E (2004) High-accuracy zenith delay prediction at optical wavelengths. Geophys Res Lett 31:L14602. doi:101029/2004GL020308

Müller J, Biskupek L, Hofmann F, Mai E (2014) Lunar Laser Ranging and relativity. In: Kopeikin S (ed) Frontiers of relativistic celestial mechanics, vol 2. De Gruyter, Berlin, pp 103–156

Nicholson A, Slojkowski S, Long A (2010) NASA GFSC Lunar Reconnaissance Orbiter (LRO) orbit estimation and prediction. In: SpaceOps 2010, Huntsville

Petit G, Luzum B (2010) IERS Conventions. IERS Technical Note 36. Frankfurt am Main: Verlag des Bundesamts für Kartographie und Geodäsie

Slojkowski S (2014) Lunar Reconnaissance Orbiter orbit determination accuracy analysis. In: 24th International Symposium on Space Flight Dynamics, Laurel

Smith D, Zuber M, Lemoine F, Torrence M, Mazarico E (2008) Orbit determination of LRO at the moon. In: 16th International Workshop on Laser Ranging, Poznan

Tooley C (2009) Lunar Reconnaissance Orbiter mission update. In: Wernher von Braun Memorial Symposium 2009, Huntsville

Vondrak R, Keller J, Chin G, Garvin J (2010) Lunar Reconnaissance Orbiter (LRO): observations for lunar exploration and science. Space Sci Rev 150:7–22. doi:101007/s11214-010-9631-5

Williams J, Boggs D, Folkner W (2008) DE421 lunar orbit, physical librations, and surface coordinates. JPL Interoffice Memorandum 335-JW,DB,WF-20080314-001

Zuber M, Smith D, Lehman D, Hoffman T, Asmar S, Watkins M (2013) Gravity Recovery and Interior Laboratory (GRAIL): mapping the lunar interior from crust to core. Space Sci Rev 178:3–24. doi:101007/s11214-012-9952-7

Part V

Usage and Applications of Reference Frames in Geosciences

Terrestrial Reference Frame Requirements for Studies of Geodynamics and Climate Change

Geoffrey Blewitt

Abstract

Scientific applications critically depend on the ITRF and impose the most stringent requirements on terrestrial reference frame accuracy and long-term stability. A recent US National Research Council report (Minster et al., Precise geodetic infrastructure: national requirements for a shared resource. The National Academies Press, Washington, DC, 2010) found that the applications demanding the highest accuracy and long-term stability were sea level, geodynamics from vertical land motion and large-scale horizontal deformation, and decadal satellite survey missions. A key recommendation was to make a long-term commitment to maintain the ITRF to ensure its continuity and stability, so as to provide a foundation for Earth system science and studies of global change. In this paper, we focus on characteristics of the ITRF that have demands placed upon them by these most stringent scientific users. We consider in detail each characteristic in terms of what the user needs, and provide examples of how such needs can be met, and identify factors that strengthen or weaken terrestrial reference frames from a user's perspective. We find the most important feature of a terrestrial reference frame is "predictability", the ability of the frame to predict future positions of stations in a multi-technique network to support science. Specifically, the key requirement of the ITRF, in order to support the most demanding scientific applications with large societal impacts, is to provide access to station coordinates that have secular predictability at the level of 1 mm per decade (0.1 mm/year).

Keywords

Climate change • Geodynamics • Reference frames • Scientific requirements

1 Introduction

Given that the emphasis of this REFAG symposium is on how to improve and develop future versions of terrestrial reference frames (TRFs), it is of useful to consider requirements and desirable characteristics from the user's perspective. Of all the users of TRFs, it is scientific users that place the most stringent demands on these characteristics (Blewitt et al. 2010; Plag and Pearlman 2009). In its report, the National Research Council in the United States discussed in detail the national requirements for precise geodetic infrastructure as a share resource (Minster et al. 2010). The report identifies the most stringent scientific applications, including the study of sea level change, geodynamics, orbit determination of decadal satellite missions, and vertical land motion. The report notes that the International Terrestrial Reference Frame (ITRF) enables coordinates from multiple techniques to be meaningfully compared anywhere on the Earth's surface (Altamimi et al. 2011), and thus forms the basis of scientific investigations using the Global Geodetic Observing System (Plag and Pearlman 2009). The ITRF is also inherited by denser, continental-scale TRFs to support

G. Blewitt (✉)
Nevada Bureau of Mines and Geology, University of Nevada, Reno, NV, USA
e-mail: gblewitt@unr.edu

T. van Dam (ed.), *REFAG 2014*, International Association of Geodesy Symposia 146,
DOI 10.1007/1345_2015_142

more targeted science (Blewitt et al. 2013; Brunini and Sánchez 2013; Bruyninx et al. 2012), thus the relationships between TRFs have a solid foundation through the ITRF.

One of most difficult problems that an accurate and stable ITRF addresses is how to tie together time series of global sea level change from different satellite altimeter missions (Nerem et al. 2010). The ITRF also addresses the problem of being able to reference tide gauge data with respect to a stable and physically meaningful origin, so that sea level can be mapped in a system consistent with altimetry, and can be compared from one decade to the next (Blewitt et al. 2010; Santamaría-Gómez et al. 2012). The interpretation of vertical land motion in terms of geodynamic processes also places stringent demands on coordinate accuracy and stability (Plag and Pearlman 2009). These processes include glacial isostatic adjustment, ice sheet mass variation, coastal subsidence, mountain growth, decade-scale hydrological loading, and elastic strain accumulation at plate boundaries. In many cases the requirements on the TRF to support these scientific applications also have societal implications, for example, the need for accurately referenced surface displacements to improve tsunami early warning systems around the Pacific Rim, the risk to infrastructure from coastal subsidence, and the depletion of water resources in central California (Amos et al. 2014).

In this paper, to consider the user's perspective, we first categorize the TRF characteristics that have user demands, and then each of these characteristics are discussed in some detail. We then consider the most stringent application, which is the determination of long-term change in sea level, to learn how those characteristics are used to meet scientific goals. Finally, we infer the key requirements on ITRF to support geodynamics and climate change, and attempt to answer the question, "What is the primary mission of a terrestrial reference frame?"

2 Terrestrial Reference Frame Characteristics That Have User Demands

What is the primary mission of a terrestrial reference frame? Whatever the answer to this question may be, it should relate to meeting the user's needs. We will keep this question in mind as we consider the desirable characteristics that a TRF should have to meet the needs of science, and we will attempt to answer it in the Sect. 4.

Let us first categorize the types of characteristics that may have user demands, and then we will look at these characteristics in some detail:

1. the associated reference system;
2. reference frame definition and inheritance;
3. realization;
4. spatial coverage;
5. temporal coverage;
6. quality; and
7. life cycle.

2.1 The Associated Reference System

Terrestrial reference frames are specific realizations of terrestrial reference systems (IERS 2010). The associated reference system comprises fundamental constants, conventions, and models. To be useful scientifically, we must consider physical characteristics of the reference system. The following provide some examples, all of which are subjects of ongoing research in the geodetic community:

1. The accurately predictable part of a station's motion, for example, solid Earth tidal displacements, may be modeled at the observation level of data analysis. The unpredictable or less-accurately predictable parts may be the target of scientific investigation, for example, co/post-seismic displacements.
2. The origin may be coincident with the entire Earth system center of mass (CM), which requires that the predictable part of the station motion model relating to mass redistribution such as ocean tides must consistently account for displacement of the Earth's geometrical figure from CM, so called "geocenter motion".
3. Scale is specified by the conventional speed of light together with timing measurements of sufficient accuracy in a consistent relativistic framework. For satellite geodesy, the so-called near-Earth approximation of the general relativistic metric defined by the International Astronomical Union (IAU) suffices (unlike interplanetary missions, for example). Scale is a function of the adopted reference gravitational potential, such as a conventional geoid for an Earth-fixed observer, or the potential at Earth's center of mass, or the solar system's center of mass. Since different reference potentials may be more convenient to perform observation modeling, the scale of the reference system should be unambiguously defined, among the various possible choices (IERS 2010).
4. The reference system must take care with any models that may affect scale, models that should be consistent with the conventional speed of light. Examples include GM, atmospheric refractivity, and system biases such as those associated with satellite reflectors and antennas.
5. The rotation of the reference system is mathematically arbitrary and thus conventional, but to be useful to geophysicists, it should be defined to be consistent with some physical principle, such as the no-net rotation of the tectonic plates, or the rotation of hot spots representing the mantle, or the integrated surface motion of the Earth (Kreemer et al. 2006).

2.2 Reference Frame Definition and Inheritance

"Reference frame definition" is the method chosen to realize the origin, scale, and orientation of the TRF, and their time evolution. Examples are:

1. Origin: One option is to use SLR for its superior orbit modeling and predictability, setting degree-1 gravity terms to zero.
2. Scale: One option is to use VLBI, given its insensitivity to GM. Another option is to use SLR, given its less sensitivity to errors in atmospheric refractivity.

"Inheritance" is often used to ensure continuity and consistency of the TRF. Examples are:

1. Maintain the same orientation as the former ITRF, so as to ensure continuity of polar motion time series and consistency in updated station coordinate time series.
2. IGS08 is aligned with ITF2008 to ensure reference frame consistency, which improves precision for daily GPS orbit alignment in the terrestrial frame (Rebischung et al. 2012).
3. Stable North America reference frame NA12 (Blewitt et al. 2013) was realized using time series in global reference frame IGS08 (Rebischung et al. 2012) to ensure reference frame consistency with ITRF2008 (Altamimi et al. 2011), while allowing for a different no-net rotation condition to meet scientific needs.

2.3 Realization

TRF realization extends the reference frame definition and inheritance by a self-consistent process that explicitly determines the coordinates of stations of each contributing technique. There are specific aspects of design in the process of TRF realization that may address user's needs:

1. The selection of space-geodetic techniques used to supply the input data and TRF stations allows for a synergistic combination, such that the TRF can better conform to the physical ideal expressed by the associated reference system. In addition, techniques such as GNSS densify the frame by orders of magnitude, facilitating user access to the frame. Thus, GNSS user solutions that use such a TRF become leveraged by SLR and VLBI (without the user having to explicitly use such data), and become more physically grounded for scientific interpretation.
2. Multi-technique site collocations and local ties between the stations are required to enable a rigorous combination of techniques. The number, spatial distribution, and variety of techniques of collocated sites have an impact on the quality of the combination, and thus the ability of GNSS to gain leverage from VLBI and SLR.
3. Relative data weights between contributing techniques, and between contributing solutions within each technique will have an impact on the quality of the combination, thus ongoing assessment is needed to determine the optimal relative weights that give the most physically meaningful and high-quality frame.
4. The selection of TRF stations and time-windows of acceptable data can affect the quality and long-term stability of the origin, scale, and orientation of the frame. In particular, if a TRF station has non-stationary variation from linear motion (such as random walk monument noise, or post-seismic deformation), its future position will be less predictable, therefore degrading the quality of the frame. Stationary processes such as flicker noise and repeatable seasonality are of less concern as they do not affect long-term predictability, however they may affect relative data weights of contributing stations.
5. The estimation of empirical station motion model parameters is required. For example, station velocities are required because secular motion from tectonics and plate boundary deformation is not sufficiently well known a priori. Moreover, GNSS time series generally suffer from discontinuities arising from equipment changes, and all techniques are generally prone to permanent displacements due to earthquakes. Note that from a user's perspective, the estimated parameters define the realized frame. Estimated parameters are conditioned by the choice of the frame origin, orientation, scale, and their evolution in time, and in turn, the user has access to these frame parameters through alignment to the predicted station positions. At present, the ITRF estimated parameters for each station include coordinates at a conventional epoch, station velocity coordinates, step coordinates, and local tie discrepancies. Currently it is not possible to physically model seasonality with sufficient accuracy (Davis et al. 2012), therefore empirical parameters such as annual and semi-annual sine and cosine amplitudes are currently being considered to enhance the sub-annual realization of ITRF. Of particular concern is the out-of-phase seasonality of the northern and southern hemisphere, which currently may bias the sub-annual realization of scale and frame origin. For example, seasonal inter-hemispheric surface mass redistribution would lead to seasonal variation in hemispheric scale and mean station position (through degree-1 deformation). Hemispheric asymmetry in the station distribution could therefore lead to anomalous seasonal variation in scale and origin.
6. Quality control (QC) is defined as the screening of input data, which in this case are the contributing geodetic solutions. Quality assessment (QA) is defined as the evaluation of the final product, which in this case are the estimated station motion parameters, and thus the realization of origin, scale, and orientation. The effectiveness of

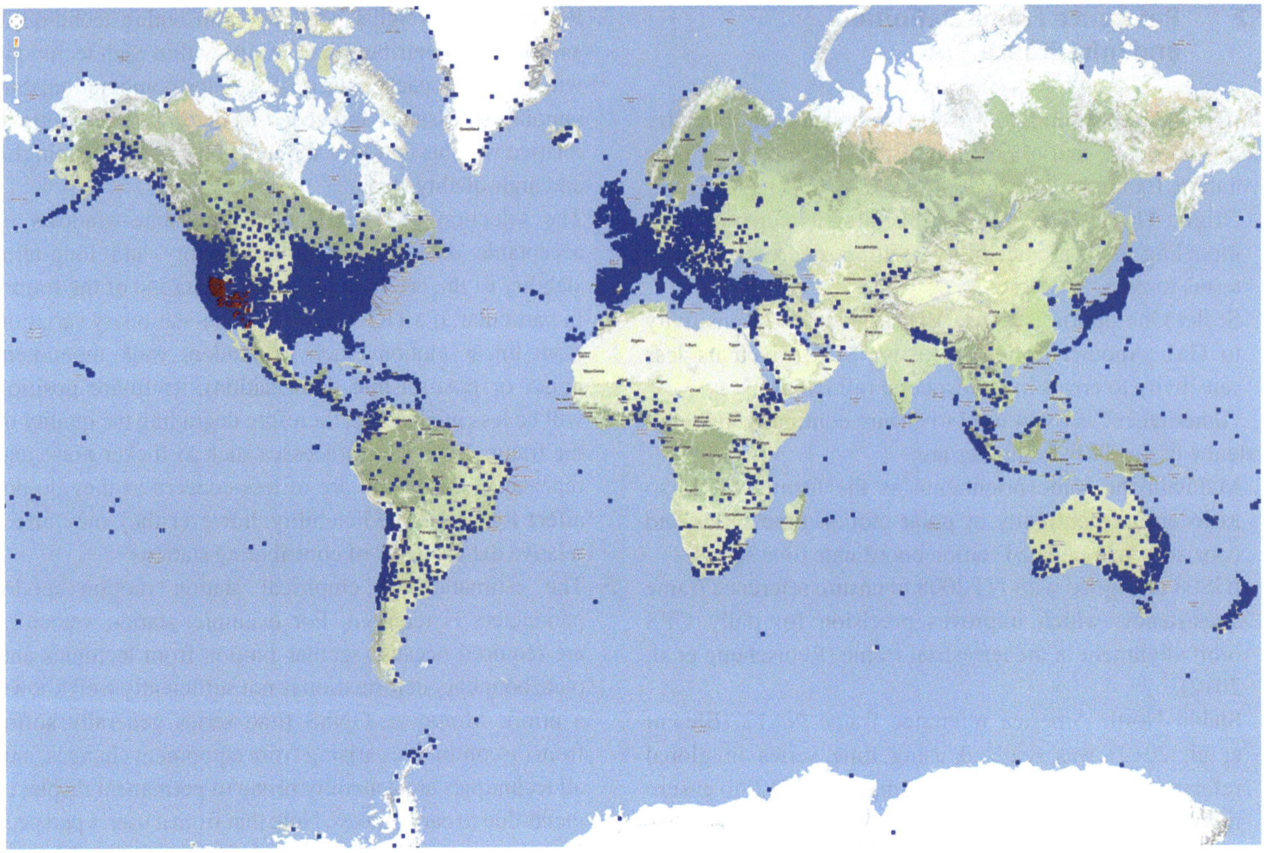

Fig. 1 Spatial coverage for geodetic GPS stations routinely processed by the Nevada Geodetic Laboratory, http://geodesy.unr.edu

QC is of paramount importance to improve the quality of the final product. QA is a somewhat more difficult process, and a conservative approach can be valuable to put bounds on the physical interpretation in scientific investigations. For example, a QA determination that the origin drift is accurate to 0.5 mm/year (Altamimi et al. 2011) has important implications for conclusions that may be drawn on global sea level rise.

2.4 Spatial Coverage

Spatial coverage of a TRF has significant impact on the usefulness of a TRF to various users. Spatial coverage is important to consider at global, continental, regional and local scales. For example:

1. Global coverage is required to accurately represent the physical Earth and its rotation, and is used for global-scale geodynamics (Plag and Pearlman 2009).
2. Continental coverage is used for plate tectonics, and to tie national reference systems to ITRF (Blewitt et al. 2013; Brunini and Sánchez 2013; Bruyninx et al. 2012)
3. Regional coverage is used to characterize plate boundary deformation.
4. Local coverage, or "footprint", is used to assess the local physical stability and monumentation of TRF stations (Minster et al. 2010).
5. TRFs naturally have a "spatial domain of applicability" relating to the extent of their spatial coverage. On the one hand, ITRF has global extent, and so is applicable globally. On the other hand, NA12 should only be used in or near North America (Blewitt et al. 2013). For frames that are not global, the user's geodetic solutions will generally degrade moving away from the region encompassed by the network.
6. In addition to spatial extent, spatial sampling also has important consequences for scientific interpretation. Hemispheric asymmetry in the SLR global network can create systematic biases in the realization of the ITRF origin and scale. The frame can also be biased because of oversampling on specific continents. Assuming that spatially-correlated systematic errors exist in any contributing network solution, one might consider a scheme to de-weighting regions that are oversampled. In a sense, this is analogous to generating normal points, but in this case, in the spatial domain.
7. Figure 1 shows the current spatial coverage for >12,000 geodetic GPS stations that are processed routinely by

the Nevada Geodetic Laboratory (http://geodesy.unr.edu). The asymmetry of the distribution is strikingly obvious.

2.5 Temporal Coverage

Temporal coverage affects the quality of the frame with time. Some aspects of temporal coverage are important to consider from the user's perspective:

1. TRFs have a start date and an end date of contributing data. The predictability of TRF station coordinates degrade at least quadratically with time, even for the ideal case of a perfectly secular frame, and so rapidly degrade outside of this time window. Therefore it is imperative to update the TRF from time to time, typically every few years, depending on the time span of contributing data. As the time span increases with each subsequent frame realization, the frame can be updated less frequently. It should be noted, however, that a significant problem remains whenever a great earthquake displaces a significant region within the TRF. To resolve this problem, temporary fixes may be required in between significant TRF updates.
2. Each station has a "time window of applicability" to the frame. At most, this window can include all possible data, from when the station began delivering data, to the general end date of all contributing data to the frame. However a common situation is that a station may only have linear behavior within a specific time window, because the station was in the region of a great earthquake, and is subject to post-seismic deformation. There could be time spans where a station was behaving poorly, perhaps because of failing equipment, or periods where there was snow and ice on an antenna. These windows of applicability need to be carefully specified to the user (Blewitt et al. 2013).
3. That each station has a time window implies that at any given epoch, there will generally be a different subset of contributing stations to the frame. Indeed, the subset at the start time of the frame may not even overlap with the subset at the end time of the frame. This situation requires that there be significant times of overlap between stations such that the frame is stable as a function of time. This is a temporal analogue to struts within a spatial structure. Naturally, stations with longer time series and less steps (discontinuities) provide more stability to the frame, provided their station coordinates are accurately predictable (i.e., secular with stationary noise). This has implications on the stewardship of equipment at fundamental stations, which have multiple collocations. For example, it may not be a good idea to swap old GPS equipment for new GNSS equipment. Rather it might be better to install the GNSS equipment in addition to the old GPS equipment.

2.6 Quality

The quality of a TRF can be characterized in several ways. Some of the following concepts were articulated by the National Research Council report (Minster et al. 2010):

1. Accuracy refers to how close a determined quantity is to the truth, and precision refers to how close the determination of a quantity can be repeated. Consistently high accuracy requires high precision.
2. Reference frame accuracy and precision refers to the determination of scale, orientation, and origin of a frame. This aspect of quality is important for the study of geodynamics and climate change, particularly in the determination of vertical land motion and sea level rise. Note that some of these parameters may not be observable to a specific technique, for example, VLBI is insensitive to the Earth center of mass. GNSS techniques determine scale with extremely high precision, but have an unknown overall bias between the satellite center of mass and electrical phase center of the transmitter, which requires calibration by SLR. Orientation must be conventionally aligned for all systems, though GNSS techniques determine the position of the rotational pole with extremely high precision.
3. Station coordinates themselves are, strictly speaking, not observable. However, it is useful to consider the concepts of "internal accuracy" and "internal precision," which refer to the determination of coordinates within a given TRF. This internal aspect of quality is important to the scientific user who studies Earth deformation. A reasonable proxy for internal precision is the standard deviation in repeated estimation of station coordinates, though it should not be assumed that this scatter is constant in time and geographic location. Internal accuracy is more difficult to assess. A reasonable proxy for internal accuracy is the degree to which different techniques agree on station coordinates, though local ties confound this assessment. It should also be recalled that all techniques use the IERS conventions (IERS 2010), hence errors in those conventions may not be well assessed by looking at differences between solutions.
4. Stability refers to predictability, that is the ability of the TRF to extrapolate station coordinates accurately into the future (and into the past, though with less application for space-based geodesy). Stability requires that the scale and origin have long-term accuracy and predictability. As a consequence, the best stability requires long-term linear behavior of the origin and scale. The useful lifetime of a TRF is related to its stability. Good stability is a result of having longer contributing time series, simple empirical station motion models (e.g., linear with few or no steps),

and lots of overlap in time windows of applicability. Good stability also requires continuous monitoring of the TRF to detect and retire TRF stations that have poorly predicted positions, for example, some stations may have unstable monumentation, or may be subject to regional instability from hydrological effects, or may have step discontinuities due to earthquakes or changes of equipment. Users that absolutely depend on TRF stability include satellite altimeter missions to measure long-term change in sea level, and scientists studying acceleration of ice sheet melting in polar regions.

5. Spatial heterogeneity of the probability distribution function (PDF) of coordinate accuracy can occur due to non-uniformity in TRF station distribution or source distribution. For example, GPS satellites do not pass near the celestial pole, hence vertical precision is weaker in polar regions. Oceans are poorly sampled compared to continents. As a result, certain hemispheres of the Earth are more poorly sampled, for example, the hemisphere centered on the Pacific Ocean. This leads to global asymmetry and geographic dependence in the level of errors in user's positions. The comments previously made with regard to spatial coverage apply to this problem.
6. Temporal heterogeneity of the PDF, or "heteroscedasticity", is a common feature of geodetic time series, with data from earlier years having more scatter and bias than more recent data. This naturally arises because new techniques typically have growing networks in earlier years. However, the opposite can occur, where in older networks the number of stations can decrease in time, or where equipment ages and starts to generate more noisy data. Recent degradation is of particular concern for the SLR network contributing to the ITRF, as this would heavily impact TRF stability, discussed above. This is because data at both the start and end of a time series have more effective leverage in the determination of the secular trend and thus the prediction of future coordinates, which is so essential for climate change science. This raises the issue of long-term commitment to essential contributing techniques in the ITRF, particularly at multi-technique stations. In any case, care should be taken when realizing a TRF to properly account for heteroscedasticity. Formal error naturally takes into account number of stations, but will not take into account geographic systematic error that can occur when a long running station in a remote region fails and is not replaced. Nor does formal error account for time-dependent noise in the data, such as the seasonal variation in scatter that is seen in GNSS time series arising from seasonal tropospheric conditions, and increase in noise due to decreased signal to noise ration when antenna elements start to fail.

2.7 Life Cycle

Reference frames have to be upgraded from time to time. As a result, a TRF can be said to have a life cycle, which can be summarized by the following iterative process:

1. *User requirements demand a new TRF.* This naturally occurs when the stability of the frame is approaching a critical point where accuracy no longer meets the user requirements. Or a new and fundamentally better frame may be demanded by new science.
2. *The reference system is upgraded.* In particular, models of the observables, station motion, source structure, etc. are typically improved with time. New models are recommended to contributing analysis centers to facilitate consistency and incremental improvement in accuracy with each frame release (IERS 2010).
3. *The TRF is designed to best meet user requirements.* For example, the end date of contributing data is specified. The list of contributing stations is reassessed based new criteria, and on the changing list of potential candidate stations. A new technique may be introduced into the ITRF combination, or how it contributes may be strengthened or relaxed. The way that the origin is realized may be revisted and a new method may be applied. The empirical station motion model may be expanded to include seasonal terms. A new scheme to determine relative weights of contributing solutions might be implemented (Altamimi et al. 2011).
4. *Contributing analysis centers reprocess data over a specified time span* (e.g., Steigenberger et al. 2006). Ideally, the specified time span is the full history to improve predictability. However there may be mitigating circumstances in practice, for example, sparseness of observations and/or high levels of uncertainty in the quality of earlier solutions.
5. *The TRF is realized and published* (Altamimi et al. 2011).
6. *The TRF is used.* For best results, this requires users to reprocess all their data using the new recommended models of the upgraded TRF, and to apply transformations into the new frame.
7. *The TRF is inherited by new frames.* For studies of tectonic plate boundary deformation and possible intraplate deformation, scientific users often prefer to refer to a frame that is co-rotating with the stable interior of the plate, for example, the NA12 frame constructed by Blewitt et al. (2013). The first step in producing NA12 was to determine a frame aligned to ITRF with a much denser network in North America than in ITRF. The second step was to apply a rotation rate to the frame to minimize the apparent rotation of stations selected to represent the stable interior of North America. Other

examples of inherited regional frames include SIRGAS in South America (Brunini and Sánchez 2013) and the EUREF Permanent Network in Europe (Bruyninx et al. 2012).

8. *The TRF degrades with time*. Actually, frame degradation begins well before the last date of contributing data. This is because it is typical to impose a lower limit on the time span of contributing stations to the input solutions. For example, contributing stations to the NA12 frame must have a minimum of 5.5 years of continuous data (Blewitt et al. 2013). Given that the last date of contributing data is 2012.1, this means that the last contributing stations started to deliver data no later than 2006.6, which represents the time of maximum number of contributing stations. Since that time, stations are not added, and can only be removed because either dataflow stopped from the stations, or because its position is no longer predictable. This reduction of stations in time causes the frame to degrade in precision since 2006.6. Moreover, since 2012.1, the lack of any contributing data causes the stability to degrade quadratically in time. At some critical point in the future, the frame will degrade beyond the threshold of user requirements. Before this happens, we need to loop back to the top of this list, and again work our way through the previous item.

3 Scientific User Demands: The Example of Sea Level Rise

Satellite altimeter measurements from the Topex/Poseidon, Jason-1, and Jason-2 missions suggest that global mean sea level is over the last two decades now rising 1 mm/year faster than what tide gauges infer over the century prior to that (Nerem et al. 2010). This result depends critically on the quality of the TRF (Minster et al. 2010). Notably, the altimeter measurements derive from three different missions that have different time spans. Therefore to infer a recent secular rate using all three missions requires that the measurements of sea surface height be completely consistent.

Consider the following chain of dependency on the TRF (Blewitt et al. 2010). The satellite radars only measure the range between sea surface and the satellite. The satellite position at any time can be inferred by precision orbit determination using three techniques: GPS, DORIS, and SLR (Cerri et al. 2010). These positions must be accurate with respect to the Earth center of mass, and with respect to absolute scale as set by the conventional speed of light, hence the need for SLR and VLBI. Bias in the radar measurement is calibrated using buoys or tide gauges with positions determined by GPS with respect to the Earth center of mass at the origin of the TRF (Watson et al. 2015). Moreover, to infer sea level change from tide gauges with respect to the Earth center of mass requires monitoring vertical land motion at the tide gauge using GPS in the TRF, or some combination of nearby GPS and local leveling (Santamaría-Gómez et al. 2012).

Considering these dependencies tells us the user requirements of ITRF for this most stringently demanding application:

1. ITRF must have continuity spanning all of the missions and going forth into the future, and must have sufficient coverage in the past to infer secular vertical land motion at tide gauges. Even then, there is the limitation that we cannot know for sure that vertical land motion has been linear prior to the space geodetic era.
2. ITRF must be stable to interpret measurements made more recently than the last contributing data to ITRF.
3. GPS, DORIS, SLR, and VLBI must contribute significantly to ITRF.
4. Local ties at collocation sites between all ITRF contributing techniques must be accurate to exploit inter-technique synergy demanded of sea level change investigations.
5. All techniques must have their data processed using IERS conventions consistent with ITRF.
6. ITRF must accurately align its origin with the Earth system center of mass.
7. ITRF must have a scale and origin that is accurate and stable in time.
8. Frame degradation due to stations that recently lack predictable frame coordinates (such as due to equipment changes or earthquakes) must be monitored and mitigated to support current missions.
9. Balanced, collocated coverage in time and geographic location is required to reduce spatio-temporal biases that can mimic decadal-scale variations in regional sea level.
10. Systematic errors of techniques contributing to ITRF must be monitored (e.g., by intercomparison) and mitigated as well as possible, for example by appropriate relative weighting of the contributing solutions, or by using specific techniques to realize specific aspects of the ITRF (e.g., SLR for origin).

Ultimately, when answering the question as to whether sea level rise is really accelerating, all of the above points must be addressed with sufficient rigor to assess the statistical significance of the results. It should also be noted that improving the TRF is a minimum requirement, in that other errors associated with altimetry must also be addressed.

4 Conclusions

Scientific users place the most stringent demands on the TRF. Such users need to interpret physical vertical station motions and large-scale horizontal deformation. The most stringent application is the determination of long-term sea level change, vertical land motion, and multi-technique satellite orbit determination. Such applications need to access a physical origin and scale with high accuracy and stability, at the level of 1 mm/decade. By meeting such a requirement, we can meaningfully connect data from satellite altimeters to GNSS buoys for radar calibration, we can consistently determine vertical land motion in a global system, and we can meaningfully compare data from different satellite missions to infer secular variation over decades.

This brings us back to our original question: "What is the primary mission of the TRF?" Considering the most stringent application discussed above, the answer to this might be "predictability". We primarily need the ability to predict accurately the coordinates of a multi-technique network of stations required by the user at any time needed, past, current, and future. To the most stringent users, this "predictability" defines the essence of a successful TRF.

Acknowledgments I am grateful for the careful reviews of Zuheir Altimimi and two anonymous reviewers, who have helped to improve the manuscript. This research was supported by NASA grants NNX12AK26G, NASA subaward 1551941 through the University of Colorado, and NSF grant EAR-1252210.

References

Altamimi Z, Collilieux X, Métivier L (2011) ITRF2008: an improved solution of the international terrestrial reference frame. J Geod 85:457–473. doi:10.1007/s00190-011-0444-4

Amos C, Audet P, Hammond WC, Burgmann R, Johanson IA, Blewitt G (2014) Uplift and seismicity driven by groundwater depletion in central California. Nature 509:483–486. doi:10.1038/nature13275

Blewitt G, Altamimi Z, Davis JL, Gross R, Kuo C-Y, Lemoine FG, Moore AW, Neilan RE, Plag H-P, Rothacher M, Shum CK, Sideris MG, Schöne T, Tregoning P, Zerbini S (2010) Geodetic observations and global reference frame contributions to understanding sea-level rise and variability. In: Church J, Woodworth PL, Aarup T, Wilson S (eds) Understanding sea-level rise and variability. Wiley-Blackwell, Chichester, pp 256–284. ISBN 978-1-443-3451-7

Blewitt G, Kreemer C, Hammond WC, Goldfarb JM (2013) Terrestrial reference frame NA12 for crustal deformation studies in North America. J Geodyn 72:11–24. doi:10.1016/j.jog.2013.08.004

Brunini C, Sánchez L (2013) Geodetic reference frame for the Americas. GIM Int 27(3):26–31, ISSN: 1566–9076

Bruyninx C, Baire Q, Legrand J, Roosbeek (2012) The EUREF permenant network (EPN): recent developments and key issues. Proceedings of the EUREF 2011 symposium. http://www.epncb.oma.be/_documentation/papers/

Cerri L, Berthias JP, Bertiger WI, Haines BJH, Lemoine FG, Mercier F, Ries JC, Willis P (2010) Precision orbit determination standards for the Jason series of altimeter missions. Mar Geod 33:379–418. doi:10.1080/01490419.2010.488966

IERS Conventions (2010) In: Petit G, Luzum B (eds) IERS Technical note: 36. Frankfurt am Main: Verlag des Bundesamts für Kartographie und Geodäsie, p 179. ISBN 3-89888-989-6

Davis JL, Wernicke B, Tamisea M (2012) On seasonal signals in geodetic time series. J Geophys Res 117:1403. doi:10.1029/2011JB008690

Kreemer C, Lavallée DA, Blewitt G, Holt WE (2006) On the stability of a geodetic no-net-rotation frame and its implication for the international terrestrial reference frame. Geophys Res Lett 33(17). doi:10.1029/2006GL027058

Minster JB, Altamimi Z, Blewitt G, Carter WE, Cazenave A, Dragert H, Herring TA, Larson KM, Ries JC, Sandwell DT, Wahr JM, Davis JL (2010) Precise geodetic infrastructure: national requirements for a shared resource. The National Academies Press, Washington, DC, p 142. ISBN 10-309-15811-7

Nerem RS, Chambers DP, Choe C, Mitchum GT (2010) Estimating mean sea level change from the TOPEX and Jason altimeter missions. Mar Geod 33(S1):435–446. doi:10.1080/01490419.2010.491031

Plag H-P, Pearlman M (2009) Global geodetic observing system. Springer, Berlin

Rebischung P, Griffiths J, Ray J, Schmid R, Collilieux X, Garayt B (2012) IGS08: the IGS realization of ITRF2008. GPS Solutions 16:483–494. doi:10.1007/s10291-011-0248-2

Santamaría-Gómez Á, Gravelle M, Collilieux X, Guichard M, Martin Miguez B, Tiphaneu P, Woppelmann G (2012) Mitigating the effects of vertical land motion in tide gauge records using a state-of-the-art GPS velocity field. Global Planet Change 98–99:6–17. doi:10.1016/j.gloplacha.2012.07.007

Steigenberger P, Rothacher M, Dietrich R, Fritsche M, Rülke A, Vey S (2006) Reprocessing of a global GPS network. J Geophys Res 111(B5). doi:10.1029/2005JB003747

Watson CS, White NJ, Church JA, King MA, Burgette RJ, Legresy B (2015) Unabated global mean sea-level rise over the satellite altimeter era. Nat Clim Change Lett. doi:10.1038/nclimate2635

The Phase 2 North America Land Data Assimilation System (NLDAS-2) Products for Modeling Water Storage Displacements for Plate Boundary Observatory GPS Stations

Zhao Li and Tonie van Dam

Abstract

In this paper, we compare the efficiency of two models to estimate the surface displacements due to continental water storage (CWS) variations over continental North America. The first model, the monthly North America Land Data Assimilation System Phase 2 Noah (NLDAS-2 Noah) is a model of CWS restricted to North America. The second data set, the Global Land Data Assimilation System Noah (GLDAS Noah), is global. To compare the models, we use coordinate time series from GPS stations within the Plate Boundary Observatory (PBO). We find that the NLDAS-2 Noah CWS estimates of vertical surface displacements are correlated with PBO height coordinate time series with an average correlation of 0.4. Of the selected 986 PBO stations, stations with their weighted root mean square (WRMS) reduced after removing the surface displacements predicted using NLDAS-2 Noah surface mass, account for 13%, 27% and 56% for the north, east and up components respectively. The highest reductions in scatter occur on coordinate time series from stations in the mountains.

Comparing NLDAS-2 Noah to GLDAS Noah, we find that the NLDAS-2 Noah model reduces the horizontal WRMS more than GLDAS for 88% and 73% of the PBO stations in the North and East components. In addition, stations in the mountains of the northwest and southeast part of the NLDAS-2 Noah spatial coverage (25% of the total stations) have their vertical scatter reduced by more than 10%. Therefore, we conclude that the NLDAS-2 Noah model better estimates the CWS induced 3-D surface displacement for PBO GPS stations in continental North America. The reasons may due to the finer spatial resolution, the updated Noah model, together with the more accurate surface forcing data of the NLDAS-2 Noah model.

Keywords

Continental water storage • GLDAS Noah • GPS time series • NLDAS-2 Noah • PBO

1 Introduction

Strong correlations exist between the continental water storage (CWS) and height changes in global positioning system (GPS) coordinate time series (van Dam et al. 2001, 2007; Tregoning et al. 2009; Fritsche et al. 2012). This environmental surface displacement could add residual signal to GPS data that is being used for geodynamic studies, e.g. tectonics, and postglacial rebound. To remove this environmental signal, CWS mass models are needed to predict surface displacements. Currently, one of the most cited models used for estimating CWS loading effects is the monthly Global Land Data Assimilation System (GLDAS) model. The model has a spatial resolution of 1° in longitude and latitude (Rui 2011). The components of water storage in GLDAS include

Z. Li (✉) • T. van Dam
Faculté des Sciences, de la Technologie et de la Communication, University of Luxembourg, Luxembourg
e-mail: zhao.li@uni.lu

T. van Dam (ed.), *REFAG 2014*, International Association of Geodesy Symposia 146,
DOI 10.1007/1345_2015_176

soil moisture (SM) and snow water equivalent (SWE). There is no groundwater change component in GLDAS.

The Plate Boundary Observatory (PBO), part of the EarthScope project, was installed to measure Earth deformation across the coterminous western United States and Alaska, primarily using permanent GPS receivers. PBO is the most precise spatial reference system realization available in United States history (Anderson et al. 2006). It consists of 1,100 continuously operating GPS stations[1]. CWS driven surface displacements introduce residual signal into these time series primarily at annual periods with significant inter-annual variability.

Under funding from the National Oceanic and Atmospheric Administration (NOAA) Climate Prediction Program of the Americas (CPPA), the National Centers for Environmental Prediction (NCEP) Environmental Modeling Center (EMC) team has developed the North America Land Data Assimilation System Phase 2 Noah (NLDAS-2 Noah) that includes improved forcing data and land surface models (LSM) simulations (Xia et al. 2012; Rui 2013). This model is provided, in near real time, at 1/8th-degree grid spacing over North America for a period extending from January 1979 to the present at hourly and monthly temporal resolutions. As NLDAS is eight times more spatially dense than the GLDAS product, we want to evaluate whether the higher spatial resolution of LDAS might improve the correlation between the CWS driven displacement and the GPS coordinate time series as compared to GLDAS Noah model.

Since our previous analysis show that under the same spatial resolution, for example, the GLDAS monthly and 3-hourly products, there is only a slight difference for the loading displacement at the weekly samples (see Fig. 2 of Li et al. 2014), and the higher temporal resolution could improve the performance by almost the same magnitude for different CWS models (see Table 2 of Li et al. 2014), while the GLDAS monthly model is currently one the most cited models for estimating the CWS loading effects, in this paper, we investigate the difference between the weekly surface displacements interpolated from the monthly GLDAS Noah and NLDAS-2 Noah CWS (SWE + SM) models. Even though the PBO network extends up into Alaska, we will restrict our comparison to sites in the continental US, as this is essentially the coverage for the NLDAS-2 Noah data set.

2 Data Processing

2.1 Farrell's Green's Function Approach

The predicted 3-D surface displacement of a point on the Earth's surface driven by changes in CWS can be determined by convolving Farrell's Green's functions (Farrell 1972) with a surface mass model over the surface of the Earth (van Dam and Wahr 1987). The basic equations can be written as:

$$\begin{aligned} d_n(\theta,\phi) &= \sum_{i=1}^{nlon}\sum_{j=1}^{nlat} \Delta P_{i,j} G_{i,j}^{n} A_{i,j} \\ d_e(\theta,\phi) &= \sum_{i=1}^{nlon}\sum_{j=1}^{nlat} \Delta P_{i,j} G_{i,j}^{e} A_{i,j} \\ d_u(\theta,\phi) &= \sum_{i=1}^{nlon}\sum_{j=1}^{nlat} \Delta P_{i,j} G_{i,j}^{u} A_{i,j} \end{aligned} \quad (1)$$

where the subscripts i and j denote a unique grid point in the GLDAS Noah or NLDAS-2 Noah data sets, $\Delta P_{i,j}$ is the CWS variation at the grid point, $A_{i,j}$ is the area of the grid point, $nlon$ and $nlat$ represent the maximum number of grid units in longitude and latitude. For GLDAS Noah grid, $nlon = 360$; $nlat = 150$, while for NLDAS-2 Noah data, $nlon = 464$; $nlat = 224$. θ and ϕ represent the co-latitude and longitude of the point on the Earth where the loading effect is being determined, $d_n(\theta,\phi), d_e(\theta,\phi), d_u(\theta,\phi)$ are the 3-D surface displacements of the given point, $G_{i,j}^{n}, G_{i,j}^{e}, G_{i,j}^{u}$ denote the Green's function for each component (Farrell 1972). The Green's functions are a function of the angular distance between the loading point and the point where the effect of the load is being calculated. We choose the Green's function derived in the center of figure (CF) frame to maintain consistency between the predicted loading and GPS coordinate time series (Dong et al. 1997; Dong et al. 2003; Blewitt 2003).

For all of the CWS data, we remove a 10-year mean that is calculated using data from 2000 to 2009 for each model. Then, the residual from this mean is convolved with the Farrell's Green's function to obtain the 3-D surface displacement. The resulting monthly surface displacements are then detrended and interpolated into weekly solutions that correspond to the GPS week.

2.2 Data Description

2.2.1 Water Storage Model

We model the CWS induced surface displacements for 986 PBO GPS stations in continental North America using both the GLDAS and NLDAS-2 Noah models. The time period for our comparison is 01/01/2000 to 31/12/2012. For the GLDAS Noah products, we use the 1-degree SM and SWE data[2]. We do not include the SWE data above the latitude of 60.5 N. This area includes Greenland and most Arctic regions. GLDAS Noah does not model snow dynamics well

[1]http://www.unavco.org/instrumentation/networks/status/pbo

[2]ftp://hydro1.sci.gsfc.nasa.gov/data/s4pa/GLDAS_V1

Table 1 Details of the CWS data in the GLDAS and NLDAS-2 Noah models

	GLDAS Noah	NLDAS-2 Noah
Data source	SM, SWE (below 60.5°N)	SM, SWE
Unit	Kg/m^2	Kg/m^2
Spatial resolution (degree)	1° × 1°	0.125° × 0.125°
Latitude extent (degree)	−59.5 to 89.5	25.063 to 52.938
Longitude extent (degree)	−179.5 to 179.5	−124.938 to −67.063
Dimension	360 (lon) × 150 (lat)	464 (lon) × 224 (lat)
Latency	1–4 months	1–2 months

in these regions (Rui 2011; Jiang et al. 2013). The summations in Eq. (1) are over the entire globe.

For NLDAS-2 Noah, we also use the SM and SWE data[3], however for this data set the spacing is at 0.125°. Table 1 shows the details of the CWS data in both models.

The top panel of Fig. 1 shows the root mean square (RMS) of the weekly vertical loading time series using NLDAS-2 Noah for each PBO (Wessel and Smith, 2013) station, while the bottom panel represents the vertical RMS of each station with respect to the elevation. We can see that the amplitude of the vertical displacement induced by NLDAS-2 Noah shows a slight increase at the higher elevations[4], among which almost all the stations above the elevation of 500 m have the scatter larger than 1 mm, and the maximum RMS reaches 4 mm. In comparison the NLDAS-2 Noah horizontal displacements for all PBO stations are small, with a maximum RMS at the coast less of than 0.7 mm (not shown).

In Fig. 2 we compare NLDAS-2 Noah surface displacements with those predicted using GLDAS Noah. Figure 2 shows the standard deviation (SD) (top) and the maximum (bottom) of the difference between the vertical loading displacement using NLDAS-2 and GLDAS Noah models. We use SD in this comparison, because we want to determine which model has the larger signal. Positive SDs indicate that the NLDAS-2 Noah predicted displacements are larger than those determined the GLDAS Noah model. Similar as Fig. 1, here we also show the relationship between SD and the elevation to better illustrate the difference between NLDAS-2 Noah and GLDAS Noah with respect to the elevation (Fig. 3). From Figs. 2 and 3, we find that large difference exists in the predicted vertical displacement, and generally the difference increases with increasing elevation. The SD and the maximum displacements reach more than 1.4 and 4 mm for the predicted verticals in the high mountains respectively. For the horizontal components, we only observe small differences between the NLDAS-2 and GLDAS Noah predictions (not shown).

2.2.2 GPS Data

The latest PBO coordinate time series in IGS08 reference frame[5] are used to evaluate the performance of the CWS models. Note that the UNAVCO announced a GPS data quality issue that the GPS coordinate time series between the dates of 01 January 2014 and 15 October 2014 contain inaccurate daily positions[6]. Here we only use the GPS data until the end of 2012, so it has no impact on our comparison results. Before comparing the GPS observations with the detrended loading results, offsets and obvious errors in the GPS coordinate time series are detected and removed. These two steps are manually done station by station. The obvious errors refer to those station coordinates deviate largely from the neighboring points, and those with uncertainty larger than 10 and 5 mm for the vertical and horizontal components respectively. After we remove the above obvious errors, we then look at the linear trend of the time series. Whenever the linear trend changes, we define the epoch as an offset, and separate the time series. Although this manually detection method is time consuming, we think that this is the most reliable way to prepare the GPS data. Finally, since the published PBO GPS time series are daily solutions, we need to first average the daily GPS data into weekly solutions. The weekly averaging is sufficient since water storage changes are primarily annual. Then a linear trend should also be removed from the weekly GPS solutions.

To evaluate the effectiveness of the two models for precisely estimating CWS in the PBO time series we compare the predicted 3-dimensional surface displacements from NLDAS-2 and GLDAS Noah models with coordinate times series from the PBO GPS sites.

3 Results

Figure 4 shows an example of the detrended weekly loading time series for station SC02 (Friday Harbor, Washington) generated from the NLDAS-2 Noah and GLDAS Noah data sets in millimeters. From the top to the bottom, the panels represent up, north, and east components respectively. The CWS monthly time series are interpolated to GPS weeks using a cubic spline interpolation. The GPS time series represented by the black curve in the figure.

In Fig. 4, we observe that the predicted peak-to-peak horizontal displacement for station SC02 from both the

[3]ftp://hydro1.sci.gsfc.nasa.gov/data/s4pa/NLDAS-2

[4]http://www.sage.wisc.edu/atlas/maps/elevation/atl_elevation_nam.jpg

[5]ftp://data-out.unavco.org/pub/products/position

[6]https://www.unavco.org/data/gps-gnss/derived-products/derived-products.html

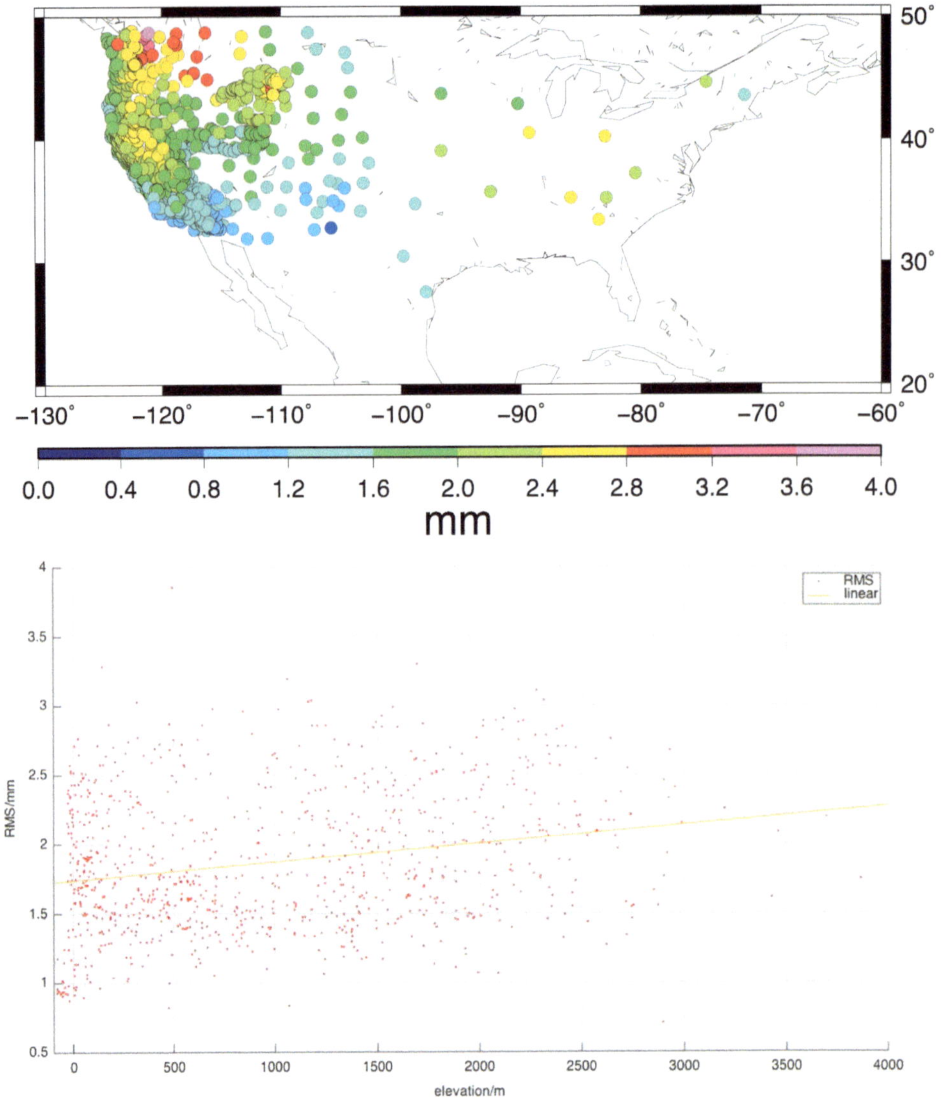

Fig. 1 (*Top*) RMS of the predicted weekly vertical displacement driven by NLDAS-2 Noah CWS. (*Bottom*) Vertical RMS of each station with respect to the elevation

NLDAS-2 and GLDAS Noah models are smaller than 2 mm. There is big discrepancy between the predicted and the horizontal GPS time series. With respect to the vertical component, both models fit the GPS height variation well, with the NLDAS-2 Noah model being slightly closer to the GPS when compared with the GLDAS Noah model.

The top panel of Fig. 5 shows the correlation between the NLDAS-2 Noah predicted vertical displacement and the GPS height time series. We find that in general the larger the RMS the higher the correlation. The average correlation between the predicted and the observed vertical component is approximately 0.4. Of all the stations, 52% have a correlation greater than 0.4. Stations located in the mountains, including the Pacific Coast Ranges, have correlations greater than 0.6.

Compared to the vertical results, poor correlations exist between NLDAS-2 Noah CWS loading and the GPS in the horizontal components. Stations with correlation coefficients larger than 0.4 account for only 5% and 12% for the North and East components respectively, most of which are located in the mountains.

The bottom panel of Fig. 5 shows the weighted root mean square (WRMS) reduction rate of the GPS coordinate time series after removing the predicted NLDAS-2 Noah CWS loading effect. The WRMS reduction rate here is a percentage, and is defined as

$$\%diff = [WRMS(GPS) - WRMS(GPS - CWS)] \times 100 / WRMS(GPS)$$

Positive values indicate that a station's WRMS is reduced.

After removing the NLDAS-2 Noah loading effects in the vertical coordinates, Fig. 5 shows that 56% of the PBO stations have their WRMS reduced. Stations with an improvement greater than 5% represent 31% of all stations. Most of these sites are located in the mountains. With respect to the

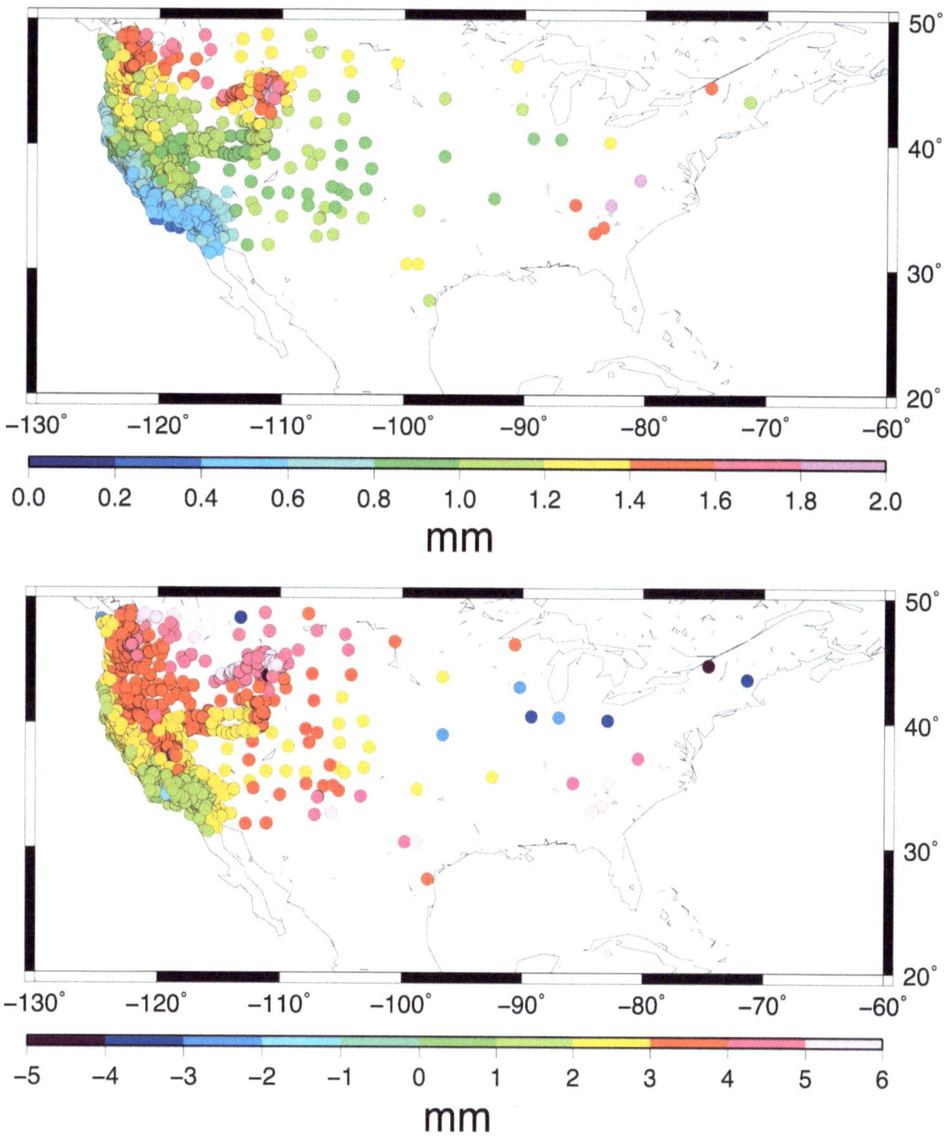

Fig. 2 Standard deviation (SD) (*top*) and the maximum differences (*bottom*) between NLDAS-2 and GLDAS Noah predicted vertical loading displacement. *White circles* indicate that the SD and the maximum difference are bigger than the maximum value on the scale. *SD* represents the scatter of the difference between NLDAS-2 and GLDAS Noah obtained displacement time series

horizontal component, only a small number of stations in the mountains have their WRMS reduced. The WRMS increases on the remaining stations when displacement from NLDAS-2 Noah are removed from the GPS horizontal coordinate time series. Specifically, 87% and 73% of the time series have their WRMS increased in the north and east respectively. This result is expected given the observed poor correlation.

4 Comparison of GLDAS and NLDAS-2 Noah Models

GLDAS Noah is one of the most common datasets used to model the CWS driven surface displacements. To determine if there are advantages to using NLDAS-2 Noah, a regional model (coterminous United States) with higher spatial resolution than GLDAS, we compare loading displacements predicted using NLDAS-2 Noah and the GLDAS Noah with observed station coordinates in GPS time series from the PBO network. From Sect. 2.2, we find only small differences between the NLDAS-2 Noah and GLDAS Noah horizontal predictions, with larger differences in the predicted vertical displacement.

Figure 6 shows the WRMS (in units of %) difference between GPS coordinate time series corrected for NLDAS-2 and GLDAS Noah CWS models. Positive values indicate that the scatter is reduced more using NLDAS-2 Noah product. From Fig. 6, we observe that 58% of the stations have their vertical WRMS improved using NLDAS-2 Noah instead of GLDAS Noah. Moreover, 25% of the improvements are more than 10%, and these stations are concentrated in the mountains of the northwest and southeast. With respect to the horizontal components, NLDAS-2 Noah could improve the WRMS reduction rate obtained

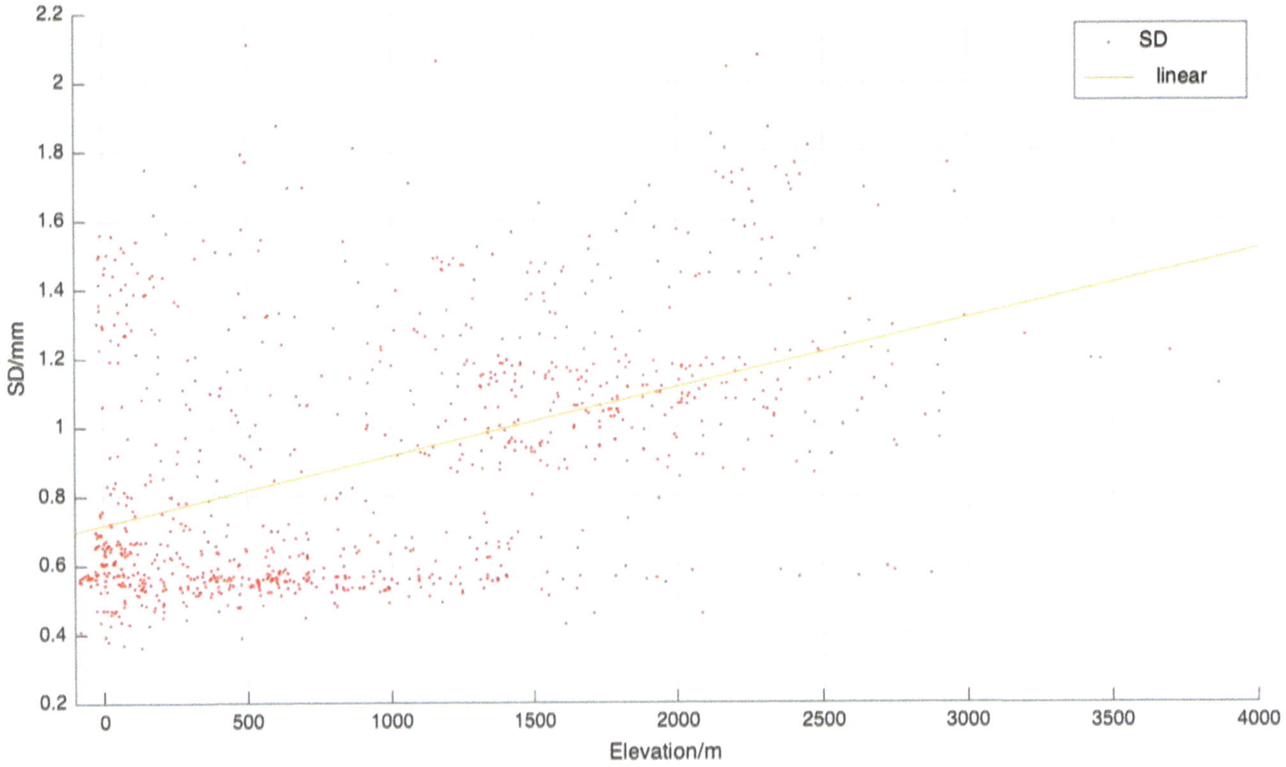

Fig. 3 Scatter of the SD for each PBO station with respect to the elevation

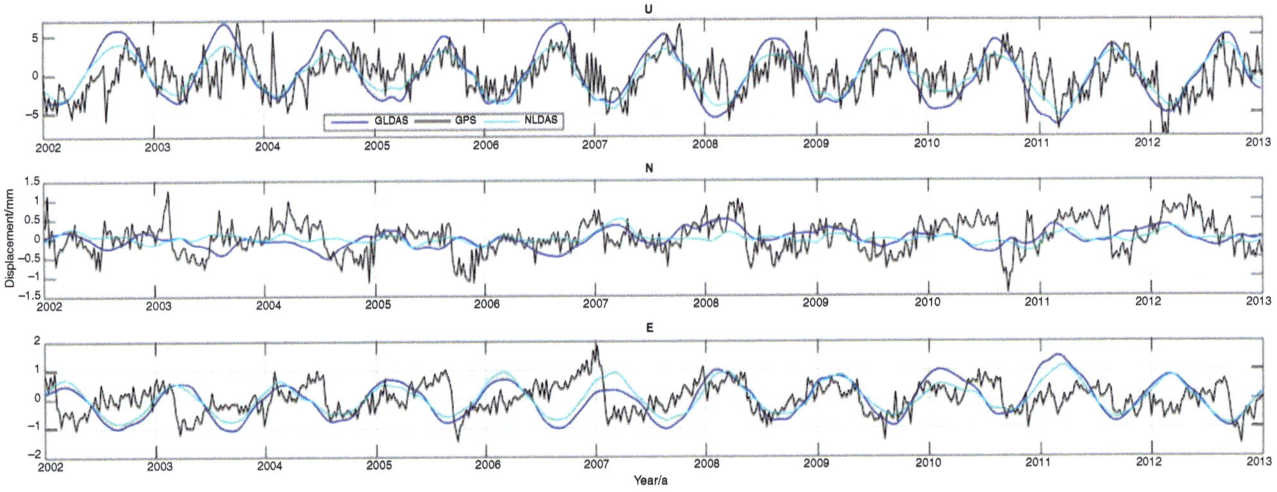

Fig. 4 3-D displacements of station SC02 generated from NLDAS-2 and GLDAS Noah models. The NLDAS-2 Noah model is shown in *cyan* (or *light blue*); GLDAS Noah is shown in *blue*

from the GLDAS Noah product to some extent for most PBO stations (88% and 73% for the North and East components respectively), although the performance is very small (see Sect. 3). The best improvement is in the North component where most stations show an improvement of more than 10%. These account for about 63% of the total stations. We partly contribute this improvement to the much higher spatial resolution of the NLDAS-2 Noah model.

Another reason is that the NLDAS-2 Noah uses an upgraded Noah version (Noah version 2.8) compared with GLDAS Noah model (Noah version 2.7.1), which includes a snow model enhancement for cold season (Livneh et al. 2010) and model parameter tuning for warm season (Wei et al. 2013). Moreover, the NLDAS-2 total column soil moisture and snow water equivalent have been comprehensively evaluated against in situ observations and

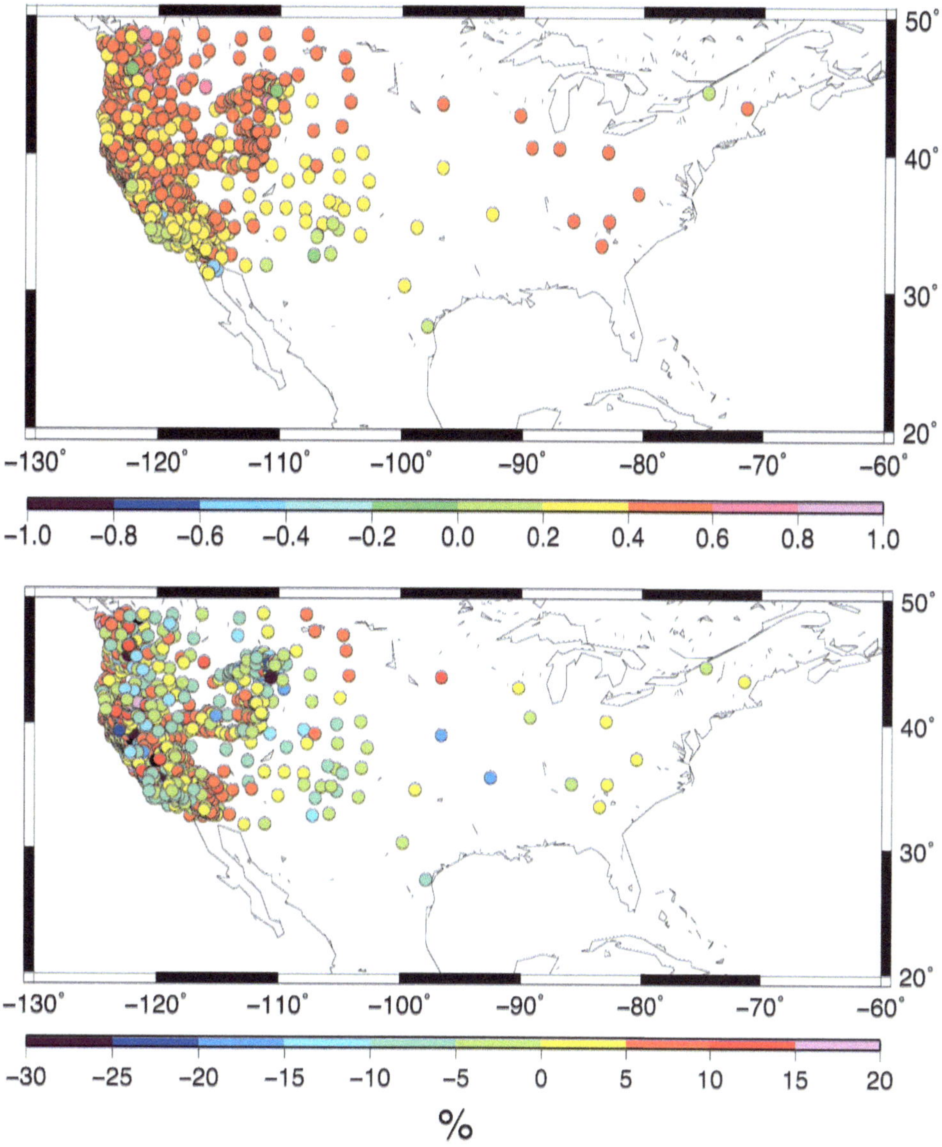

Fig. 5 *Top*: Correlation between NLDAS-2 Noah predicted loading and the GPS height time series. *Bottom*: WRMS reduction (%) after removing the NLDAS-2 Noah predictions from the GPS height

satellite retrievals (Xia et al. 2014; Livneh et al. 2010) in the continental United States. Therefore, the overall performance is that Noah version 2.8 has better performance than Noah version 2.7.1. Besides the model version differences, the surface meteorological forcing data driving the Noah model are different, with NLDAS-2 surface forcing data maybe more accurate. Due to these reasons, we demonstrate that NLDAS-2 Noah is better than GLDAS Noah in modeling the CWS driven surface displacement for PBO GPS stations over the continental North America.

Since our focus here is to compare the CWS models only, we do not evaluate the impact of atmospheric mass on our comparison result. However, in our previous studies, we did this kind of analysis. Our results show that after considering the impacts of non-tidal atmosphere and ocean loading effects, the performance of CWS model in correcting the GPS height could improve by at least 10% globally, but the characteristics of the comparison results between different CWS models would not change (Li et al. 2014). Therefore, we expect that the performance of NLDAS-2 Noah model would be better in reducing the GPS height if removing the effects of non-tidal atmospheric and oceanic mass, but our comparison result between NLDAS-2 and GLDAS would not change.

5 Conclusions

We model the 3-D surface displacement induced by CWS from the regional NLDAS-2 Noah model with high spatial resolution for the continuous PBO GPS stations over the continental North America. We find that the CWS induced displacements in the horizontal are small and have a poor correlation with the GPS coordinate time series. Correspondingly,

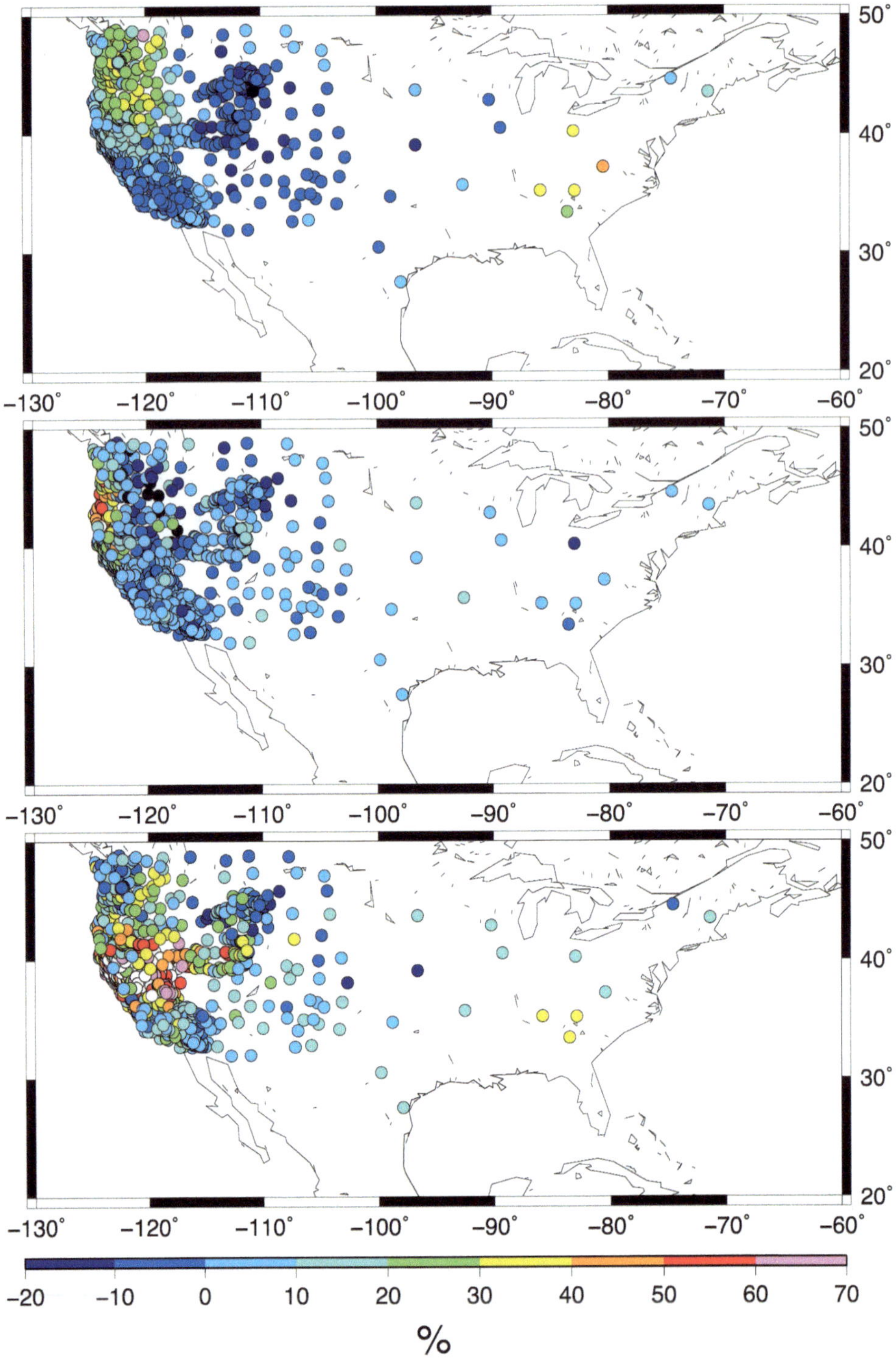

Fig. 6 WRMS difference between NLDAS-2 and GLDAS Noah models. *White* and *black dots* indicate stations' WRMS difference exceed the maximum and the minimum scale respectively. From *top* to *bottom* are the Up, East and North components

only 13% and 27% of the selected stations, mainly those located in the mountains, have their WRMS reduced in the North and East components after removing the NLDAS-2 Noah CWS loading effects from the GPS observations.

For the vertical coordinate, we find that the magnitude of the NLDAS-2 Noah predicted vertical displacement for the PBO stations increases with increasing elevation, and that the maximum RMS reaches 4 mm in the mountains. We also find that much higher correlations between the predicted vertical height changes and the GPS height coordinates. Stations with vertical WRMS reduced when the NLDAS-2 Noah CWS signal is removed represent 56% of the total; stations with the scatter reduced the greatest are concentrated in the mountains.

Compared with the GLDAS Noah product, we find that NLDAS-2 Noah could improve the WRMS reduction rate of the horizontal GPS time series obtained from the GLDAS Noah for most stations (88% and 73% of the stations in the north and east components respectively). With respect to the vertical component, 25% of the stations have their WRMS reduced by more than 10% using NLDAS-2 Noah product. Hence, we conclude that the NLDAS-2 Noah estimates of CWS are better for modeling the CWS induced 3-D surface displacement for PBO GPS stations over the continental North America. The reason may due to the finer spatial resolution, the updated Noah model, together with the more accurate surface forcing data of the NLDAS-2 Noah model.

NLDAS-2 Noah corrections for the PBO sites can be accessed from the PBO H2O data portal located at http://xenon.colorado.edu/portal/.

Acknowledgements We thank NASA for making the GLDAS and NLDAS-2 Noah data freely available. We thank UNAVCO for providing the latest PBO GPS coordinate time series. Figures are plotted with the GMT and MATLAB software.

References

Anderson G, Hodgkinson K, Herring T, Agnew DC (2006) Plate boundary observatory data management system critical design review version 1.2

Blewitt G (2003) Self-consistency in reference frames, geocenter definition, and surface loading of the solid. Earth J Geophys Res 108, B22103. doi:10.1029/2002JB002082

Dong D, Dickey JO, Chao Y, Cheng MK (1997) Geocenter variations caused by atmosphere, ocean and surface ground water. Geophys Res Lett 24(15):1867–1870

Dong D, Yunck T, Heflin M (2003) Origin of the international terrestrial reference frame. J Geophys Res 108, B42200. doi:10.1029/2002JB002035

Farrell WE (1972) Deformation of the earth by surface loads. Rev Geophys Space Phys 10(3):751–797

Fritsche M, Döll P, Dietrich R (2012) Global-scale validation of model-based load deformation of the Earth's crust from continental water mass and atmospheric pressure variations using GPS. J Geodyn 59–60:133–142

Jiang WP, Li Z, van Dam T, Ding WW (2013) Comparative analysis of different environmental loading methods and their impacts on the GPS height time series. J Geod. doi:10.1007/s00190-013-0642-3

Li Z, van Dam T, Collilieux X, Altamimi Z, Rebischung P, Nahmani S (2014) Quality evaluation of the weekly vertical loading effects induced from continental water storage models. Accepted, IAG proceeding, 2014, Potsdam

Livneh B, Xia Y, Mitchell KE, Ek MB, Lettenmaier DP (2010) Noah LSM snow model diagnostics and enhancements. J Hydrometeorol 11:721–738

Rui H (2011) Readme document for Global Land Data Assimilation System Version 1 (GLDAS-1). http://disc.sci.gsfc.nasa.gov/services/grads-gds/gldas

Rui H (2013) Readme document for North America Land Data Assimilation System Phase 2 (NLDAS-2). http://ldas.gsfc.nasa.gov/NLDAS-2/NLDAS-2news.php

Tregoning P, Watson C, Ramillien G, McQueen H, Zhang J (2009) Detecting hydrologic deformation using GRACE and GPS. Geophys Res Lett 36, L15401. doi:10.1029/2009GL038718

van Dam TM, Wahr J (1987) Displacements of the Earth's surface due to atmospheric loading: effects on gravity and baseline measurements. J Geophys Res 92:1281–1286

van Dam T, Wahr J, Milly PCD, Shmakin AB, Blewitt G, Lavallée D, Larson KM (2001) Crustal displacements due to continental water loading. Geophys Res Lett 28(4):651–654

van Dam T, Wahr J, Lavallée D (2007) A comparison of annual vertical crustal displacements from GPS and gravity recovery and climate experiment (GRACE) over Europe. J Geophys Res 112, B03404. doi:10.1029/2006JB004335

Wei H, Xia Y, Mitchell KE, Ek MB (2013) Improvement of the Noah land surface model for warm season processes: evaluation of water and energy flux simulation. Hydrol Process 27:297–303

Wessel P, Smith WH (2013) The generic mapping tools technical reference and cookbook (version 4.5.11). http://www.soest.hawaii.edu/gmt/gmt/pdf/GMT_Docs.pdf, GMT online document 250 pp

Xia Y, Mitchell K, Ek M, Sheffield J, Cosgrove B, Wood E, Luo L, Alonge C, Wei H, Meng J, Livneh B, Lettenmaier D, Koren V, Duan Q, Mo K, Fan Y, Mocko D (2012) Continental-scale water and energy flux analysis and validation for the North American Land Data Assimilation System project phase 2 (NLDAS-2): 1. Intercomparison and application of model products. J Geophys Res 117, D03109. doi:10.1029/2011JD016048

Xia Y, Sheffield J, Ek MB, Dong J, Chaney N, Wei H, Meng J, Wood EF (2014) Evaluation of multi-model simulated soil moisture in NLDAS-2. J Hydrol 512:107–125

Non-parametric Estimation of Seasonal Variations in GPS-Derived Time Series

Janusz Bogusz, Marta Gruszczynska, Anna Klos, and Maciej Gruszczynski

Abstract

In this research, we focus on determining the quasi-annual changes in GNSS-derived 3-dimensional time series. We use the daily time series from PPP solution obtained by JPL (Jet Propulsion Laboratory) from more than 300 globally distributed IGS stations. Each of the topocentric time series were stacked into data sets according to year (from January to December) and then decomposed and approximated with a Meyer wavelet. This approach allowed investigating changes of the amplitudes in time. An observed quasi-annual signal for a set of European stations prompted us to divide the stations into different sub-networks called clusters. For Up component seven clusters were established. The signals were then averaged within each cluster and median quasi-annual signal was revealed. The vast majority of the GNSS time series is characterized by vertical changes of 3 mm with their maximum in Summer. The maximum vertical amplitude was at the level of 14 mm with the minimum equal to −13 mm, giving the peak-to-peak position changes up to 27 mm.

Keywords

Clustering • GPS • PPP • Seasonal variations • Wavelet decomposition

1 Introduction

The seasonal variations in a GNSS station's position may arise from gravitational excitation, thermal changes, hydrological effects, or other errors. These, when superimposed, may introduce seasonal oscillations that are not entirely of geodynamical origin, but still have to be included in time series modelling (Dong et al. 2002). Displacements of the surface of the Earth caused by the atmosphere, soil moisture, snow and ocean mass changes can reach several millimetres at some locations (e.g. Mangiarotti et al. 2001; van Dam et al. 2001; Poutanen et al. 2005; Tregoning and Watson 2009). These variations with different periods all affect the reliability of the station's velocity (Blewitt and Lavallée 2002; Bos et al. 2010), which, in turn, can disturb the quality of kinematic reference frames or rightness of geodynamical interpretations. As shown before by a number of authors (e.g. Chen et al. 2013), the commonly modelled annual signal is not time-constant either in amplitude or in phase due to season-to-season variability. However, some components of the seasonal changes observed by satellite navigation techniques are due to artefacts such as the aliasing of periodic signals (e.g. Dong et al. 2002; Penna and Stewart 2003) or mismodelling at sub-daily periods (e.g. Watson et al. 2006; King et al. 2008; Bogusz and Figurski 2012). In this research, we focus on the amplitudes and phases of seasonal changes from globally distributed IGS (International GNSS Service) stations and present results for European stations. The determination of the quasi-annual signals was performed by dividing the stations characterized by similar temporal patterns into different clusters. As a result, the median quasi-annual curves are presented for each cluster.

J. Bogusz • M. Gruszczynska (✉) • A. Klos • M. Gruszczynski
Faculty of Civil Engineering and Geodesy, Military University of Technology, Warsaw, Poland
e-mail: marta.gruszczynska@wat.edu.pl

T. van Dam (ed.), *REFAG 2014*, International Association of Geodesy Symposia 146,
DOI 10.1007/1345_2015_191

2 Data

We used position time series obtained from the Jet Propulsion Laboratory (JPL) using the GIPSY-OASIS software in a Precise Point Positioning mode. The details concerning the processing of the GNSS observations can be found at https://gipsy-oasis.jpl.nasa.gov. The pre-analysis consisted of removing outliers and offsets. Short data gaps (up to 40 days) were interpolated using a white noise assumption. Having introduced a 6.5-year threshold in time series length, the number of 320 globally distributed GPS permanent stations were taken here.

3 Stacking and Smoothing

Within this research the significant seasonal peaks around 1 cpy and its overtones in the power spectral densities (using the Lomb-Scargle method) of the IGS time series were found. Other researchers have previously discovered similar results (e.g. Ray et al. 2008; Collilieux et al. 2007). To focus on the annual signal, the daily topocentric time series (North, East and Up) were stacked here as proposed in Freymueller (2009).

The North, East and Up time series of each station were sorted by day of year (from January to December) into specific years. Observations from the day of year were stacked in order to emphasize the annual signal in the observations. It this way, we created the matrices of $356 \times n$ size, where n is a number of years in the time series. Then for each day of the year, i.e. each row of the matrix, the values of the weighted median and the weighted median absolute deviation (WMAD) were calculated providing the averaged representative quasi-annual signal for each of the three components for each station. These series will be referred to as "stacked" and "smoothed", respectively.

One of the method for determination of the seasonal signals is approximation with a set of sinusoidal functions (e.g. Kenyeres and Bruyninx 2009; Bogusz and Figurski 2014). However, many seasonal signals are not purely sinusoidal (e.g. snow loading and precipitation can vary from year to year) and are not time-invariant (Santamaría-Gómez et al. 2011). For that reason, in this research we performed wavelet decomposition on the smoothed data. The wavelet transform allows us to decompose the original time series into a number of new time series, each with a different degree of resolution. We used a symmetric and orthogonal Meyer's wavelet (Meyer 1990) for this task, because it is compact in

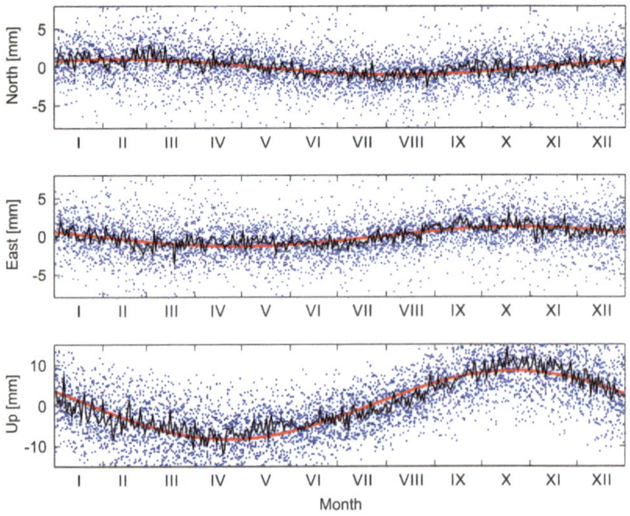

Fig. 1 The stacked and smoothed data for the BRAZ (Brasilia, Brazil) IGS station using 19.5 years of observations. The *blue dots* represent the original (yearly stacked) data; a *black curve* represents the daily medians; a *red curve* represents a wavelet quasi-annual approximation. The North component is shifted in phase when being compared to East and Up. The amplitude in the Up direction is twice as large in amplitude for North or East

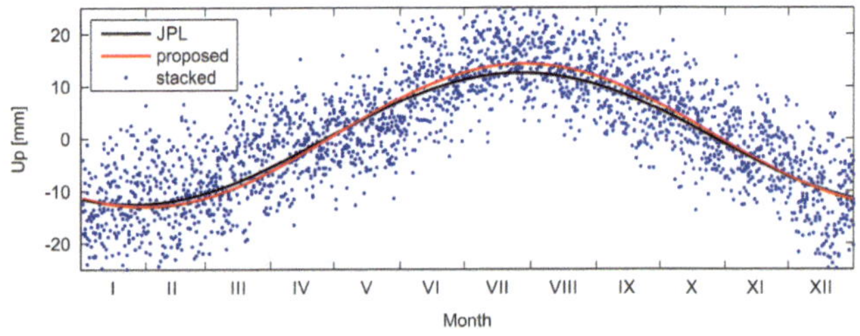

Fig. 2 The comparison of JPL-derived sine wave determined for annual signal (a *black curve*) with the non-parametric approach so the quasi-annual curve proposed in this research (a *red curve*) for the NOVM (Novosibirsk, Russian Federation) station

the frequency domain. Seasonal signals dominate the spectrum near the lowest (annual) harmonics (e.g. Blewitt and Lavallée 2002; van Dam et al. 2007; Ray et al. 2008; Amiri-Simkooei 2013), so for further analysis we took the quasi-annual approximation from wavelet decomposition (Fig. 1).

Then, we compared our approach to the one provided by the JPL, available at http://sideshow.jpl.nasa.gov/post/tables/table4.html. Although the vast majority of amplitudes determined with wavelets on the smoothed data is almost the same as the ones from the JPL sine waves, some of them show the evident change during year. Figure 2 shows stacked and smoothed 1-year signal for the NOVM (Novosibirsk, Russian Federation) station presented with blue dots. A black curve

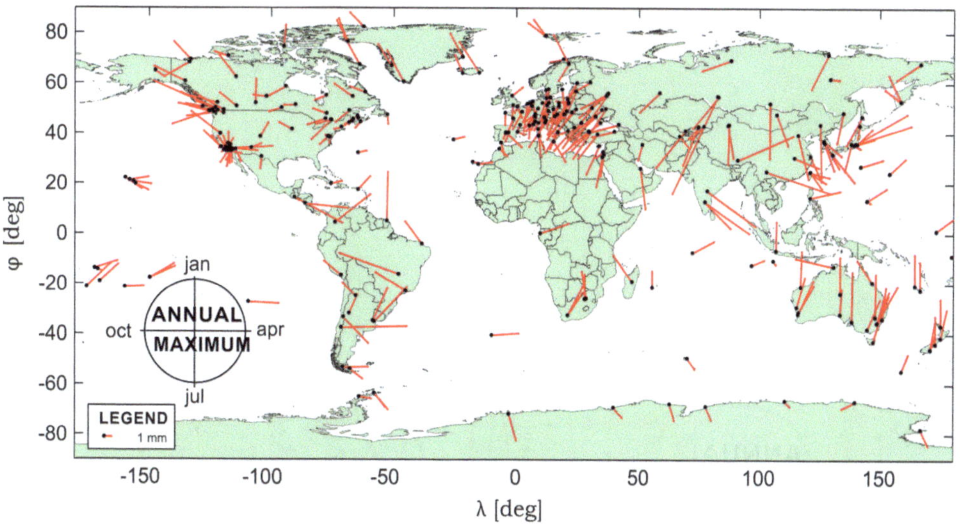

Fig. 3 The amplitudes and phases of quasi-annual curves in the vertical direction. The length of the *arrow* means the value of the amplitude, while the azimuth-like angle stays for phase of the quasi-annual maximum

indicates a sine parametric function used to estimate the seasonal components by the JPL. This curve is being compared with our non-parametric approach presented with a red line. It can be easily noticed, that both of the estimated seasonal variations are quite similar to each other, however, the proposed approach shows the amplitude changing in time. This change is at the level of 2 mm, but when taking into consideration the growing demands of the models that value seems to be significant. The quasi-annual curves determined here from the JPL PPP solutions for the vertical component show similar amplitudes and phases when the stations close to one another are being analyzed (Fig. 3). The maxima for the vast majority of stations are directed "latitudinal", which means that they have their maxima in Spring or Autumn. The stations situated in North America have their quasi-annual maxima during August and November being close to August on the East coast of America and to November as you move toward to the West Coast. For these stations, the quasi-annual amplitudes are about 3 mm. All stations situated on the ocean islands have their quasi-annual maxima close to April with the median amplitude of 2 mm. Stations situated in South America are also characterized by a "latitudinal" direction of quasi-annual maxima. Here, two of the IGS stations (BRAZ, Brasilia, Brazil and BOGT, Bogota, Colombia) have prominent annual amplitudes of about 7 mm. All stations situated in Antarctica have their quasi-annual maxima in July with amplitudes close to 1.5 mm. The quasi-annual signal estimated for stations situated in South–East Asia have their maximum in May. The amplitudes are quite similar from East to West part of Asia. The largest amplitude of 14 mm was noticed for NOVM (Novosibirsk, Russian Federation) station. The Western part of Asia is characterized by maxima close to July with the median amplitudes of 6 mm.

Nearly one hundred of permanent IGS stations are situated in Europe. They are discussed in details in the next section of the paper.

4 Clustering

The idea of clustering for investigating the annual signal was successfully introduced by Tesmer et al. (2009) for homogeneously reprocessed VLBI and GPS height time series or Poutanen et al. (2005). We investigate the magnitude of seasonal oscillations for the 90 of IGS permanent stations located in Europe. The longest time series were 22 years of available data (e.g. GRAZ, Graz, Austria), while the shortest duration of our observational series was 6.8 years (e.g. ROAP, San Fernando, Spain). The analysis of the quasi-annual curve determined with wavelet decomposition for each station allows us to sort the IGS stations into various sub-networks, called clusters, whose annual signals display similar characteristic. We defined the X parameter, which specifies the maximum acceptable phase difference for stations classified within a cluster. The next parameter, Y, is defined as the maximum distance between any two stations within a cluster. Stations were examined whether their vertical components meet the criterion of the maximum phase difference, $X = 30$ days, and in the maximum distance

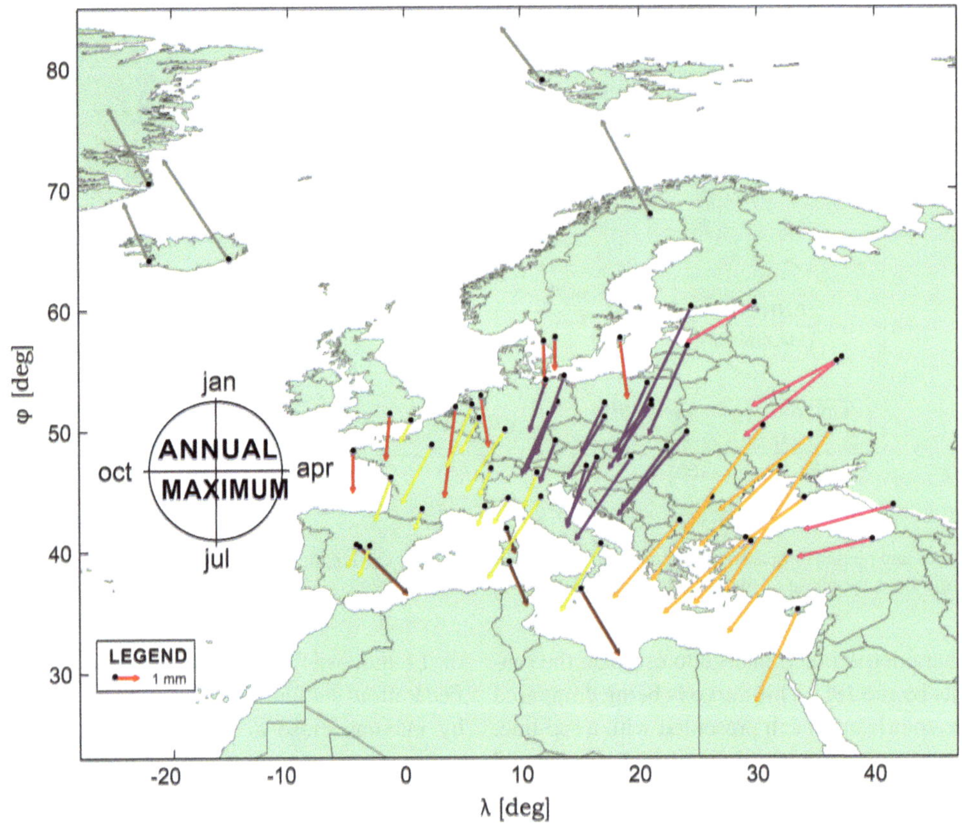

Fig. 4 The quasi-annual signal in vertical component of 65 European permanent stations. The different clusters are marked with different colours. They strictly show different characteristics of annual signals

between stations, $Y = 2{,}000$ km. Sixty-five of the ninety European IGS stations were grouped in this way into seven clusters (Fig. 4). Each cluster consists, on average, of 9 stations, however, some of 5. Such differences in the size of clusters indicate a local and regional nature of the phenomena that generates seasonal signal in the GPS time series. For the remaining 25 stations, we found no similarities in their quasi-annual signal with any nearby station. This result might be caused by local effects as snow or radome effects or by numerical artefacts as multipath or mismodelling in short periods. Therefore, these stations were excluded from clustering.

For the majority of the clusters, the quasi-annual signal in the vertical component has a minimum in Winter and a maximum in Summer (Fig. 5, Table 1). The clusters differ in terms of their quasi-annual amplitude of 1 or 2 mm, and phase of more than 30 days. The E6 cluster has a clear maximum in Spring, whereas the E1 cluster has its maximum in Autumn. Both of previously mentioned include stations located near to the sea/ocean. Worth noting is the fact that stations situated in Eastern Europe (especially the E7 cluster) have larger annual amplitudes when being compared to the remaining clusters.

5 Discussion

Some part of the annual signal in the GPS coordinates reflects the real geophysical effects. Dong et al. (2002) estimated their amplitudes into 4 mm for atmospheric mass changes, 2–3 for ocean non-tidal loading, 3–5 for snow mass, 2–7 for soil moisture and 0.5 mm for bedrock thermal expansion. Freymueller (2009) underlined that the seasonal signal has nothing in common with sine function of annual plus semi-annual curve and revealed that non-parametric approach is more suitable for GPS-derived time series. Extended studies have been conducted by Tesmer et al. (2009), who has performed a cluster analysis on the basis of globally distributed stations. These authors were using similar methods based on the moving average for "mean year" determination.

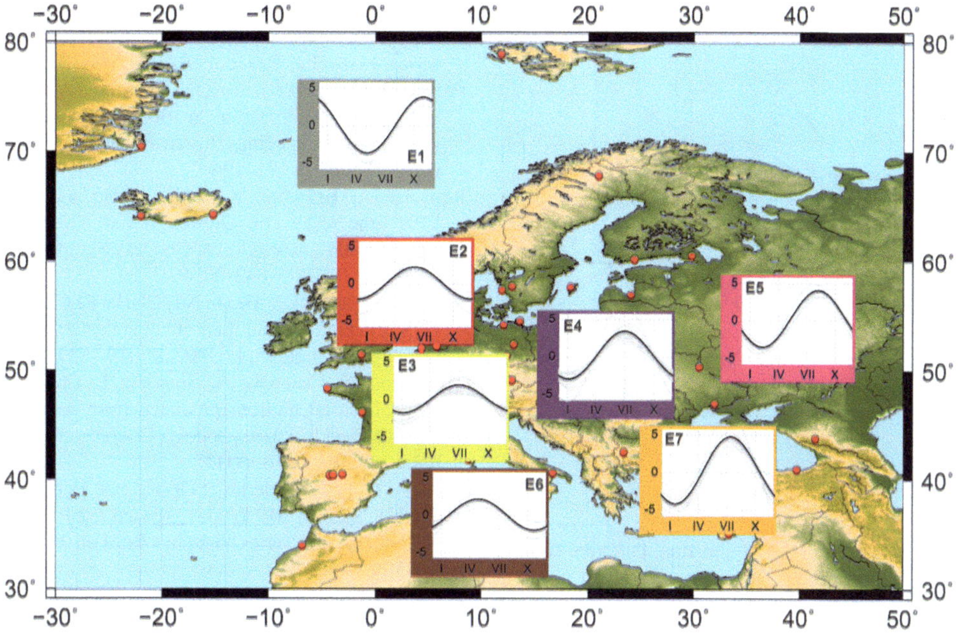

Fig. 5 The mean quasi-annual signal for individual European clusters that stations were divided into in this research. Each cluster has its description from E1 to E7

Table 1 All clusters with the names of stations included within it; the amplitudes of quasi-annual oscillation is given both with phase in days and months

Cluster	Station included	Amplitude (mm)	Phase (days)	Phase (months)
E1	HOFN KIRU NYA1 NYAL REYK SCOR	3.67	334	November
E2	BRST DLFT HRM1 ONSA SPT0 VIS0 WSRT	2.10	182	July
E3	BZRG FFMJ GENO GRAS HERT KOSG LROC MADR MAT1 MEDI OPMT TITZ TLSE YEBE ZIMM	1.81	211	July
E4	BOGI BOR1 GRAZ JOZ2 LAMA LEIJ METZ MTBG PENC POTS RIGA SASS SULP UZHL WARN WROC WTZR	3.24	211	July
E5	MDVJ SVTL TRAB ZWE2 ZECK	3.86	248	September
E6	AJAC CAGL NOT1 RABT VILL	2.03	146	May
E7	ANKR BUCU CRAO GLSV ISTA KHAR MIKL NICO POLV SOFI TUBI	4.56	224	August

The method proposed in this paper deals with weighted median and wavelet decomposition for quasi-annual curves estimation. In this way a greater reliability of non-parametric seasonal variations model by assuming year-to-year changes is ensured.

Secondly, the method of clustering is an effective algorithm to describe the spatial phenomena whereas geophysical studies are conducted. We proposed realistic model, in which phase shift is important as well as amplitude value. We found a good consistency in the quasi-annual signal for nearby stations. Figure 6 presents all quasi-annual curves estimated with wavelet decomposition for the 65 European IGS stations plotted together. The vast majority of the stations have their minima during the Winter with maxima in Summer. The remaining stations with the maxima in Spring and Autumn are those situated near the sea/ocean. Although the quasi-annual curves determined here look at a first glance like perfect sine functions, they are in fact not sinusoidal. The amplitudes of the maxima differ from one another at the level of one-tenth of mm. Our results show that a station's location (near or distant to the ocean) impacts the annual signal. Here, the seasonal amplitudes for the Up component may arise from atmospheric and hydrospheric changes (see van Dam et al. 1997).

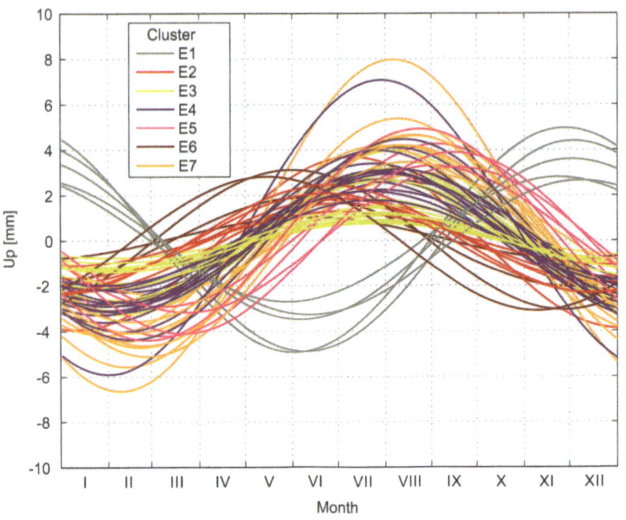

Fig. 6 The quasi-annual curves in vertical direction. Different *colours* means cluster number that the certain station was classified to (as in Fig. 3)

Acknowledgments This research was financed by the Military University of Technology, Faculty of Civil Engineering and Geodesy Young Scientists Development funds, grant No. 739/2015. JPL repro2011b time series accessed from ftp://sideshow.jpl.nasa.gov/pub/JPL_GPS_203_Timeseries/repro2011b/raw on 2014-11-10.

References

Amiri-Simkooei AR (2013) On the nature of GPS draconitic year periodic pattern in multivariate position time series. J Geophys Res Solid Earth 118:2500–2511. doi:10.1002/jgrb.50199

Blewitt G, Lavallée D (2002) Effect of annual signals on geodetic velocity. J Geophys Res 107(B7):2145. doi:10.1029/2001JB000570

Bogusz J, Figurski M (2012) GPS-derived height changes in diurnal and sub-diurnal timescales. Acta Geophys 60:295–317. doi:10.2478/s11600-011-0074-5

Bogusz J, Figurski M (2014) Annual signals observed in regional GPS networks. Acta Geodyn Geomater 11:125–131. doi:10.13168/AGG.2014.0003

Bos M, Bastos L, Fernandes RMS (2010) The influence of seasonal signals on the estimation of the tectonic motion in short continuous GPS time-series. J Geodyn 49:205–209. doi:10.1016/j.jog.2009.10.005

Chen Q, van Dam T, Sneeuw N, Collilieux X, Weigelt M, Rebischung P (2013) Singular spectrum analysis for modeling seasonal signals from GPS time series. J Geodyn 72:25–35. doi:10.1016/j.jog.2013.05.005

Collilieux X, Altamimi Z, Coulot D, Ray J, Sillard P (2007) Comparison of very long baseline interferometry, GPS, and satellite laser ranging height residuals from ITRF2005 using spectral and correlation methods. J Geophys Res 112, B12403. doi:10.1029/2007JB004933

Dong D, Fang P, Bock Y, Cheng MK, Miyazaki S (2002) Anatomy of apparent seasonal variations from GPS-derived site position time series. J Geophys Res. doi:10.1029/2001JB000573

Freymueller JT (2009) Seasonal position variations and regional reference frame realization. In: Bosch W, Drewes H (eds) GRF2006 symposium proceedings. International Association of Geodesy Symposia, Berlin, pp 191–196

Kenyeres A, Bruyninx K (2009) Noise and periodic terms in the EPN time series. In: Drewes H (ed) Geodetic reference frames. International Association of Geodesy Symposia, Berlin, pp 143–148. doi:10.1007/978-3-642-00860-3_22

King MA, Watson CS, Penna NT, Clarke PJ (2008) Subdaily signals in GPS observations and their effect at semiannual and annual periods. Geophys Res Lett 35, L03302. doi:10.1029/2007GL032252

Mangiarotti S, Cazenave A, Soudarin L, Cretaux JF (2001) Annual vertical crustal motions predicted from surface mass redistribution and observed by space geodesy. J Geophys Res 106(B3):4277–4291. doi:10.1029/2000JB900347

Meyer Y (1990) Ondelettes et Opérateurs. Hermann, Paris

Penna NT, Stewart MP (2003) Aliased tidal signatures in continuous GPS height time series. Geophys Res Lett 30(23):2184. doi:10.1029/2003GL018828

Poutanen M, Jokela J, Ollikainen M, Koivula M, Bilker M, Virtanen H (2005) Scale variation of GPS time series. In: Sansò F (ed) A window on the future of geodesy, IAG General Assembly in Sapporo, Japan 2003, IAG Symposia 128, Springer, pp 15–20. doi:10.1007/3-540-27432-4_3

Ray J, Altamimi Z, Collilieux X, van Dam T (2008) Anomalous harmonics in the spectra of GPS position estimates. GPS Solution 12:55–64. doi:10.1007/s10291-007-0067-7

Santamaría-Gómez A, Bouin MN, Collilieux X, Woppelmann G (2011) Correlated errors in GPS position time series: implications for velocity estimates. J Geophys Res 116(B1), B01405. doi:10.1029/2010JB007701

Tesmer V, Steigenberger P, Rothacher M, Boehm J, Meisel B (2009) Annual deformation signals from homogeneously reprocessed VLBI and GPS height time series. J Geod 83:973–988. doi:10.1007/s00190-009-0316-3

Tregoning P, Watson C (2009) Atmospheric effects and spurious signals in GPS analyses. J Geophys Res 114, B09403. doi:10.1029/2009JB006344

van Dam TM, Wahr J, Chao Y, Leuliette E (1997) Predictions of crustal deformation and geoid and sea level variability caused by oceanic and atmospheric loading. Geophys J Int 99:507–517. doi:10.1111/j.1365-246X.1997.tb04490.x

van Dam T, Wahr J, Milly PCD, Shmakin AB, Blewitt G, Lavallée D, Larson KM (2001) Crustal displacements due to continental water loading. Geophys Res Lett 28:651–654. doi:10.1029/2000GL012120

van Dam T, Wahr J, Lavallee D (2007) A comparison of annual vertical crustal displacements from GPS and gravity recovery and climate experiment (GRACE) over Europe. J Geophys Res 112, B03404. doi:10.1029/2006JB004335

Watson C, Tregoning P, Coleman R (2006) The impact of solid Earth tide models on GPS coordinate and tropospheric time series. Geophys Res Lett 31, L08306. doi:10.1029/2005GL025538

Part VI

Georeferencing in Practice

A Two-Frame National Geospatial Reference System Accounting for Geodynamics

Nic Donnelly, Chris Crook, Richard Stanaway, Craig Roberts, Chris Rizos, and Joel Haasdyk

Abstract

This paper presents a high level proposal for how a Local Reference Frame (LRF) could be implemented alongside the International Terrestrial Reference Frame (ITRF) as part of a two-frame national geospatial reference system. By accounting for both local and global geodynamic effects using time-dependent transformations, the LRF can minimize the complexity that results when objects that are fixed with respect to the ground have continuously time-varying coordinates in a global frame.

The role of the national geospatial reference system, of which the reference frame is a core component, has changed. Whereas traditionally a national geodetic datum of the highest available precision has been required for accurate surveying and positioning, GNSS-derived positioning now provides easy access to precise global reference frames such as ITRF. However, the exponential growth of spatial data sets has created a need for a geospatial reference system providing coordinates that are "ground-fixed". That is, the system provides coordinates that can be used to locate and relate physical features, and to align spatial data sets acquired at different times. This requires the definition of a LRF and reference epoch, with clear traceability to a global reference frame such as ITRF.

The ITRF has long been adopted as the most precise means of accessing a LRF using Precise Point Positioning (PPP) and global post-processing services, or for applications where the highest precision is required. However, transformation to the local frame has not always been carried out robustly, whether due to a lack of officially defined transformations or failure of systems to utilize time-dependent transformation parameters. Formal recognition of ITRF within a national spatial referencing system will support the increasingly broad range of users and applications utilizing high precision ITRF coordinates derived from absolute positioning, including rapidly emerging real-time PPP services and geodetic imaging techniques such as Lidar and InSAR.

While some of the implementation details will differ to reflect the local tectonic and legislative environment, the suggested framework could be used by any jurisdiction

N. Donnelly (✉)
Cooperative Research Centre for Spatial Information, University of New South Wales, Sydney, NSW, Australia

School of Civil and Environmental Engineering, University of New South Wales, Sydney, NSW, Australia

National Geodetic Office, Land Information New Zealand, Wellington, New Zealand
e-mail: ndonnelly@linz.govt.nz

C. Crook
National Geodetic Office, Land Information New Zealand, Wellington, New Zealand

R. Stanaway • C. Roberts • C. Rizos
School of Civil and Environmental Engineering, University of New South Wales, Sydney, NSW, Australia

J. Haasdyk
Cooperative Research Centre for Spatial Information, University of New South Wales, Sydney, NSW, Australia

Survey Infrastructure and Geodesy, NSW Land and Property Information, Bathurst, NSW, Australia

T. van Dam (ed.), *REFAG 2014*, International Association of Geodesy Symposia 146,
DOI 10.1007/1345_2015_188

considering an updated approach to defining the reference frame in its national geospatial reference system.

Keywords
Datum • Deformation Model • Reference frame • Transformation

1 Introduction

The *national geospatial reference system* (NGRS) for a country is the geodetic infrastructure which supports positioning to the highest levels of precision and robust management of spatial data. It includes survey marks and their coordinates, infrastructure and data providing connections to the reference frame, as well as tools, recommendations and standards to assist with its use (Johnston and Morgan 2010). At its core is the authoritative reference frame, supported (and in many cases mandated) by government for use in diverse spatial applications.

Prior to the advent of GNSS, accurate spatial positioning could only be achieved by measuring to nearby marks which defined the geodetic datum or by astronomical observations. GNSS-derived positions based on global reference frames have replaced this function, allowing ubiquitous and very accurate positioning without the need for intervisible local marks (although these are still useful for supporting some surveying techniques). GNSS technology dominates positioning to the extent that it is now very rare for a position to be calculated that does not utilize GNSS, albeit that the use may be indirect (for example, when positioning relative to an existing coordinate calculated from GNSS observations). It matters little whether the GNSS-derived position is direct or indirect; the outcome is that coordinates are natively in terms of a global reference frame, often with a high level of precision.

Geodetic imaging describes any technique which uses massive point clouds (or pixels) to produce an image which is georeferenced to a level typically associated with geodetic coordinates (0.1 m or better). Examples include precise photogrammetry, pictometry, Lidar, Terrestrial Laser Scanning (TLS) and InSAR. The points generated by these images are often defined in terms of a global reference frame, due to the combination of the imaging sensor with a GNSS receiver. The use of geodetic imaging is increasing rapidly and the massive volume of data generated greatly exceeds that from direct GNSS positioning. Reference systems need to support this technology.

At the same time the need for coordinate reference systems for GIS has greatly increased. These systems require coordinates that represent physical features and their relative real-world spatial relationships. A requirement of GIS is that the coordinate can be used to relocate the feature in the future if needed and that the coordinates provide good relative precision, both horizontally and vertically. Most GIS systems assume static relationships between these features and do not account for movement or deformation over time. In the absence of any localized deformation, spatial data acquired over long periods of time over a common area should be spatially aligned in a LRF. The requirement for accurate calculations based on coordinates (for example, distances at geodetic type accuracies) is of much lower priority for a LRF. There is, however, a requirement for moderate local accuracy, particularly to determine which features are coincident, overlapping, or within a specified distance of each other. For many users, these needs are better met by the traditional geodetic datum of well-placed marks with fixed coordinates, subject to understanding the limitations of relative precision that ensues from this where there is non-negligible deformation. Thus a LRF in which coordinates are generally constant over time remains very useful.

For many applications, heights related to the gravity field are of most interest, since these will reliably represent fluid flow. Traditionally, this has led to local vertical reference frames based on mean sea level and precise levelling. Increasingly, the trend is for modern vertical reference frames to be defined in terms of the geoid or quasigeoid and accessed using GNSS-derived ellipsoidal heights and a geoid model. For example, New Zealand implemented a quasigeoid-based vertical reference frame in 2009 (Amos 2009). This paper does not further consider the vertical reference frame, except to note that analysis is required to determine how deformation modelling should be incorporated into a vertical reference frame, if at all, given that many engineering applications require that vertical deformation is visible in measurements. These decisions about handling deformation will impact on the relationship between the vertical and geometric reference frames.

Overall, a national geospatial reference system consisting of both global and local reference frames may better meet current and future requirements than a system consisting of a single frame.

2 Classifying Reference Frames

The consistent use of clearly defined, unambiguous terminology greatly assists with understanding the concepts associated with the geospatial reference system. Without a sound understanding of the concepts, there is a significantly increased risk that geospatial datasets are incorrectly managed. In particular, many users are now managing datasets of sufficiently high precision that they must correctly account for geodynamic effects. Unfortunately, some of the terminology commonly used to describe elements of the national geospatial reference system adds to user confusion.

The term *reference frame*, as distinct from *datum*, correctly describes contemporary geospatial reference systems, which are usually based on a realisation of the International Terrestrial Reference System (ITRS). The term *datum* describes the relationship between a reference system and reference frame (origin, axes orientation and scale) (Drewes 2009). Thus *datum* was appropriate to use prior to the availability of the ITRF, when the datum was typically fixed to the Earth's surface by fixing the coordinates of at least two stations to define the origin, orientation and scale. In the case of ITRF (and any LRF aligned to ITRF), station coordinates are not fixed; they may change due to improved observations and/or land movement. The fact that the coordinates of features may change with time is a new concept for many users, so in addition to being technically correct, the use of *reference frame* highlights that coordinate behaviour may be different to that associated with traditional geodetic datums.

More generally, there is confusion resulting from conflicts between the terminology used by the International Organization for Standardization (ISO) and that in long-standing usage by the International Astronomical Union (IAU) and International Association of Geodesy (IAG) through the International Earth Rotation and Reference Systems Service (IERS). The IAU/IAG distinguish between the *reference system*, being a set of conditions that need to be met to define spatial references, and the *reference frame*, being the realization of that reference system by precisely determining coordinates at physical points (Petit and Luzum 2010). For the wider geospatial community, the definitions of the ISO standard 19111: Geographic Information – Spatial Referencing by Coordinates are more widely used. This standard uses the term *datum* in preference to *reference frame* and *coordinate reference system* to describe how coordinates are expressed in terms of a datum/reference frame (for example: geocentric or geographic) (ISO 2007). The *coordinate reference system* of ISO is not the same as the *reference system* of the IERS. This paper utilizes the terminology of the IERS, but until such time as there is agreement on preferred terminology between the geodetic and wider geospatial communities, it is important that the meaning of these terms is clearly defined within each national geospatial reference system.

Frequently, the terms *dynamic*, *semi-dynamic* and *static* are used to describe reference frames (or geodetic datums). Use of these terms causes confusion for reference frame managers and users alike. For example, when a reference frame is described as *dynamic*, this means that the coordinates for a ground-fixed feature are time-varying within that frame. Thus it is not the reference frame, but the coordinates which are "dynamic". Even when referring to coordinates, the term "dynamic" is not rigorously correct, as technically this implies force in the coordinate movement, which is not necessarily the case. "Kinematic" is a more appropriate term, as it implies nothing about the cause of the motion. Thus ITRF, which is sometimes described as a "dynamic datum", is in reality a static reference frame with kinematic coordinates for ground-fixed physical features.

Similarly, "semi-dynamic" has been used to describe a reference frame where coordinates change with time, but only periodically, with the period between updates being determined subjectively by the reference frame manager. Coordinate updates may be triggered either by localized deformation, or secular movement over a sufficiently long period that discrepancies with global frames become problematic. Otherwise, the coordinates in a semi-dynamic frame are static, and a time-dependent model is used to ensure that consistent coordinates can be calculated from observations made at various times. Conceptually, there is no clear factor that differentiates a "semi-dynamic" from a "static" frame. In fact, coordinates in traditional static geodetic datums have also been updated as required to reflect localized deformation, although the official coordinates usually represent only the current state. For example, coordinates affected by the 1987 Edgecumbe earthquake in New Zealand were updated, even though the geodetic datum at the time was static. Similarly, there are static reference frames that make use of time-dependent transformations, so this characteristic is not unique to semi-dynamic frames. An example is the Geocentric Datum of Australia 1994 (GDA94), which utilizes a time-dependent 14-parameter transformation to transform coordinates from ITRF (Dawson and Woods 2010).

A better way to classify reference frames is to consider the body onto which they are fixed. The concept of an *Earth-fixed* frame is that it is fixed to the whole solid Earth at a depth where no tectonic movement occurs. In practice, this may be difficult or impossible to achieve as even deep beneath the surface there are geodynamic processes which would likely compromise attempts to define a stable reference frame. Thus the Earth-fixed frame may be realized using the no-net-rotation condition, as is the case for ITRF (Altamimi et al. 2011). The concept of a *plate-fixed* frame is that it is fixed to (and therefore moves with) a tectonic plate (or plates in the case of a country which straddles a plate boundary). Again this concept may be challenging to implement in practice in those regions where there is non-uniform plate motion. In effect, the reference frame is deforming with the plates, which needs to be accounted for using a deformation model. Despite the implementation issues, these terms are useful as they relate explicitly to the reference frame and the body to which it is fixed, rather than trying to describe the behaviour of coordinates of features within the frame.

Reference frames can also be classified as global or local. This terminology is reasonably intuitive, and correlates well with the use of "global" to describe the likes of the Global Positioning System, which is familiar to both spatial professionals and the general populace. The term "local" used here describes a geographical area, no larger than continental-scale, over which the relevant authorities have determined that a single reference frame is required. It could vary from a few square kilometres for a small island nation to covering an entire continent, as would be the case for Australia. In many parts of the world, such as Europe and South America, regional reference frames provide a continent-wide frame on which the local frames of individual countries (if required) are based to ensure consistency across borders. Some countries may decide that the regional reference frame is sufficient to act as the local frame. The discussions relating to plate-fixed local frames generally also apply to a plate-fixed regional reference frame.

3 The Two-Frame System: Global and Local

3.1 Global Frame

The first frame in the proposed two-frame system is the Earth-fixed, global reference frame. This should be the most recent ITRF (currently ITRF2008 (Altamimi et al. 2011)), to ensure consistency with data and products used to compute high-precision positions, such as the precise orbit and clock products of the International GNSS Service (IGS). This also ensures that coordinates calculated using techniques such as Precise Point Positioning (PPP), or derived from such techniques (as may be the case for a geodetic imaging point cloud), are immediately in terms of the official global reference frame.

As new (improved) ITRF realisations are released, these should be incorporated into the national geospatial reference system, replacing the previous realisation. The exact timing of the adoption of a new ITRF would be determined by the national geodetic agency and would likely not take place until organisations such as the IGS were providing products in terms of the new ITRF. The current stability of successive ITRFs means that for many applications, the impact of adopting a new ITRF would be negligible. However, the adoption of any new ITRF is important if the national geospatial reference system is to support the highest-precision geodetic applications.

The adoption of the ITRF as an official reference frame for a jurisdiction formalizes the long-standing practice of carrying out processing in the ITRF, before transforming to a LRF if necessary. Making it an official frame within the country recognizes that there are some computations that cannot be accurately carried out in a local frame. One example is GNSS processing, which must be carried out in a global reference frame to remove biases due to plate rotation and enable use of precise global data products, such as orbit and clock parameters. Another is precise engineering design in a deforming area, where LRF coordinates may not sufficiently represent the physical kinematic reality. Precise engineering requirements could provide a trigger to update the LRF, where precision tolerances for large scale engineering projects are exceeded.

Adoption of the ITRF will make it easier to support applications related to navigation, such as automated aircraft guidance systems, where the use of the ITRF-aligned WGS84 is mandated (ICAO 2002). It will also be easier to support geodetic imaging techniques generating large sets of ITRF-aligned coordinates. In the near future, the volume of data being collected using these systems and techniques is likely to be so much greater than the volume of data in existing datasets that it may be more efficient to bring these existing datasets to the epoch of the geodetic imaging dataset, rather than transforming the geodetic imaging dataset to the local frame. Another possibility is to provide large geodetic datasets in the local frame (data are transformed once) whilst retaining the original in ITRF for future data combinations or transformations. Duplication of data is perhaps less of an issue than repeated transformations of data both in terms of risk and computing cost.

There are also non-technical advantages. Defining a global frame officially within the country emphasizes to decision-makers and the spatial community that the global reference frame is of critical importance to GNSS-derived positioning. In some jurisdictions, this may make it easier to justify contributions to the infrastructure and analysis required to develop and maintain the ITRF. Direct linkage of the LRF to ITRF is also of value at jurisdictional boundaries.

In the context of a national geospatial reference system, the global frame may be a specific ITRF-aligned realization computed by the reference frame manager, or adopted from an organization such as the IGS. It is not necessarily the exact reference frame published by the IERS. The key requirements are that the global frame used within a jurisdiction is as consistent as practicable with the official ITRF and that details of the procedure used to generate coordinates and velocities is well-documented and publically accessible.

3.2 Local Frame

The second frame in the proposed two-frame system is the plate-fixed, local reference frame. Where possible, it is fixed to the stable portion of a tectonic plate at a particular epoch, which becomes the reference epoch for the frame. In some countries, such as New Zealand, there is no stable plate but the LRF could still be precisely fixed to the deforming tectonic plates at a chosen reference epoch.

There are two approaches that could be used to define a LRF. The first is to explicitly define a set of coordinates using a suitably precise GNSS campaign, as undertaken in Europe for EUREF. The second approach is to define the LRF implicitly through its relationship to the global frame via transformation and/or deformation models, as done in New Zealand. Regardless of which approach is chosen, permanent GNSS stations and other precise geodetic observations are then used to monitor and update the relationship between the local and global frames. In this discussion we assume the second approach has been taken, although similar outcomes are achieved with either approach.

Use of a local frame removes uniform plate rotation and various non-uniform deformation effects. Within this frame the coordinates of fixed features are stable – and the velocities are minimized (near-zero) – see Fig. 1. The accuracy with which a coordinate identifies a ground-fixed point over time is determined by the extent to which plate motion and deformation is accounted for in the time-dependent models used to transform between the global and local frames. Since official time-dependent models are specified by the national geodetic agency, new versions of the time-dependent models can be produced to respond to local requirements. For example, after the 2010 and 2011 Canterbury earthquake sequence in New Zealand, there was strong demand from spatial professionals for LRF coordinates to be updated to reflect the post-earthquake relationships among physical features. However, there has been little demand for the reference epoch of the secular deformation component to be updated. Thus a new version of the deformation model was developed that updated coordinates to reflect earthquake movements but did not update coordinates to reflect the 10 years of secular motion (Crook and Donnelly 2013). This flexibility ensures that the LRF is responsive to user needs.

Within the local frame, any non-zero velocities or coordinate changes for ground-fixed physical features are indicative of land movement not accounted for in the time-dependent transformation models.

Provision of a LRF addresses the current difficulties with using kinematic coordinates for some applications. Some of these problems are likely to reduce in the near future, as widely used GIS software better incorporates time-dependent transformation models. But for some applications, there are legal requirements that are difficult to change in the short term. For example, property boundaries are often described by physical relationships and/or fixed coordinates in terms of the LRF. While it may not be technically difficult to utilize kinematic coordinates in terms of a global reference frame, the legislative change that could be required is unlikely to occur quickly, given that there are unlikely to be advantages from making such a change for this application.

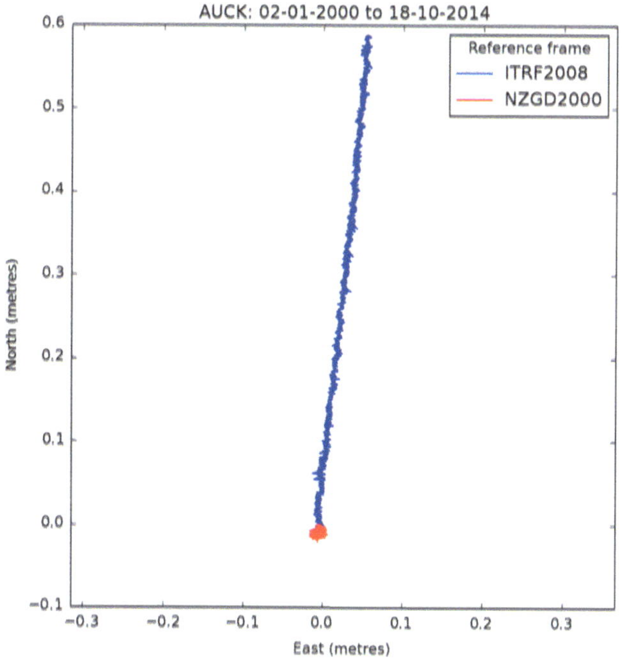

Fig. 1 Horizontal daily time series of the IGS station AUCK in terms of ITRF2008 (global frame), in *blue*, and New Zealand Geodetic Datum 2000 – NZGD2000 (local frame), in *red*. In this case the daily ITRF2008 coordinate is calculated by aligning to the official IERS coordinates of a regional subset of ITRF2008 stations. The trajectory in terms of the local frame is almost static

4 Time-Dependent Transformation Models

In a two-frame system, the term *time-dependent transformation model* refers to any model which describes the relationship between the global and local frames, enabling coordinates to be transformed between the two frames. Reference frame transformation models, plate motion models (PMMs) and deformation models are all examples of models which could be required in the two-frame system. A key feature of the transformation approach is that it would be possible for users requiring different precisions to apply different layers / portions of the transformation model(s), based on their accuracy requirements and/or computational resources. The full transformation path is shown in Fig. 2. Note that the order in which the reference frame transformation, PMM and

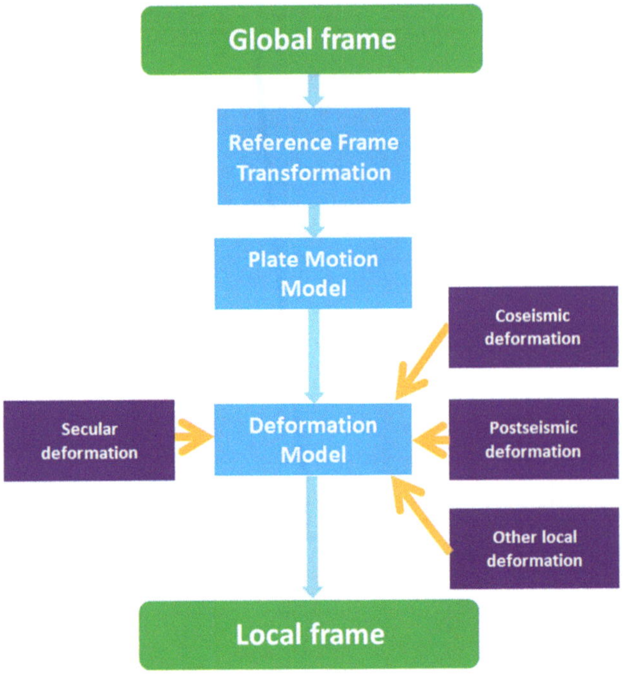

Fig. 2 Transformation path between the global and local frames. This figure assumes that the PMM and deformation model are defined in terms of the local frame, so the reference frame transformation is carried out first

deformation model should be applied depend on how these have been defined.

The reference frame transformation model is required where the local frame is aligned to a different ITRF realisation. For example, New Zealand's local frame is currently aligned to ITRF96, so a 14-parameter transformation is required when transforming coordinates from ITRF2008 (three translations, three rotations, one scale parameter and their time-dependent equivalents). The coordinate epoch is differenced from the reference epoch for the transformation to appropriately calculate the time-dependent parameters which apply over that period.

The PMM describes the impact of rigid plate motion on the coordinates being transformed. It can be defined using three rotation-rate parameters. As with the reference frame transformation, the coordinate epoch is differenced from the reference epoch for the transformation to appropriately calculate the specific rotation to apply to the coordinate over that period. In stable countries such as Australia, the PMM would remove almost all of the effect of land movement (Stanaway et al. 2014).

The deformation model accounts for the changes in the relative position of features due to geophysical processes. It may consist of a number of submodels, each of which relates to a particular deformation event (or type of deformation).

For example the New Zealand deformation model comprises submodels for a secular deformation component and for a number of earthquakes. Each submodel may include multiple components to represent different types of deformation associated with the event. Thus an earthquake event might include a component for coseismic deformation and one or more postseismic components.

Figure 3 shows a representation of part of the deformation model used in New Zealand. Two submodels are shown, one for the 2007 George Sound earthquake and one for the 2009 Dusky Sound earthquake. The George Sound submodel has a single component modelling the coseismic deformation. However the Dusky Sound submodel has two components, to model both coseismic and postseismic deformation.

The deformation model may also include some rigid plate movement where it is not possible or practical to reliably separate this component into a PMM. For example, Tregoning et al. (2013) identify potential rigid plate movement in the far north of New Zealand, but it may be simpler to implement a single secular deformation model than to divide a region up into discrete microplates or crustal blocks, each defined by a PMM. However, a limitation of a single deformation model is how the model is interpolated across active faults and plate boundaries. An advantage of a polygon based model is that the boundaries of crustal blocks can be defined along active bounding faults. A deformation model can be overlain on the PMM to define intraplate and interseismic plate boundary deformation (back-slip) (Stanaway et al. 2015). Note that the deformation model excludes highly localized monument damage or disturbance. It also does not include smaller deformation events (either tectonic or human induced) which are either not significant, or for which there is currently insufficient data to model reliably.

5 Versioning of Models

The time-dependent transformation models are versioned, which leads to a versioned realization of the LRF. That is, coordinates in the local frame may change if a new version of the deformation model is published. For example, a new version of the deformation model might be released within a few days of a significant earthquake, based only on CORS and seismometer data. An improved version might be released a year later, based on a much wider range of data, such as campaign GNSS and InSAR, or incorporating postseismic deformation.

6 Referencing Coordinates

A fully referenced dataset must have information to enable the coordinates to be reliably updated when a new version of one of the time-dependent transformation models is released.

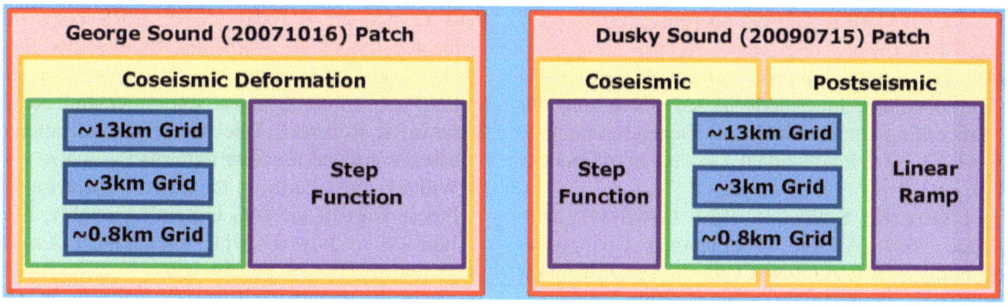

Fig. 3 Representation of part of the NZGD2000 deformation model, available from Land Information New Zealand (2015)

This could either be accomplished by retaining a primary copy of the raw untransformed ITRF data, or more simply by recording the following key metadata items:

1. Reference frame
2. Transformation model version(s) used to calculate coordinates
3. Coordinate epoch

The coordinate epoch is the epoch at which the coordinate was observed. This enables improvements to models of continuous deformation to be applied to coordinates. To illustrate, consider New Zealand's original deformation model, NZGD2000v20000101, which modelled the secular deformation of the country. In 2013 an improved secular deformation model, using data over a much longer time span, was published as part of NZGD2000v20130801 (Crook and Donnelly 2013).

Consider a set of LRF coordinates transformed from ITRF2008 coordinates at epoch 2010.0 using NZGD2000v20000101. The ITRF2008 coordinates were not retained. To improve the precision of these LRF coordinates using the new version NZGD2000v20130801, NZGD2000v20000101 must first be used to recalculate the coordinates at epoch 2010.0. NZGD2000v20130801 is then used to calculate improved coordinates at the reference epoch. Without knowledge of the coordinate epoch, it is not possible to update coordinates when models of continuous deformation are improved.

7 Time-Dependent Models for Trajectory Estimation

The trajectory of a ground fixed feature in the global reference frame can be estimated using the PMM and/or deformation model (as applicable). This enables ITRF coordinates for the feature to be calculated for any desired epoch, which is useful for visualization or analysis of datasets collected or archived at various epochs. It is also necessary for CORS-NRTK and GNSS RTK and post-processing if the reference coordinates are epoch fixed.

8 Concluding Remarks

The official inclusion of two reference frames into a national geospatial reference system would overcome some of the key challenges to implementing a reference frame that meets the needs of a diverse range of users and applications. Critical to the success of a two-frame system is the provision of time-dependent transformation models and the coordinate metadata required to utilize them. Both New Zealand and Australia are likely to implement some form of two-frame system, with some of the concepts, particularly relating to deformation models and descriptions of reference frames, already being used in New Zealand to make improvements to the existing local frame.

Many of the implementation details of the two-frame system are still to be determined. In many cases these will be strongly influenced by local circumstances, but the framework outlined here is sufficiently flexible to be implemented in diverse tectonic and regulatory settings. What is clear is that any modern reference frame will require regular updates and successfully propagating these updates into the geospatial community will be critical to the success of the geospatial reference system.

Acknowledgements This work has been supported by the Cooperative Research Centre for Spatial Information, whose activities are funded by the Australian Commonwealth's Cooperative Research Centres Programme.

References

Altamimi Z, Collilieux X, Metivier L (2011) ITRF2008: an improved solution of the international terrestrial reference frame. J Geod 85:457–473

Amos M (2009) New Zealand Vertical Datum 2009. N Z Surveyor 300:5–16

Crook C, Donnelly N (2013) Updating the NZGD2000 deformation model. In: Proceedings of the 125th New Zealand Institute of Surveyors Conference, Dunedin, 29–31 August 2013. http://www.nzisltsurveybook.org.nz/sites/all/files/joint%20proceedings%20nzis%20sircnz_V1.pdf. Accessed 20 Aug 2011

Dawson J, Woods A (2010) ITRF to GDA94 coordinate transformations. J Appl Geod 4:189–199

Drewes H (2009) Reference systems, reference frames, and the geodetic datum – basic considerations. In: Sideris M (ed) Observing our changing earth. Proceedings of the 2007 IAG General Assembly, Perugia, Italy, July 2–13, 2007, International Association of Geodesy Symposia, vol 133, pp 3–9

ICAO (2002) World Geodetic System – 1984 (WGS-84) manual, 2nd edn, Doc 9674, AN/946, International Civil Aviation Organization. http://www.icao.int/NACC/Documents/Meetings/2014/ECARAIM/REF08-Doc9674.pdf. Accessed 4 July 2015

ISO (2007) ISO19111:2007 geographic information – spatial referencing by coordinates. International Organization for Standardization

Johnston G, Morgan L (2010) The status of the national geospatial reference system and its contribution to global geodetic initiatives. In: Proceedings of XXIV FIG World Congress, 11–16 April 2010, Sydney, Australia. http://www.fig.net/pub/fig2010/papers/ts06c%5Cts06c_johnston_morgan_4116.pdf. Accessed 4 July 2015

Land Information New Zealand (2015) NZGD2000 deformation model. http://www.linz.govt.nz/data/geodetic-system/datums-projections-and-heights/geodetic-datums/new-zealand-geodetic-datum-2000. Accessed 4 July 2015

Petit G, Luzum B (eds) (2010) IERS conventions (2010), IERS technical note 36. Verlagdes Bundesamts für Kartographie und Geodäsie, Frankfurt am Main, Germany. http://www.iers.org/IERS/EN/Publications/TechnicalNotes/tn36.html. Accessed 4 July 2015

Stanaway R, Roberts C, Blick G (2014) Realisation of a geodetic datum using a gridded absolute deformation model (ADM). In: Rizos C, Willis P (eds) Earth on the edge: science for a sustainable planet. Proceedings of the IAG General Assembly, Melbourne, Australia, June 28 to July 2, 2011. International Association of Geodesy Symposia, vol 139, pp 259–256

Stanaway R, Roberts C, Rizos C, Donnelly N, Crook C, Haasdyk J (2015) Defining a local reference frame using a plate motion model and deformation model. In: Proceedings of IAG Commisison 1 Symposium 2014: Reference Frames for Applications in Geosciences (REFAG2014), 13–17 October, Kirchberg, Luxembourg

Tregoning P, Burgette R, McClusky SC, Lejeune S, McQueen H, Watson CS (2013) A decade of horizontal deformation from great earthquakes. J Geophys Res 118. doi:10.1002/jgrb.50154

List of Reviewers

Graham Appleby
Sabine Bachmann
Yoaz Bar-Sever
Oliver Baur
Dirk Behrend
Geoffrey Blewitt
Mathis Bloßfeld
Johannes Boehm
Qiang Chen
Ludwig Combrinck
John Dawson
Paul Denys
Athanasios Dermanis
Denise Dettmering
Nicolas Donnelly
Shu Fengchun
Mathias Fritsche
Daniel Gambis
Pascal Gegout
John Gipson
Richard Gross
Chris Hartnady
Thomas Herring
Thomas Hobiger
Guorong Hu
Lennard Huisman
Christopher S. Jacobs
Rolf Koenig
Sebastien Lambert
Martin Lidberg
Chopo Ma
Daniel MacMillan
Zinovy M. Malkin
Benjamin Männel
Stavros Melachroinos
Angelyn W. Moore
Michael Moore
Faramarz Nilfouroushan
Tobias Nilsson
Jean-Mathieu Nocquet
Erricos C. Pavlis
Chris Pearson
William Terry Petrachenko
Leonid Petrov
Lucia Plank
Markku Poutanen
Paul Rebischung
John C. Ries
Eldar Rubinov
Alvaro Santamaria-Gomez
Elifuraha Saria
Tilo Schöne
Manuela Seitz
Sarah Stamps
Felix Norman Teferle
Daniela Thaller
Joao Agria Torres
Paul Tregoning
Tonie van Dam
Jan Weiss
Simon David Paul Williams
Richard Wonnacott
Youlong Xia

T. van Dam (ed.), *REFAG 2014*, International Association of Geodesy Symposia 146,
DOI 10.1007/1345_2015

Author Index

T. van Dam (ed.), *REFAG 2014*, International Association of Geodesy Symposia 146,
DOI 10.1007/1345_2015

MIX
Papier aus verantwortungsvollen Quellen
Paper from responsible sources
FSC® C105338

If you have any concerns about our products,
you can contact us on
ProductSafety@springernature.com

In case Publisher is established outside the EU,
the EU authorized representative is:
Springer Nature Customer Service Center GmbH
Europaplatz 3, 69115 Heidelberg, Germany

Printed by Libri Plureos GmbH
in Hamburg, Germany